Krassimir Russew • Liljana Stojanova

Glassy Metals

Springer Reference

Krassimir Russew
Institute of Metal Science, Equipment
and Technologies with
Hydroaerodynamics Center
"Acad. A. Balevski"
Bulgarian Academy of Sciences
Sofia, Bulgaria

Liljana Stojanova
Institute of Metal Science, Equipment
and Technologies with
Hydroaerodynamics Center
"Acad. A. Balevski"
Bulgarian Academy of Sciences
Sofia, Bulgaria

ISBN 978-3-662-56906-1 ISBN 978-3-662-47882-0 (eBook)
DOI 10.1007/978-3-662-47882-0

Springer Heidelberg New York Dordrecht London

Softcover reprint of the hardcover 1st edition 2016

Printed on acid-free paper

Springer-Verlag GmbH Berlin Heidelberg is part of Springer Science+Business Media (www.springer.com)

This book is addressed to research scientists, lecturers at universities and academic institutes, engineers, as well as to students especially PhD students. It is based on the author's efforts and experience as researchers at the Institute of Metal Science, Equipment and Technologies "Acad. A. Balevski" with Hydroaerodynamics Centre, Bulgarian Academy of Sciences, as well as being guest scientists and grantees of the Alexander von Humbolth Foundation (K. Russew) and DAAD Foundation (L. Stojanova) at the Max-Planck-Institut für Metallforschung, Stuttgart, Germany, NWO Foundation, the Netherlands (K. Russew), and EU COST Fellowship of the European Union (K. Russew and L. Stojanova).

Abbreviations

CBMS	Chill block melt spinning
DSC	Differential scanning calorimeter
DTA	Differential thermal analyzer
FVM	Free volume model
FVT	Vogel-Fulcher-Tammann
GFA	Glass forming ability
KAJM	Kolmogorov-Avrami-Johnson-Mehl
KWW	Kohlrausch-Wiliams-Watts
LTM	Late transition element
LVDT	Linear voltage displacement transformer
OM	Optical microscopy
PFC	Planar flow casting
RE	Rare earth element
SEM	Scanning electron microscope
TEM	Transmission electron microscope
TM	Transition metal
TMA	Thermomechanical analysis
TMS	Thermomechanical system
TTT	Temperature-time-transformation
XRD	X-ray diffraction

Contents

Production, Properties and Applications of Glassy Metals

1

Abstract

A brief overview about the production properties and applications of amorphous metallic alloys is given. The possibility of synthesis and production of bulk amorphous alloys is discussed. The most important properties of application of metallic glasses are discussed. The ways for control change of physical and application properties of metallic glasses are presented. The main author's objectives of this book are presented.

Keywords

Amorphous metals • Glassy alloys • Properties • Applications

Introduction

Amorphous metallic alloys (glassy metals) represent new class materials in the contemporary materials science. They can be produced from the melt by its cooling with sufficiently high cooling rate in order to avoid the crystallization in the temperature range between liquidus temperature and the temperature of glass transition T_g. The critical cooling rate for the most glassy metals is usually very high – about 10^6 K/s. Due to this circumstance at least one of the dimensions of the produced amorphous alloy should be not more than approximately 0.02–0.05 mm. Recently, the discovery has been made [1] that a series of metallic alloys can be produced in amorphous state by moderate or even low cooling rates of 1–100 K/s. This is of substantial scientific and practical importance because these metallic alloys can form metallic glasses by casting in bulk machine details sizing from several mm to several cm. To these bulk amorphous metallic alloys belong, for example, the alloy compositions Al-RE-TM (RE = rare earth element, TM = transition metal) [2–5], Mg-RE-TM [6], Zr-Al-TM [7, 8], Zr-Al-Cu-Ni [9], and Zr-Ti-Cu-Ni [10]. A very important attainment represents the development of

K. Russew, L. Stojanova, *Glassy Metals*,
DOI 10.1007/978-3-662-47882-0_1

ferromagnetic multicomponent $(Fe,Co,Ni)_{70}Zr_{10}B_{20}$ bulk amorphous metallic alloys [11] that can be produced in the form of ingots of several cm diameter. These ingots can be rolled to thin transformer core lamellae for cores of high filling degree, in the temperature range between the crystallization temperature and the glass transition temperature of the glassy alloy. A general feature of metallic glasses is, in most cases, the combination of excellent and unique application properties. This in mind, as well as the technological production scheme of the metallic glasses, linked with great energy savings, is the reason for the great interest in their practical application.

For the first time, the metallic glasses were obtained in 1960. The basic methods for their production are described in a series of synopsis articles [12–17]. Since then and till now, the scientific and applied research in the field of (metallic) materials rapidly solidified from the melt follows rapid exponential increase. The industrial production of these materials takes place in many countries, on the first place in the USA, Europe, Japan, and Russia. Amorphous metallic alloys are produced in the form of thin ribbons, by width of up to 500 mm and practically unlimited length up to several hundred meters or even several km. Their shape is especially advisable for application in electronics and electrotechnics for manufacturing of magnetic cores for power supply and small-sized high-frequency transformers, magnetic heads, magnetic shields, etc. The energy losses by alternating magnetization are from 2 to 10 times lower as compared to the conventional transformer cores produced from Fe-Si iron sheet. Some metallic glasses possess high coefficient of magnetostriction, what makes them suitable for manufacturing of highly efficient tensosensors. The above-discussed applications are more extensively discussed in ▶ Chap. 13 of this book as well as in synopsis articles [18–51]. Rapidly solidified metallic alloys are extremely homogeneous. By precisely chosen chemical composition, they are very suitable for application as soldering alloys [52–57].

Most metallic glasses possess high mechanical strength combined with good plasticity. They are suitable for armoring elements for car tires, composite materials, and constructive elements in device building [58–60]. Some amorphous metallic alloys possess very low coefficient of thermal expansion in a wide temperature range. They are suitable to be used as Invar alloys [61, 62]. The above stated is only a brief survey of the glassy metals fields of application.

Due to the fact that metallic glasses represent systems, very far from thermodynamic equilibrium, their properties substantially depend both on the conditions of their production by rapid solidification from the melt and on the conditions of their subsequent heat treatment and/or thermomagnetic treatment. Although no phase transformations (except their crystallization) are observed by heat treatment of glassy metals, very fine structural changes take place by temperatures lower than their onset temperature of crystallization. These processes are called structural relaxation, which final stage is the crystallization, bringing the material into its thermodynamic equilibrium state for given temperature and pressure.

In this way rapidly quenched from the melt amorphous metallic alloy pass over a series of structural states with definite physicochemical and application properties. This circumstance gives the opportunity for a very precise control of their properties, depending on the requirements of their application [61–73]. This is why the fundamental study of structural relaxation in its connection to the application properties is of great importance from both scientific and practical point of view.

During the heat treatment of metallic glasses, they undergo fine changes of atomic ordering leading to decrease of the internal free energy of the system. Although not leading to the final thermodynamic equilibrium state of the system, some a kind of intermediate metastable state is reached, which by moderate temperatures and pressure can remain unchanged for practically unlimited time duration. Extremely sensitive to the structural relaxation are the coefficients of atomic transport, especially the viscosity η. It is well known that under isothermal heating, the viscosity of metallic glasses can increase with more than 5 orders of magnitude. As it will be shown later on, this is due to the fact that viscosity is proportional to the concentration of structural defects and in this way it depends exponentially on the amount of the so-called free volume of amorphous metallic alloys. The free volume, on the other hand, is the basic parameter, which endures changes (annihilation, production) as a result of thermally activated structural relaxation of amorphous materials. This is why one of the most important characteristics of metallic glasses is the temperature dependence of their coefficient of viscous flow – the viscosity. It determines not only the possibility itself to produce glassy metals via rapid quenching from the melt but corresponds directly to the changes of their physicochemical properties during the structural relaxation. The above discussion explains the importance of getting extensive and reliable information concerning the viscosity of amorphous metallic alloys – its magnitude and temperature dependence, as well as about the mechanism of structural relaxation.

The topics included in this book treat first of all some fundamental problems of structural relaxation of glassy metals and the possibility to study it on the basis of the characteristic parameters of their viscosity and its relation to the rheological behavior of metallic glasses. The rheological behavior is presented by the viscosity temperature dependence, the relaxation of bend stresses, the thermal expansion and specific heat anomalies, the density changes, and crystallization behavior. It is taken into account that the study of structural relaxation of amorphous materials has been an object of numerous studies of many researchers and is in the stage of its ripeness. The basic natural regularities are already established. Nevertheless, there exist a number of not very-well-studied and clarified problems, especially by the relaxation processes taking place under nonisothermal conditions around the glass transition temperature. Structural changes caused by the relaxation processes are directly reflected by changes in the physical and application properties of amorphous metallic alloys. Most of the structural relaxation studies, however, are dealing with a single, definite property of the material and not with a wider spectrum of physical properties giving with this respect the possibility to embrace them by a more general model about the mechanism of relaxation processes. This state of the problems determines the main objectives and the scope of this book:

- To make a short review on the basic methods for production of ribbon-like and bulk amorphous metallic alloys clarifying the influence of the technological parameters of their production upon the geometric characteristics of the obtained metallic glasses.
- To present the known experimental methods described in the scientific bibliography as well as those developed by the authors for viscosity measurements under nonisothermal conditions
- To present their capability for determination of the viscosity temperature dependence and the influence of different factors on it, such as variations in alloy chemical composition, alloy thermal prehistory, the varying production conditions, and other structure determining factors.
- To present the author's model contributions for extending and going deeper into the theoretical ideas about the temperature dependence of viscosity of deeply undercooled melts in the framework of Free Volume Model (FVM) of Cohen and Turnbull [74, 75], further developed by Spaepen [76].
- To show the correctness of the author's FVM extensions for nonisothermal conditions on the basis of numerous experimental proofs, obtained by the authors, showing that the FVM extended to describe the relaxation phenomena under nonisothermal conditions, does not contain internal contradictions by describing the nonreversible structural relaxation of metallic glasses and their rheological behavior.
- To present the author's FVM contributions about the temperature dependence of isothermal and nonisothermal relaxation of bend stresses in ribbon-like amorphous metallic alloys, about the anomalies of their the coefficient of thermal expansion α_l, about their specific heat Δc_p, the density changes, etc.
- To show that it is possible to describe the results of different independent experiments with a single common set of FVM model parameters for a definite amorphous metallic alloy. Such kinds of experiments are the direct measurement of viscous tensile creep (viscosity measurements) under continuous heating, the relaxation of bend stresses, the anomalies of thermal expansion, the anomalies of specific heat, and the peculiarities of the density changes under heating.
- To show that under heating the structural relaxation of amorphous metallic alloys before reaching the glass transition consists of fully reversible low-temperature structural relaxation and fully irreversible high-temperature structural relaxation. To show also that these two kinds of relaxation can be quantitatively separated from each other. The irreversible structural relaxation can be quite adequately described by the FVM.
- To show that a quantitative relation between the viscosity temperature dependence and the kinetics of crystal growth in amorphous metallic alloys exists.

In this way, it becomes once again obvious that the maxim of A. Einstein "Die *meisten Grundideen der Wissenschaft sind an sich einfach und lassen sich in der Regel in einer für jedermann verstaendlichen Sprache wiedergeben*"[1] is quite valid for FVM also.

The scope of the proposed book includes also a brief discussion about some possibilities for practical application of different kinds of amorphous metallic alloys.

The authors have been working together upon the topics, an object of the present book, more than two decades. This is why it should be clear that the book is mainly focused upon their own theoretical and experimental results. This long-term cooperation determines also the fact that both authors should undertake together the responsibility for slipping into possible shortcomings and faults.

Substantial parts of the studies have been carried out during the research stays of one of the authors (K. Russew) in F. R. Germany as an Alexander von Humboldt and Max-Planck Gesellschaft grantee at the Max-Planck-Institut fuer Metallforschung, Stuttgart, as well as in the Netherlands at a TU-Delft as a grantee of NWO. The fruitful long-term cooperation with Prof. Dr. Ferdinand Sommer, Prof. Dr. Arie van den Beukel, Prof. Dr. Jilt Sietsma, and Dr. Paul de Hey is gratefully acknowledged. The authors appreciate also the long-term research cooperation with Prof. Dr. G. Konczos, Dr. L. Lovas, Prof. Dr. I. Bakonyi, Prof. Dr. L. K. Varga, Dr. E. Fazakas, etc., from the Research Institute for Solid State Physics and Optics of the Hungarian Academy of Sciences.

References

1. Greer AL (1993) Nature 366:303
2. Inoue A et al (1990) Mater Trans Jpn Inst Met 31:425
3. Inoue A et al (1993) Mater Trans Jpn Inst Met 34:351
4. He Y et al (1994) Philos Mag Lett 70:371
5. Inoue A et al (1991) Mater Trans Jpn Inst Met 32:609
6. Inoue A et al (1992) Mater Trans Jpn Inst Met 33:937
7. Zhang T et al (1991) Mater Trans Jpn Inst Met 32:1005
8. Inoue A et al (1993) Mater Trans Jpn Inst Met 34:1234
9. Kim YJ et al (1994) Appl Phys Lett 65:2136
10. He Y et al (1996) J Non Cryst Solids 205–207:602
11. Yang B, Liu CT, Nieh TG (2006) Appl Phys Lett 88:221911–1
12. Giessen B et al (1972) In: Beer SZ (ed) Physics and chemistry of liquid metals. Marcel Dekker, New York, p 633
13. Jones H (1973) Rep Prog Phys 36:1425
14. Jones H et al (1973) J Mater Sci 8:705
15. Davies HA (1978) In: Cantor B (ed) Proceedings of conference on rapidly quenched metals RQ3, vol 1. The Metals Society, London, p 1
16. Liebermann HH (1979) IEEE Trans Mag Magn MAG-15:1393

[1]The most important scientific ideas are simple by themselves and could be presented principally in a way understandable for everyone.

17. Duwez P et al (1960) J Appl Phys 31:1136
18. Werner FE (1981) In: Marder AR, Stephenson ET (eds) Energy efficient electrical steels. TMS-AIME, Warrendale, p 1
19. Douglas J et al (1982) EPRI J 7(1):14
20. De Cristofaro N et al (1982) In: Masumoto T, Suzuki K (eds) Proceedings of conference in rapidly quenched metals RQ4, vol 1. Japan Institute of Metals, Sendai, p 238
21. Datta A, De Cristofaro N (1982) Magnetic amorphous metal alloys. US Patent 4321090 A, 23 Mar 1982
22. Yamamoto H et al (1983) Pulse arc welding method and device in which pulse current and background current have a constant current characteristic. US Patent 4,409,465 A, 11 Oct 1983
23. Boll R et al (1981) IEEE Trans Magn MAG-17:3053
24. Mischler WR, Rosenburry GM et al (1981) In: Proceedings of IEEE PES winter meeting, Atlanta, 1981, paper 81 WM 189-0
25. Pry RH et al (1958) J Appl Phys 29:532
26. Egami T et al (1997) In: Proceedings of annual conference on magnetism and magnetic materials, San Francisco, Dec 1977
27. Schwarzschield BM (1982) Phys Today 35(2):20
28. Warlimont H, Boll R (1982) J Magn Magn Mater 26:97
29. Hermansfeldt W (1979) In: Heavy ion fusion staff, Heavy on fusion program, University of California, Lawrence Berkeley Lab, Sept 1979
30. Hilzinger HR (1985) In: Steeb S, Warlimont H (eds) Proceedings of conference on rapidly quenched metals RQ5, vol 2. North Holland, Amsterdam, p 1695
31. Makino Y (1985) In: Steeb S, Warlimont H (eds) Proceedings of conference on rapidly quenched metals RQ5, vol 2. North Holland, Amsterdam, p 1699
32. Russew K et al (1988) Influence of the isothermal heat treatment on the soft magnetic characteristics of the amorphous alloy $(Co,Fe)_{75}(Si,B)_{25}$. In: Proceedings of 1st international conference on rapidly quenched metals, Varna, 1987, Akademie der Wissenschaften der DDR, VEB Kongress- und Werbedruck, Oberlungwitz, p 306
33. Sakakima H et al (1982) In: Masumoto T, Suzuki K (eds) Proceedings of conference rapidly quenched metals RQ4, vol 2. Japan Institute of Metals, Sendai, p 929
34. Boll R (ed) (1977) Weichmagnetische Werkstoffe, Vacuumschmelze GmbH, Hanau. Verlag Siemens AG, Berlin
35. Holmes L (1982) Electr Power 28(2):180
36. Boll R et al (1981) Siemens Forschungs- und Entwicklungsberichte 10(2):83
37. Hilzinger HR et al (1979) Phys Stat Sol A 55:73
38. Mohri K (1984) IEEE Trans Magn MAG 20:942
39. Scarzello JF et al (1977) IEEE Trans Magn MAG-13:574
40. Scarzello JF, Usher G (1977) IEEE Trans Magn MAG-13:1101
41. Foerster F (1955) Z Metallkd 46:358
42. Luz H (1968) Elektronik 17(8):247
43. Takeuchi G et al (1984) IEEE Trans Magn MAG-20:1723
44. Fearon E (1974) Open-strip ferromagnetic marker and method and system for using same. US Patent 3,790,945 A, 5 Feb 1974
45. Becker K, Hoffmann K (1978) Diebstahlschutz-alarmanlage. Patentschrift DE 2 709 522 A1, 7 Sept 1978
46. Fearon E (1983) Antitheft system. Patent WO 1,983,002,027 A1, 9 Jun 1983
47. Vazkuez N et al (1983) Phys Stat Sol A 80:195
48. Mohri K et al (1988) IEEE Trans Magn MAG-14:1071
49. Meydan T et al (1981) IEEE Trans Magn MAG-17:3376
50. Webb DC et al (1979) IEEE Trans Magn MAG-15:1410
51. Török E et al (1979) J Magn Magn Mater 10:303
52. Yao YD et al (1982) J Appl Phys 53:2258

53. Szewieczek D et al (1985) In: Steeb S, Warlimont H (eds) Proceedings of conference on rapidly quenched metals RQ5, vol 2. North Holland, Amsterdam, p 1791
54. De Cristofaro N et al (1985) In: Steeb S, Warlimont H (eds) Proceedings of conference on rapidly quenched metals RQ5, vol 2. North Holland, Amsterdam, p 1715
55. DeCristofaro NJ et al (1978) Weld J 57(7):35
56. Archangelskiy VM et al (1983) Stahl 7:69 (in Russian)
57. Cline CF, Hopper RW (1977) Scr Metall 11:1137
58. Kank S et al (1977) IEEE Trans PHP PHP-13:318
59. Naka M et al (1978) J Non Cryst Sol 30:29
60. Kikuchi M et al (1983) Sci Rep RITU A 31(1):99
61. Schneider I et al (1979) Neue Huette 24(10):379
62. Drijver JV et al (1978) Nederlands Tijdschrift voor Natuurkunde 44(1):10
63. Onn DG (1981) J Appl Phys 52:1788
64. Matey JR et al (1977) J Non Cryst Solids 23:129
65. Ravex A et al (1981) Physica 107B:397
66. Riley JD et al (1979) Phys Rev B 20:776
67. Nagel SR et al (1976) Phys Rev B 13:3284
68. Amamou A, Krill G (1980) Solid State Commun 33:1087
69. Chen HS (1983) In: Luborsky FE (ed) Amorphous metallic alloys. Butterworths, London, p 169
70. Egami T (1981) Ann NY Acad Sci 371:238
71. Van Den Beukel A et al (1984) Acta Metall 32:1859
72. Huizer E et al (1987) Acta Metall 35:2843
73. Warlimont H et al (1985) In: Steeb S, Warlimont H (eds) Proceedings of conference on rapidly quenched metals RQ5, vol 1. North Holland, Amsterdam, p 619
74. Cohen MH, Turnbull D (1959) J Chem Phys 31:1164
75. Turnbull D, Cohen MH (1970) J Chem Phys 52:3038
76. Spaepen F (1980) In: Balian R, Kléman M, Poirer JP (eds) Physics of defects, Les Houches 1980, session XXXV. NorthHolland, Amsterdam, p 135

2 Most Important Methods for Production of Amorphous Metallic Alloys

Abstract

The most frequently used methods for production of amorphous metallic alloys via rapid solidification from the melt, chill block melt spinning, and planar flow casting are presented and discussed. The influence of technological production parameters are taken under consideration. The dependence of the geometrical dimensions of ribbonlike glassy metals upon the melt ejection pressure and peripheral velocity of the quenching substrate (disk) is theoretically subjected to analysis and experimental proof. The authors' experience upon the production of bulk amorphous alloys is presented and discussed. A brief description about other methods for production of amorphous metallic materials is presented. An overview about the bibliography in the form of more than 90 cited references is proposed.

Keywords

Amorphous metallic alloys • Glassy metals • Production • CBMS method • PFC method

The materials, which could be easily obtained in a vitreous state (the natural silicates, some chalcogenides, polymers, etc.), are characterized with high melt viscosity η of 10^2–10^3 Pa s. order of magnitude. The glass forming is additionally enhanced by the rapid increase of viscosity along with increasing the melt undercooling. By conventional glasses, low or moderate rates of melt cooling (by order of magnitude 0.01–5 K/s) are sufficient to avoid the crystallization, due to the fact that rates of crystal nucleation and growth are extremely low. On the contrary, the viscosity of molten metallic alloys is very low – of 10^{-2}–10^{-3} Pa s order of magnitude. This is why extremely high cooling rates, about 10^6 K/s or more, are needed in order to avoid their crystallization and to obtain them in a vitreous state [1–4]. Because of this circumstance, at least one of the dimensions of the produced amorphous alloy should be in the range of hundreths of millimeters.

K. Russew, L. Stojanova, *Glassy Metals*,
DOI 10.1007/978-3-662-47882-0_2

Such very high melt cooling rates are realized by using of well-known production methods: chill block melt spinning (CBMS) [5–7] and planar flow casting (PFC) [8, 9]. Principally, each metallic melt can be obtained in glassy state, provided that it is cooled with sufficiently high cooling rate. In the case of metallic glasses of practical interest (the metallic glasses on the basis of transition metals Fe, Co, and Ni), the needed critical cooling rate is of 10^5–10^6 K/s order of magnitude. Obviously, in order to reach such high cooling rates, the product of rapid quenching should be practically a 2D object, i.e., the thickness must be within 0.01–0.05 mm. The length and the width of the amorphous ribbon can be theoretically unlimited.

The methods for rapid quenching from the melt of such flat objects are described in details by many authors [10–15]. The method of P. Duvez is described in [15], "hammer and evil" is discussed in [16], the method "two hammers" in [17], and the "two shafts method" is described in [18]. All these methods are suitable only for laboratory use and are of importance only for production of amorphous metallic samples of irregular shape and small dimensions for studying their properties. The production of continuous fibers and ribbons became possible after inventing the method of Pond [19], in 1958. The process consists of launching of a thin jet of molten material upon the inner surface of a concave quenching disk. The obtained fiber/ribbon is of unlimited length. This method, however, is unsuitable for production of metallic ribbons. Variant of this method is later on developed by Pond and Maddin [5], for production of amorphous metallic fibers.

The possibility for production of practically infinite in length amorphous ribbon has been demonstrated in 1972 by Polk and Chen [20]. They developed the two shafts method for rapid quenching from the melt in the laboratories of Allied Chemical Corp. According to this method, a jet of molten metallic alloy is launched into the slot between two quickly rotating cooling shafts. This approach has been used for experimental proof of the possibility for vitrification of a series of metallic alloys on Fe–Ni and Ni basis. Although useful, this method is difficult for technological control and did not found any industrial application.

In 1975, Bedell and Wellslager [21] also coworkers of Allied Chemical Corp., applied elements of Pond's method directing a jet of molten metallic alloy against the peripheral surface of copper quenching disk of 10 cm diameter, rotating with 1800 m/s peripheral speed. In this way, the CBMS method has been developed. This method allows to achieve very high quenching rates, due to the natural adhesion between the metallic melt and the quenching disk. Bedell et al. [22, 23] have shown later on that the contact between the metallic ribbon and the quenching disk can be improved by using of additional flexible pressing copper foil. In this way, a comparably simple and industrially applicable method for production of thin, narrow amorphous metallic ribbons has been developed. The length of these ribbons is practically infinite. This method has acted a key role by taking the decision from Allied Chemical Corporation to produce and offer for commercial dissemination amorphous metallic alloys.

A method for production of wires of amorphous metallic alloys is developed by Kavesh [24, 25]. The wires of infinite length are with circular cross section. The process is as follows: molten metallic jet is ejected into liquid cooling bath, flowing parallel and along the direction of melt ejection. By using this method, amorphous wires of the Pd–Cu–Si alloy have been produced. Their diameter has been approximately 0.380 mm. Amorphous wires from Fe-based metallic compositions were also produced, with a diameter approximately 0.1 mm. Due to different technological hindrances, this method did not find any broad industrial application till now.

Regardless of the big breakthrough of Bedell's method in the direction of industrial production of metallic glasses, it has become very soon clear that it is impossible to obtain long amorphous ribbons wider than 5 mm. For the production of wider ribbons, it is necessary to work with ejected stream of rectangular cross section. This is impossible to achieve with a nozzle of circular opening. It is known also, that due to the low viscosity and the high surface tension of metallic melts, the molten metallic jet of rectangular cross section quickly loses shape stability building knots. This problem is critical for the production of wide amorphous ribbons and has been successfully solved by Narasimhan [26]. He has developed the so-called PFC method. According to this method, the molten metallic melt is ejected under pressure through a rectangular nozzle positioned in a close proximity (the distance nozzle-quenching surface is about 0.2–0.5 mm) to the rotating copper disk. In this way, the width of the produced ribbon equals the width of the rectangular nozzle. Nowadays, it is possible to produce amorphous ribbons of more than 500 mm width.

Maringer and Mobley [27] have developed in 1975 a method for effecting of ultrahigh melt cooling rates through extracting thin metallic ribbons from the melt. The essence of this method is to bring a rapidly rotating in the vertical plane cooling disk into contact to the melt surface partially immersed into the melt bath. As a result, a thin metallic layer solidifies onto the disk edge. Depending on the chemical composition of the molten alloy, microcrystalline or amorphous metallic ribbons are obtained in this way. The first aim of this project has been to realize an inexpensive method the production of steel staple fibers for armoring of fireproof concrete. The resulted composite materials possessed very high exploitation characteristics. Since 1974, the companies *National Standard* and *Ribbon Technology*, USA, are applying the method for production of staple fibers from low-carbon and stainless steel. The length and the diameters of these fibers are usually 2.5–5 cm and 0.02–0.05 mm, respectively. In 1978, the production volume has reached about 1000 tons. The further refining of this method led to the successful production of amorphous wires and ribbons. The advantage of the method of Mobley and Maringer [27], as compared with the CBMS and PFC methods, consists in the fact that the molten metal is in a free state in a bath and there is no need to use an expensive special crucible with round or rectangular nozzle.

All above-described methods for production of metallic glasses resulted in obtaining of a final product in the form of ribbon/wire. This product is characterized with a very great ratio surface/volume. This is conditioned by the necessity of very high melt cooling rates. Unfortunately, such product form is very often inconvenient for application, especially when the product should possess complicated threefold configuration with relatively big size in all three directions. This is why a great attention has been paid [28–33] to the possibility to develop methods for obtaining of amorphous metallic alloys in a bulk form. To these methods belong the plasma spraying, the cold compacting, the explosive compacting, the hot extrusion, and the dynamic compacting. These methods can be most conveniently applied, when the initial product is in the form of amorphous powder with approximately spherical particle form. The existing methods for production of amorphous powders could be separated in two groups [33]. To the first group belong the methods by which the powder particles represent solidified melt droplets (dispersive methods). The second group includes methods based on the cutting/milling of amorphous ribbons/wires into tiny particles [34]. By the ultrasound gas dispersion, narrow molten metal jet encounters with a great velocity a gas stream [35]. The gas stream in this case plays a double role – to disperse the melt into small droplets and to cool them down up to solidifying. This method uses gas stream pulsing with frequency of about 80 kHz. By using He under pressure of about 8 MPa, an amorphous powder of Cu–Zr alloy is produced, with particle size ranging from 0.05 to 0.125 mm, respectively. The cooling rate attained is about 10^5 K/s.

By the water dispersion method, the working scheme is practically thc same as by gas dispersion [36, 37]. In this case, however, the dispersing and cooling medium is water. This method makes possible to attain high cooling rate. The shortcoming of the method is that it is difficult to obtain small enough droplets with even spherical shape. In most cases, the particles are agglutinated to each other and are only partially amorphous. By this method, amorphous–crystalline Fe–B–Si alloys have been produced containing up to 80 % amorphous phase. By the gas–liquid melt dispersion methods [38, 39] upon the molten metallic jet, streams of gas and water under high pressure are focused. The gas stream causes mainly the melt separation into droplets, while the water stream ensures the cooling and solidifying of the droplets. In this way, under conditions of gas streams under pressure of 4 MPa and water stream under pressure of 1.4 MPa Cu–Zr and Fe–B–Si amorphous powders have been attained. The cooling rate of the particles of about 0.02 mm size is estimated to be approximately 10^6 K/min.

The method of centrifugal spraying is developed by the *Pratt & Whitney Company* [40]. By this scheme of work, molten metallic jet is ejected to rotating with about 2500 revolutions per min cooling and dispersing disk. The centrifugal forces direct the melt to the disk periphery, where the molten jet is disintegrated into tiny droplets by the surface tension forces. The droplets obtained are additionally blown and cooled by gaseous He. The cooling rate attained is 10^5–10^6 K/s. The droplet size is about 0.03 mm. They are with almost perfect spherical shape. In this way, series of amorphous powders from Ni–B–Si and Fe–B–Si have been produced.

The two-shaft dispersing equipment is developed by Singer [41]. The production of powder is realized via molten jet ejection into the slot between two rotating with 1000–5000 revolutions/min against each other's shafts. The rotating shafts are covered with a thick grease to prevent the molten jet from solidification within the slot between the shafts. The jet is disintegrated to separate droplets, which fall into a water bath and are cooled down by a cooling rate of approximately 10^5 K/s. With the aid of such equipment, amorphous/crystalline powders from Pd–Cu–Si, Ni–Si–B, and Co–Si–B alloys have been obtained. The method has found a limited application because of the fact that the powder particle size is greater than 0.04 mm. A modification of this method uses a solid substrate for rapid cooling of molten droplets [42].

There exist also other methods for production of amorphous metallic powders. Here belong the electro-erosion and the electro-hydrodynamic method. Their productivity, however, is very low – 3.5 g/h and 20 g/day, respectively. This is why these methods are suitable only for laboratory use. The method for electro-hydrodynamic disintegration can produce very fine powder particles with a size of 0.000001 mm. In this way, very fine amorphous powders of Fe–Ni–P–B alloy have been obtained.

Mechanical impact for production of amorphous metallic powders is also possible. The plastic deformation, mechanical impact, and neutron bombardment can lead to destroying of the crystalline lattice. One of the most widely used methods of this kind is the prolonged mechanical processing in special ball mills. Amorphous powders from Gd–Co alloy after processing from 15 min to 15 h have been produced in this way. Along with increasing the duration of ball milling, the vitrification of the powder reaches up to 100 %.

The above-described methods do not exhaust the subject, but in consideration are taken mainly those methods, which can/could find an industrial application. The most prospective methods are the methods for rapid quenching from the melt CBMS and PFC. One of the objectives of this book is to present more thoroughly the methods CBMS and PFC.

The Chill Block Melt Spinning (CBMS) Method

The chill block melt spinning is a process for continuous casting of narrow ribbonlike amorphous alloys directly from the melt. The process includes ejection of molten metallic jet trough a round nozzle, against rapidly moving cooling substrate. This process is an object of increased interest during the last decades. The same is valid also for the PFC method. Both methods are based on the usage of a single massive rapidly rotating quenching disk as a substrate [43, 44].

Figure 2.1 shows schematically the working principle of CBMS method. It is believed that the most promising application of the amorphous metallic ribbons is as soft magnetic materials in electronics and electrotechnics, where they are used mainly in laminated state as transformer cores. In this case, the requirement exists for a high package density of the lamellae. This is why the soft magnetic ribbons should be as even as possible – with possibly constant width and thickness. In order to fulfill this requirement, one needs to know the dependence of the geometric characteristics of the ribbon on the technological parameters of their production.

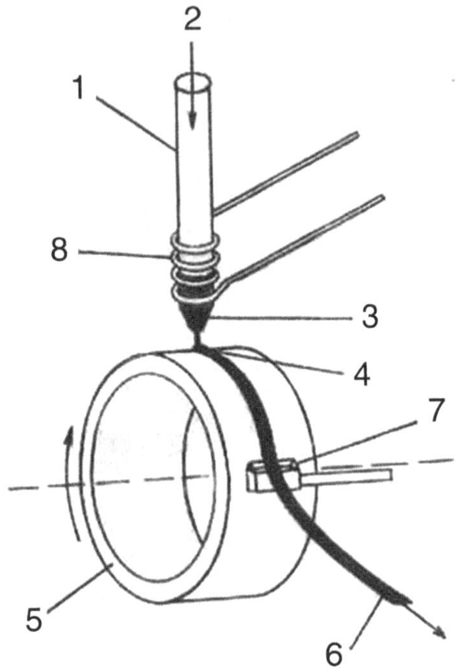

Fig. 2.1 Scheme of the equipment for chill block melt spinning CBMS: *1* cylindrical silica glass crucible, *2* ejection gas pressure (N_2, Ar, or He), *3* molten metallic alloy, *4* stationery melt paddle in contact with the cooling substrate, *5* quenching Cu disk, *6* amorphous ribbon, *7* unsticking spoiler, *8* HF inductive heater

Influence of the Production Parameters upon the Geometric Characteristics of the Amorphous Ribbons by the CBMS Method

The controllable production parameters, which first of all determine the geometric ribbon characteristics, are the volumetric melt flow rate through the crucible orifice Q_f, the peripheral rotating velocity V_s, and the angle θ between the crucible axis and the tangent to the cooling surface. When the molten metal comes into contact with the cooling surface of the rotating (copper) disk, a situation of simultaneous heat and momentum (the product mass × velocity in a definite direction) transfer arises. It is quite difficult to give an exact description of this complicated phenomenon (with exception of several special cases [45, 46]. This is why one has to imagine that one of the two transfer processes dominates during the ribbon formation.

The first model is based on the idea that the heat transfer is dominating, i.e., it is supposed that the heat transfer coefficient h_i is infinitely big. This model leads to the widely used relation of Kavesh [47], for the dependence of ribbon thickness $\bar{t}$ and width w, respectively, on the technological parameters Q_f and V_s:

$$\bar{t} = C\frac{Q_f^{1/4}}{V_s^{3/4}}, \tag{2.1a}$$

$$w = C^{-1}\frac{Q_f^{3/4}}{V_s^{1/4}}, \tag{2.1b}$$

where C is a proportionality coefficient.

The second model is based on the assumption that h_l equals zero and that the thickness of the ribbon equals the thickness of melt dragged from the stationary paddle. The thorough analysis of this model, carried out by Anestiev and Russew [44], leads to the following relation at $\theta = \pi/2$:

$$\bar{t}_{\pi/2} = 1.412\left(b\eta_k^{1/2}\right)\frac{Q_f^{1/4}}{V_s^{3/4}}, \tag{2.2a}$$

$$w_{\pi/2} = 0.708\frac{1}{\left(b\eta_k^{1/2}\right)}\frac{Q_f^{3/4}}{V_s^{1/4}}, \tag{2.2b}$$

where b is dimensionless constant, taking into account the specific physical properties of a definite metallic melt (kinematic melt viscosity η_k, surface tension, etc.). The relation between the kinematic and the dynamic viscosity η_k and η_m, respectively, is given by $\eta_k = \eta_m/\rho^l$, where ρ^l is the melt density.

As is seen, the second model proposes practically the same relations as the model of Kavesh, based on the heat transfer as governing the ribbon formation.

Russew et al. [48] have carried out thorough experimental study on the influence of peripheral rotation speed V_s of the quenching disk and the melt overheat ΔT_m upon the ribbon geometric characteristic thickness and width, respectively, by using the amorphous metallic alloy $Fe_{40}Ni_{40}Si_6B_{14}$ obtained by CBMS as an example. The melt spinning is carried out in air environment and copper quenching disk of 150 mm diameter, silica glass crucibles with 0.7 mm orifice, Ar ejection pressure 15 KPa, $\theta = \pi/2$.

All production parameters are kept constant, except the peripheral disk speed V_s, and the melt overheat ΔT_m, which varied from 20 to 50 m/s, and from 40 to 400 K, respectively. The measured ribbon thicknesses and widths, respectively, are given in Table 2.1. The results are interpreted according to Eq. (2.2a, b). The values of volumetric melt velocity Q_f is calculated according to the well-known Bernoulli equation, not taking into account the existence of local hydraulic hindrances by the melt flow through the orifice:

$$Q_f^{id} = \pi r^2\left(2\Delta p/\rho^l\right)^{0.5}, \tag{2.3}$$

where r is the orifice radius, Δp is *the ejection pressure*, and ρ^l is the melt density.

Obviously, Q_f does not depend on the melt overheating ΔT_m. At a given value of $\rho^l = 7.16 \times 10^3$ kg/m^3, one obtains for Q_f^{id} the value of 7.88×10^{-7} m^3/s. The real value of Q_f^{id} obtained from the slope of the straight line shown in Fig. 2.2 is 4.62×10^{-7} m^3/s. Hence, the ratio Q_f/Q_f^{id} equals 0.6. This value is quite realistic and is in good agreement with the experimentally observed deviations from Eq. (2.3), caused by the existing hydraulic hindrances.

Figure 2.3 shows the dependence of ribbon thickness t by 40 and 400 K melt overheating, respectively, on the parameter $(V_s)^{-0.75}$. Two clearly separated linear dependences with different slopes are observed. According to Eq. (2.2a), these

Table 2.1 Geometric characteristics of amorphous alloy $Fe_{40}Ni_{40}Si_6B_{14}$ ribbons, produced by the CBMS method, depending on the melt overheat ΔT_m and the peripheral speed V_s of the rotating quenching disk

V_s, m/s	ΔT_m, K	t, 10^{-6} m	w, 10^{-6} m
20	40	42	600
20	100	39	660
20	400	32	690
30	40	30	580
30	400	24	680
40	40	23	480
40	300	22	660
40	400	18	690
50	40	20	520
50	400	15	680

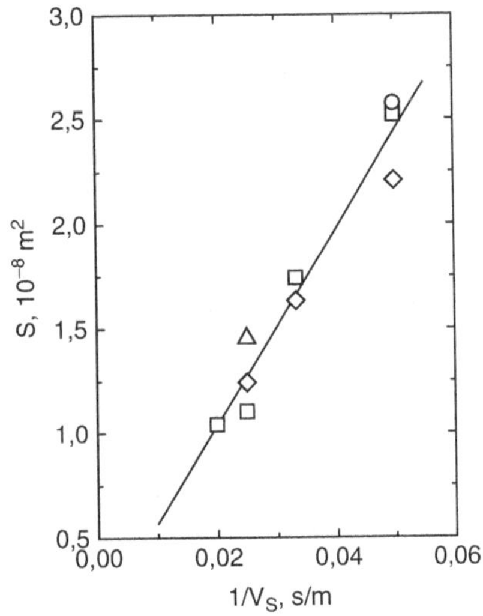

Fig. 2.2 Dependence of the ribbon cross-section area $S = t \times w$ of all amorphous ribbons, by different melt overheating ΔT_m, as a function of the $(V_s)^{-1}$ parameter. Overheating: □, 40 K; ○, 100 K; Δ, 300 K; ◇, 400 K

slopes should be equal to $b(\eta_m/\rho^l)^{0.5}(Q_{fr})^{0.25}$. It follows from the values of the two slopes that the melt viscosity η_m at T = 1630 K (ΔT_m = 400 K) is 1.7 times lower than at T = 1270 K (ΔT_m = 40 K). On the assumption that the temperature dependence of η_m is of Arrhenius-type:

$$\eta_m = \eta_o^m \exp\left(\frac{Q_m}{RT}\right), \tag{2.4}$$

where Q_m is the activation energy of viscous flow at temperatures higher than the alloy melting temperature T_m, η_o^m is pre-exponential factor, R is the gas constant, and the value of Q_m should be equal to 24.3 kJ/mol. This value is in excellent agreement with the value of 22.3 kJ/mol for the activation energy of melt viscous flow of $Fe_{40}Ni_{40}P_{14}B_6$. It has been determined by Anderson et al. [49]. By using the data of Chhabra et al. [50] for the melt viscosity of pure metals, the mean value of

Fig. 2.3 Dependence of ribbon thickness t, produced at melt overheating of 40 and 400 K, as a function of the parameter $(V_s)^{-0,75}$: ○, overheating of 40 K; □, overheating of 400 K

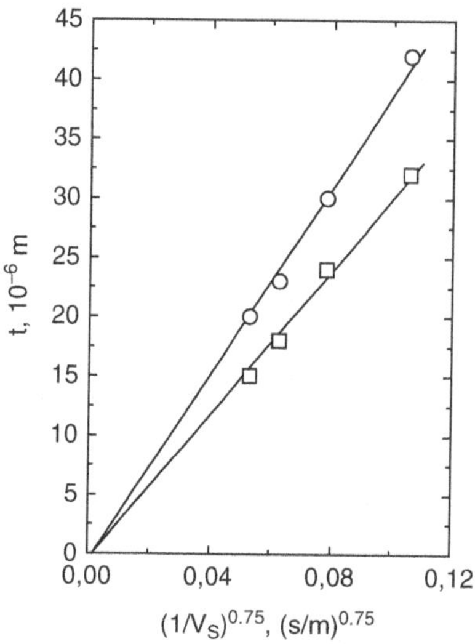

the pre-exponential viscosity factors of molten η_o^{Fe} and η_o^{Ni} were used to "construct" most probable Arrhenius temperature dependence of molten $Fe_{40}Ni_{40}P_{14}B_6$ alloy:

$$\eta_m = 4,25 \times 10^{-3} \exp\left(\frac{24300}{RT}\right), \tag{2.5}$$

where η_m is given in Pa s and R is in J/mol K. Equation (2.5) is "constructed" aiming to present a link between viscosity data around the glass transition temperature T_g and the viscosity at the melting temperature T_m with the equation of Vogel–Fulcher–Tammann. It follows from Eq. (2.5) that the melt viscosity of $Fe_{40}Ni_{40}P_{14}B_6$ alloy at T_m is 0.053 Pa s, which is in excellent agreement with the data in [49] for the melt viscosity of the metallic alloy $Fe_{40}Ni_{40}P_{14}B_6$ in the melting point – 0.035 Pa s. From the above discussed follows that Eqs. (2.2a), (2.2b) realistically describe the influence of CBMS production parameters (Q_{fr}, V_s, and ΔT_m vs melt viscosity) upon the geometric characteristics (ribbon thickness, width) of narrow metallic glass ribbons.

Planar Flow Casting (PFC) Method

Figure 2.4 shows schematically an equipment for production of wide amorphous metallic ribbons according to the PFC (planar flow casting) method. An analysis of this process is performed by Anestiev and Russew [51].

The main difference between the *CBMS* and *PFC* methods consists in form of crucible orifices and the distance between the orifice and the rotating quenching disk. In the first case, the orifice is with round cross section and a diameter between 0.5 and 1.5 mm. This maximal diameter of 1.5 mm is determined by the

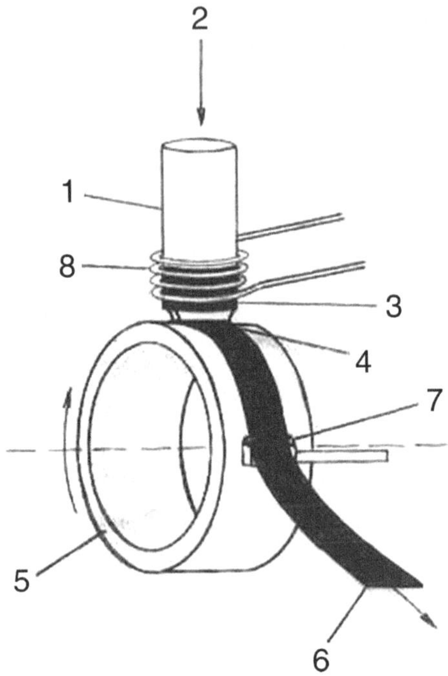

Fig. 2.4 Scheme of an equipment for melt spinning according to the PFC method: *1* cylindrical silica glass crucible with a rectangular orifice, *2* ejection pressure (N_2, *Ar* or *He*), *3* metallic melt, *4* stationery melt paddle in contact with the surface of the rotating quenching disk (Cu), *5* quenching disk, *6* amorphous ribbon, *7* unsticking spoiler, *8* HF inductive heater

requirement of the surface tension to be able to withstand successfully to the own melt weight trying to cause a spontaneous melt flow through the orifice before applying any ejection pressure.

Relation Between the Production Conditions and the Geometric Characteristics of Wide Amorphous Ribbons by the PFC Method

In the case of the PFC method, the orifice is flat, with rectangular cross section. The orifice length can be theoretically unlimited, while the orifice width should be within 0.5–1 mm. The distance between the orifice and the rotating quenching disk varies between 0.2 and 0.4 mm. By using greater distance, the flat molten jet builds knots, trying to bring the melt to a cylindrical form, due to the surface tension forces. The very short distance between the orifice and the quenching disk determines that the ribbon width corresponds practically to the length of the orifice S and does not change during the process. In this case, the dependence between the processing conditions – volumetric melt flow rate Q_f and disk peripheral velocity V_s and the ribbon geometric characteristics – thickness t, is determined by the mass balance equation:

$$t = \frac{1}{S}\frac{Q_f}{V_s}. \quad (2.6)$$

Usually the volumetric melt flow rate is calculated according to the Bernoulli equation:

$$Q_f = dS\sqrt{\frac{2\Delta p}{\rho^l}}, \tag{2.7}$$

where ρ^l is the melt density and Δp is the ejection pressure. Unfortunately, the ribbon thickness t, calculated according to the equation

$$t = \frac{\alpha d}{V_s}\sqrt{\frac{2\Delta p}{\rho^l}}, \tag{2.8}$$

with α as a correction coefficient, is approximately 40–50 % greater than the really observed thickness value. This circumstance can be explained, taking into account the resistance met by the melt flow into the narrow slot of the orifice. The examination of the process shows, that by the movement of the molten metal through the orifice, it is forced "to squeeze" along the slot, which causes loss of pressure. Due to the high value of the Reynolds number $Re \approx 10^3$ [52], the melt flow through the orifice can be considered as a flow of an ideal liquid. By using the equation for preservation of energy and impulse of an ideal liquid [53], together with the equation of mass balance, one obtains a system of three equations with three unknowns: the melt velocities V_l and V_2 at the entrance and the outlet of the rectangular orifice, respectively, and the local loss of pressure H:

$$p_1 + \frac{\rho^l V_1^2}{2} = p_2 + \frac{\rho^l V_2^2}{2} + \rho^l gH, \tag{2.9a}$$

$$\rho^l V_2^2 S_2 - \rho^l V_1^2 S_1 = S_1 p_1 - (S_2 - S_1)\left(p_1 + \frac{\rho^l V_1^2}{2}\right) - S_2 p_2, \tag{2.9b}$$

$$V_1 S_1 = V_2 S_2. \tag{2.9c}$$

In these three equations, S_l and S_2 are the cross-section areas of the crucible and the rectangular orifice, respectively, and p_l и p_2 are as shown in Fig. 2.5.

From the system of Eqs. (2.9a), (2.9b) (2.9c), one obtains for V_2:

$$V_2 = \sqrt{\frac{2\Delta p}{\rho^l\left[2 - (S_2/S_1) - (S_2/S_1)^2\right]}}. \tag{2.10}$$

As far as $S_2/S_1 << 1$, the last two terms in the denominator in Eq. (2.10) could be ignored. In this way, one obtains for melt outlet velocity:

$$Q_f = \alpha dS\sqrt{\frac{\Delta p}{\rho^l}}. \tag{2.11}$$

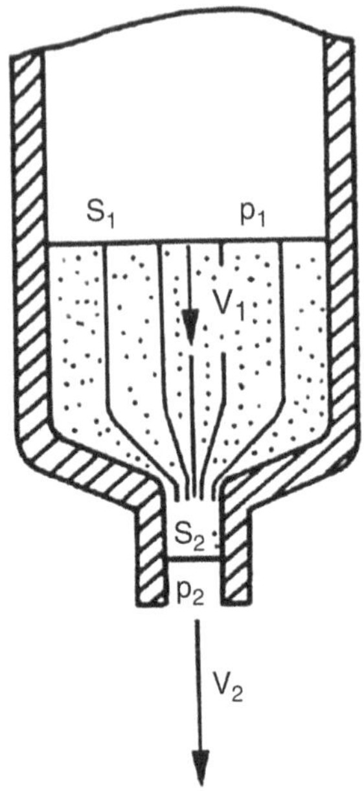

Fig. 2.5 Scheme of an incompressible liquid through a narrow flat orifice

Respectively, for the ribbons, one obtains:

$$t = \frac{\alpha d}{V_s}\sqrt{\frac{\Delta p}{\rho^l}}. \tag{2.12}$$

The recognizing of pressure loss leads to relation, which predicts ribbon thickness of about 30 % lower than calculated thickness according to Eq. (2.8), even in the case, when no correction coefficient α is introduced. An additional pressure loss influencing the relation between Q_f, t, and Δp could arise provided that the melt flow comes in contact with the front lip of the orifice, parallel to the moving substrate surface. This should be possible if the slot between the moving substrate and the front orifice lip is filled with molten metal. The analysis of the possible reasons for such a situation shows that it could be possible if:

(a) The molten metal wets well the material of the crucible and the orifice. However, as far as this material is commonly chosen to be wet very poorly by the metallic melt, this possibility should be rejected.
(b) The processing conditions are chosen in such way (too high ejection pressure Δp, too low peripheral quenching disk velocity V_s) that the quenching disk is not able to transport the whole amount of molten metal in the form of ribbon. Under such conditions, it is not possible to produce a ribbon of satisfactory quality.
(c) The distance h between the orifice and the quenching disk is so small, that the boundary layer $\delta(d)$ in the stationary melt paddle becomes thicker than h.

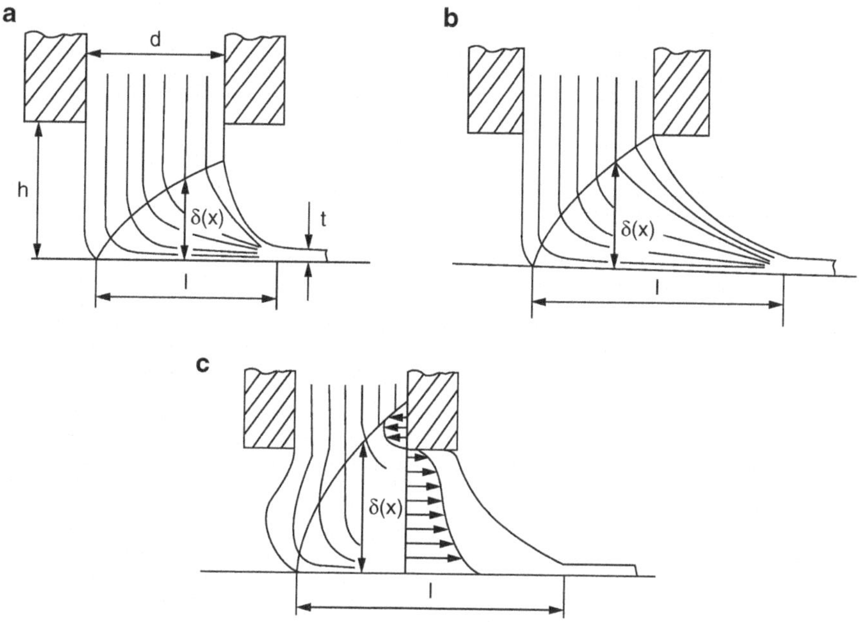

Fig. 2.6 Possible shapes of the stationary paddle at different casting conditions by the PFC method. (**a**) $\delta(d) < h$, (**b**) $\delta(d) = h$, (**c**) $\delta(\partial) > h$

The boundary layer *δ(d)* is shown arising by the drag forces of the rotating disk under the orifice at a distance $x = d$ (see Fig. 2.6a–c) if

$$h \geq \delta(d). \tag{2.13}$$

The slot could not be filled with molten metal, and there is no need to take into consideration any additional loss of pressure. The expression (2.13) can be used for defining the conditions at which Eq. (2.12) can be considered to give a correct description of the PFC process [54]. According to Eq. (2.10), the relations $\delta = \delta(x)$ and $t = t(l, V_s)$ are defined as follows:

$$\delta(x) = 7.4\sqrt{\frac{x\eta_k}{V_s}}; \quad t = 1.62\sqrt{\frac{l\eta_k}{V_s}}, \tag{2.14}$$

where η_k is the kinematic viscosity of the metallic melt and l is the length of the stationary paddle. Taking into account Eq. (2.14), one obtains for the criterion needed:

$$h \geq \frac{4.57\alpha}{V_s}\sqrt{\frac{d^3 \Delta p}{\rho l}}. \tag{2.15}$$

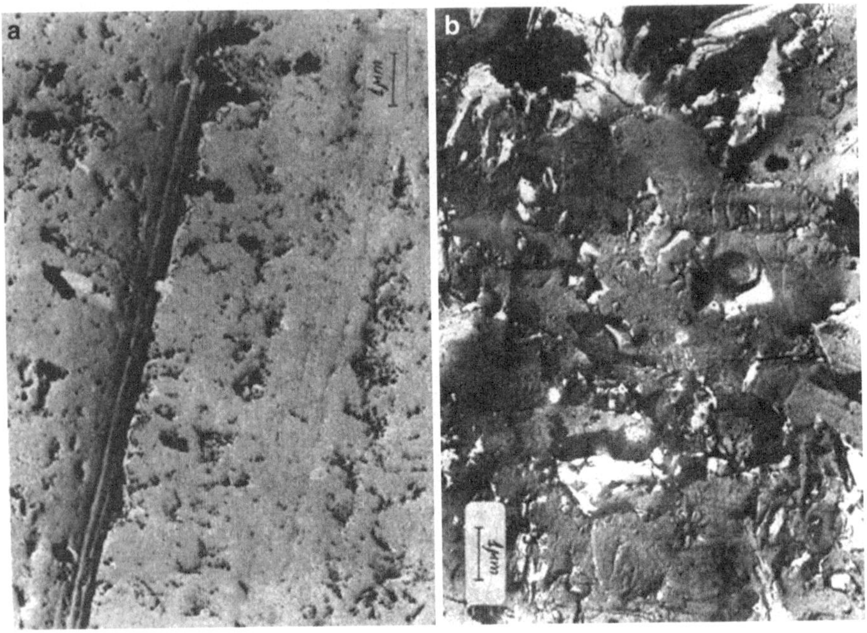

Fig. 2.7 (**a**) Relief of the orifice front lip surface parallel to the moving quenching substrate - lack-carbon replica EM image; (**b**) Relief of the inner surfaces of silica glass orifice lips, which have been undoubly in contact of the flowing metallic melt - lack-carbon replica EM image

It can be defined with the aid of the technological parameters d, h, and Δp и V_s only. The limiting condition (2.15) has been experimentally proven by Kopasz et al. [55], by the production of five amorphous metallic ribbons from $Co_{75}Fe_5Si_{13}B_7$ alloy, at $h = 0.25$ mm, $V_s = 25$ m/s and $\Delta p = 4.2\times10^{-2}$ MPa. The ribbon thickness has been within 0.031–0.033 mm.

Figure 2.7a shows the relief of the inner surfaces of silica glass orifice lips, which have been in contact with the flowing metallic melt during the ribbon production by PFC method. Figure 2.7b shows the relief of the orifice front lip surface parallel to the moving quenching substrate. Obviously, the flowing through the orifice metallic melt has not been in contact with the orifice front lip. Introducing the values of the production technological parameters into Eq. (2.15) and assuming that the characteristic contact length l equals 3 mm, one obtains that the value of the quantity h should be equal or greater than 0.137 mm in order to avoid the contact between the melt in the stationary paddle and the front orifice lip. The experimental facts are in agreement with the determined in [50] criterion Eq. (2.15).

Examining the dependence of ribbon thickness t on the technological PFC parameters Δp and V_s, Kopasz et al. [55] have derived the following relation:

$$t_{\exp}(\mu m) = \frac{43.7}{V_s(m/s)}\sqrt{\Delta p(mbar)} + 0.6. \qquad (2.16)$$

Introducing in Eq. (2.12) the values $\alpha = 0.7$, $d = 0.5$, and $\rho^l = 7 \times 10^3$ kg/m^3, for the wide ribbon thickness, one obtains:

$$t_{theor}(\mu m) = \frac{42.2}{V_s(m/s)}\sqrt{\Delta p(mbar)}. \quad (2.17)$$

The model presented by Eq. (2.17) is in very good agreement with the experimental evidence.

Production of Bulk Amorphous Metallic Alloys by Rapid Cooling from the Melt and Study of Their Structure

During the past several years, advances have been made in this field as a result of the discovery and development of several families of alloys with substantially improved glass-forming ability [56, 57]. Whereas the "conventional" ribbonlike amorphous metallic alloys were generally formed by cooling the melt at rates of 10^5–10^6 K /s, the more recently developed alloy compositions require cooling rates of only 1–100 K/s or less. As such, the new materials can be cast from the molten state into glassy objects with dimensions up to several centimeters as compared with maximum thickness or diameter of 0.01–0.1 mm for rapidly quenched ribbons and powders. These new alloys are referred to as "bulk glassy alloys" or "bulk metallic glasses." Among them, one can mention, for example, Al-RE-TM (RE = rare earth, TM = transition metal) [58–60], Mg-RE-TM [61], Zr-Al-TM [62, 63], multicomponent alloys as Zr–Al–Cu–Ni [62], Zr–Ti–Cu–Ni [63], Fe–Ga–Al–P–B–C [64], etc. These bulk amorphous alloys are produced in the form of ingots with diameter of several centimeters. With the aid of rolling in the temperature range between the glass transition temperature and the onset temperature of crystallization, lamellae for transformer cores with very high degree of filling can be produced.

The bulk metallic glasses are an object of increasing interest as well, because they can be used as precursors for production of nanocrystalline metallic alloys. For this purpose, one has to find out the suitable conditions to carry out the crystallization process [65]. In this way, it is possible to obtain nanocrystalline alloys possessing very good mechanical [66–68], soft magnetic [69–72], hard magnetic [73, 74] alloys with high magnetostriction in weak magnetic fields [75, 76] and excellent catalytic properties [77–79], which cannot be achieved by using only pure amorphous or multiphase crystalline alloys. The beginning of this new stage in the development of glassy metals started with the discovery of the first metallic alloys with chemical composition, allowing their production in amorphous state not only in the form of ribbons but also in the form of bulk ingots with dimensions up to 3 mm ($Pd_{76}Cu_6Si_{18}$ and $Pd_{40}Ni_{40}P_{20}$) with low critical cooling rate of about 10^3 K/s [80–82].

In the period between 1990 and 2000, as a result of the purposeful endeavor of A. Inoue et al. in the Sendai University, Japan, a series of bulk amorphous metallic alloys were discovered, with a critical cooling rate lower than 10^3K/s. One can mention here the multicomponent metallic systems Mg-Ln-TM [83, 84], Zr-Al-TM [85, 86], Zr-(Ti, Nb, Pd)-Al-TM [87], Zr-Ti-TM-Be [88, 89], Nd-Fe-Al [90–92], Pr-Fe-Al [93], Fe-(Co, Ni)-(Zr, Nb, Ta)-B [94, 95], Fe-(Co, Ni)-(Zr, Nb, Ta)-(Mo, W)-B [96–99], where Ln = metal from the group of lanthanides and TM = transition metal. The maximal thickness of the amorphous ingots depending on the different type is up to 30 mm for Zr-Al-TM, 25 mm for Zr-Ti-TM-Be, 40–72 mm for the system Pd–Cu–Ni–P, 6 mm for Fe–Co–Zr–Mo–W–B, and 25 mm for the system Pd–Ni–Fe–P.

Successful attempts for production of bulk amorphous metallic alloys have been carried out by the authors of this book also, with the alloy compositions Pd–Ni–P and Pd–Cu–Ni–P [100, 101]. They are objects of the following presentation.

The preparation of bulk amorphous metallic alloys through rapid cooling of liquid $Al_{7.5}Cu_{17.5}Ni_{10}Zr_{65}$, $Pd_{40}Cu_{30}Ni_{10}P_{20}$, and $Pd_{40}Ni_{40}P_{20}$ alloys has been studied. Two methods for rapid cooling have been used – vacuum suction casting into thick-wall copper tubes of outer diameter of 10 mm and inner diameter of 3 mm and rapid quenching of the alloy melts contained in thin-wall silica glass tubes of inner diameter of 4 mm into cold water. Bulk amorphous samples of $Pd_{40}Cu_{30}Ni_{10}P_{20}$ and $Pd_{40}Ni_{40}P_{20}$ and microcrystalline samples of $Al_{7.5}Cu_{17.5}Ni_{10}Zr_{65}$ have been prepared by the first method of quenching. By the second method, bulk amorphous samples of $Pd_{40}Cu_{30}Ni_{10}P_{20}$ alloy and microcrystalline samples of $Pd_{40}Ni_{40}P_{20}$ alloy were obtained. Ribbonlike amorphous samples of the abovementioned alloys were produced using the classical chill block melt spinning (CBMS) method. The structure of the rapidly cooled bulk samples of the different master alloys has been proved for traces of crystallinity by using differential scanning calorimetry (DSC), X-ray diffractometry, scanning electron microscopy (SEM), and metallographic observations.

The master alloy ingots were prepared by simultaneous arc melting of pure elements or their intermetallic compounds on a water-cooled copper mold under vacuum of 10^{-5} Torr and by using pure Mg or Zr as a getter. Special attention was paid to the oxygen content of Zirconium, which was less than 40 ppm. The $Pd_{40}Ni_{40}P_{20}$ master alloy ingot was prepared from the pure elements Pd and Ni of not less than 99.95 purity grade and Ni_2P powder. The $Pd_{40}Cu_{30}Ni_{10}P_{20}$ master alloy ingot was prepared by alloying of Cu and Ni of 99.999 purity grade and the intermetallic compound Pd_7P_3. The last one was synthesized by diffusional alloying of powder mixtures of Pd and red phosphorus which were consolidated at room temperature under high mechanical pressure. The consolidated samples were sealed into silica glass ampoules under vacuum of 10^{-5} Torr and subjected to 100 h heat treatment by stepwise increasing of the temperature up to 973 K in the last 25 h of heat treatment. The ingots obtained were remelted three times in BN crucibles under protective Ar atmosphere in a HF furnace in order to reach a high degree of homogeneity. In the case of $Pd_{40}Ni_{40}P_{20}$ and $Pd_{40}Cu_{30}Ni_{10}P_{20}$ ingots, a refinement of the melts by using B_20_3 flux treatment was also carried out in order to eliminate the heterogeneous nucleation sites of crystallization. Due to the great affinity of Zr to oxygen, the application of such B_20_3 flux treatment was not possible.

Immediately after the last remelting at temperatures higher than the starting temperature of crystallization, a thick-wall copper tube of 10 mm outer diameter and 5 mm inner diameter was immersed quickly into the melt, and the melt was sucked into the tube. In order to avoid melt contamination with dissolved copper, a small piece of BN hollow cylinder was attached to the tip of the copper tube.

By using two vacuum valves, the first one connecting the copper tube with an evacuated closed volume of 10^{-3} Torr vacuum, approximately ten times bigger than the volume of the casting space in the copper tube, and the second one, connecting the evacuated closed volume with the rotary vacuum pump, it was possible to perform an instantaneous generation of vacuum within the casting space. At a closed first valve and after evacuation of the closed volume, the second valve was closed also. Just before starting the casting process, the BN tip of the casting tube was quickly immersed into the melt, the first valve was opened, and the melt was sucked out into the casting space of the copper tube. It should be pointed out that the inner surface of the copper tube was beforehand carefully cleaned up mechanically, followed by ultrasonic treatment in acetone bath.

The melt rises quickly upward solidifying at the inner tube surface building a solid layer of approximately 1 mm thickness, while the melt in the close proximity of the tube axis remains hot enough and liquid to be able to “climb” further upward. Immediately after finishing the casting procedure, the copper tube was additionally quenched into water. By this method of casting, amorphous or microcrystalline hollow cylinders of outer diameter of 5 mm and inner diameter of approximately 3 mm are formed.

Experiments for preparation of bulk $Pd_{40}Cu_{30}Ni_{10}P_{20}$ and $Pd_{40}Ni_{40}P_{20}$ bulk amorphous samples through water quenching of molten master alloys in sealed evacuated silica glass tubes of 4 mm inner diameter were also carried out. A sample obtained through vacuum suction casting and a piece of the copper casting tube are shown in Fig. 2.8.

In Fig. 2.8, X-ray diffraction pattern of five samples of $Pd_{40}Cu_{30}Ni_{10}P_{20}$ bulk amorphous alloy is shown. The bulk samples are obtained through rapid water quenching of its melt in a silica glass capillary. The absence of sharp diffraction peaks due to traces of crystallinity is obvious as well as the uniformity of all different samples.

As a conclusion, it can be pointed out that a partially successful attempt for preparation of bulk amorphous metallic alloys has been carried out. Using the method of suction casting in thick-wall copper tubes, bulk glassy alloy of the $Pd_{40}Cu_{30}Ni_{10}P_{20}$ and $Pd_{40}Ni_{40}P_{20}$ systems can be easily obtained. The first alloy can be also obtained into a glassy state after rapid quenching into water. The attempts to prepare bulk samples of $Al_{7.5}Cu_{17.5}Ni_{10}Zr_{65}$ by using the above methods of casting remained unsuccessful as yet.

More details and methods for preparation of bulk amorphous metallic alloys of various chemical compositions can be found in the reference book of A. Inoue [70].

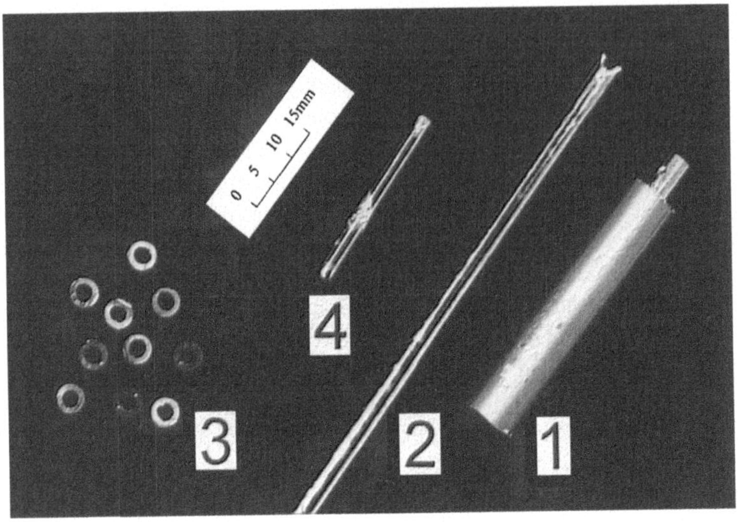

Fig. 2.8 *1* Thick-wall copper tube for preparation of bulk amorphous and/or microcrystalline samples by using vacuum suction casting; *2* hollow cylinder of the bulk amorphous $Pd_{40}Ni_{40}P_{20}$, prepared through "suction casting"; *3* small rings cut off using circular diamond saw from the hollow cylinder shown in *2* in order to be used as bulk samples for density change measurements as a function of the relaxation heat treatment conditions; *4* dense cylindrical sample of bulk $Pd_{40}Cu_{30}Ni_{10}P_{20}$ amorphous metallic alloy obtained through water quenching of a silica glass capillary of 4 mm diameter filled with the molten alloy

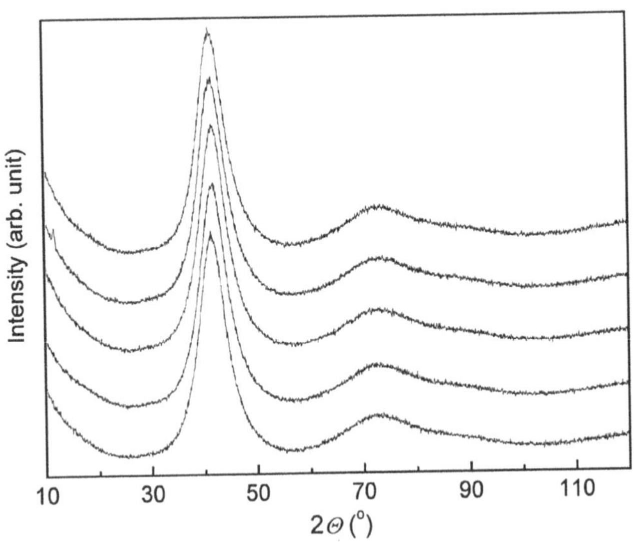

Fig. 2.9 X-ray diffraction pattern of five samples of $Pd_{40}Cu_{30}Ni_{10}P_{20}$ bulk amorphous alloy

References

1. Davies HA (1983) In: Luborsky FE (ed) Amorphous metallic alloys. Butterworths, London, p 13
2. Hafner J (1981) In: Guentherodt H-J, Beck H (eds) Metallic glasses. Springer Verlag, Berlin, p 93
3. Giessen BS, Whang SH (1980) J Phys C8:95
4. Sommer F (1981) Zs Metallkde 72:219
5. Pond R, Maddin P (1969) Trans Met Soc AIME 245:2475
6. Chen HS, Miller CE (1970) Rev Sci Instrum 41:237
7. Anestiev L (1991) Mater Sci Eng A131:115
8. Huang SC, Fiedler HC (1981) Mater Sci Eng 51:39
9. Takayama S, Oi T (1979) J Appl Phys 5C(7):4962
10. Giessen B et al (1972) In: Beer SZ (ed) Physics and chemistry of liquid metals. Marcel Dekker, New York, p 633
11. Jones H (1973) Rep Prog Phys 36:1425
12. Jones H et al (1973) J Mat Sci 8:705
13. Davies HA (1978) In: Cantor B (ed) Proceedings of the conference on rapidly quenched metals RQ3, vol 1. The Metals Society, London, p 1
14. Liebermann HH (1979) Trans Mag MAG 15:1393
15. Duvez P et al (1960) J Appl Phys 31:1136
16. Miroshnichenko IS (1982) Quenching from the melt. Metallurgy, Moscow, p 168, in Russian
17. Esslinger P et al (1965) Zs Wirtschaft Fertigung 60:1156
18. Chen HS et al (1970) Rev Sci Instrum 41:1237
19. Pond RB (1958) Metallic filaments and method of making same. US Patent 2,825,108, 4 Mar 1958
20. Polk DE et al (1974) J Non-Cryst Solids 15:165
21. Bedell JR, Wellslager JA (1975) Elevation of melt in the melt extraction production of metal filaments. US Patent 3,863,700, 4 Feb 1975
22. Bedell JR (1975) Extended retention of melt spun ribbon on quenching wheel. US Patent 3,862,658,A, 28 Jan 1975
23. Bedell JR, Rothmayer NY et al (1978) Chill roll casting of continuous filament. US Patent 4,077,462, 7 Mar 1978
24. Kavesh S (1974) Liquid quenching of free jet spun metal filaments. US Patent 3,845,805, 5 Nov 1974
25. Kavesh S (1978) In: Gilman JJ, Leamy HJ (eds) Metallic glasses. ASM, Metals Park, p 36
26. Narasimhan MC (1979) Continuous casting method for metallic strips. US Patent 4,142,571, 6 Mar 1979
27. Mobley CE, Maringer RE (1975) An improved method of formation of filament directly from molten material. US Patent 3,861,450, 21 Jan 1975
28. Beddow JK (1980) The production of metal powder by atomization. Heyden and Sons LTD, London
29. Grant NJ (1985) In: Steeb S, Warlimont H (eds) Proceedings of the conference on rapidly quenched metals RQ5, vol 1. North Holland, Amsterdam, p 3
30. Lawley A (1977) Int J Powder Met Powder Tech 13:169
31. Lubanska H (1970) J Metals 22:45
32. Klar E et al (1072) In: Burke JJ, Weiss V (eds) Powder metallurgy for high-performance applications. Syracuse University Press, Syracuse, p 57
33. Mehrotra SP et al (1980) Trans Indian Inst Metals 33:361
34. Lawley A (1981) J Metals 33:13
35. Dain RJ (1980) Atomizer for making powder. US Patent 4,191,516, 4 Mar 1980
36. Yamaguchi T (1981) Amorphous electronic materials and their application. Science Forum Press, Tokyo, p 78

37. Miller S et al (1979) Scr Metall 13:673
38. Anand V et al (1980) In: Reston R, Mehrabian R, Kear B, Cohen M (eds) Proceedings of the 2nd conference on rapid solidification processing: principles and technologies, Baton Rouge LA, Claitor's Publ Division, p 273
39. Thompson CC (1981) Gas manifold for particle quenching. US Patent 4,284,394, 18 Aug 1981
40. Yamaguchi T (1978) Appl Phys Letters 33:498
41. Singer A et al (1980) Powder Metall 2:81
42. Singer ARE et al (1977) In: Hausner HH, Taubenblat PV (eds) Modern developments in powder metallurgy. Metal Powder Industries Federation, Princeton, p 127
43. Ishii H (1982) In: Masumoto T, Suzuki K (eds) Proceedings of the conference on rapidly quenched metals RQ4, vol 1. Japan Institute of Metals, p 238
44. Anestiev L, Russev KA (1987) On the fluid dynamics of formation of thin metal ribbons by rapid quenching from the melt. Mater Sci Eng 95:281
45. Strange EH, Pim CA (1908) Process of manufacturing thin sheets, foil, strips, or ribbons of zinc, lead, or other metal or alloy. US Patent 905,758, 1 Dec 1908
46. Pond RB (1959) Method of forming round metal filaments. US Patent 2,879,566 A, 31 Mar 1959
47. Kavesh S (1978) In: Gilman JJ, Leamy HJ (eds) Proceedings of the symposium on metallic glasses, Niagara Falls, ASM, Metals Park, p 36
48. Russew K, Stojanova L, Lovas A (1994) Effect of processing conditions on the ribbon geometry and viscous flow behavior of $Fe_{40}Ni_{40}Si_{6}B_{14}$ amorphous alloy. Int J Rapid Solidification 8:147
49. Anderson PM III, Lord AE Jr (1980) Mat Sci Eng 43:267
50. Chhabra RP, Sheth DK (1990) Z Metallkde 81:264
51. Anestiev L, Russew K (1988) On the dependence between the casting conditions and ribbon dimensions by PFC method. In: Proceedings of the first international conference on rapidly quenched metals, Varna, 1987, Akademie der Wissenschaften der DDR, VEB Kongress- und Werbedruck, Oberlungwitz, p 63
52. Takayama S, Oi T (1979) J Appl Phys 50:4962
53. Antony T, Cline H (1978) J Appl Phys 49:829
54. Aleshko PI (1977) Mechanics of fluid and gas. High School Publ. House, Kharkov, p 320 (In Russian)
55. Kopasz C et al (1983) In: Proceedings of the EPS conference soft magnetic materials 6, vol 7E. Eger, Hungary, p 101
56. Greer AL (1993) Nature 366:303
57. Greer AL (1995) Science 267:1947
58. Inoue A et al (1990) Mater Trans Jpn Inst Met 31:425
59. Inoue A et al (1993) Mater Trans Jpn Inst Met 34:351
60. He Y et al (1994) Philos Mag Lett 70:371
61. Inoue A et al (1991) Mater Trans Jpn Inst Met 32:609
62. Inoue A et al (1992) Mater Trans Jpn Inst Met 33:937
63. Zhang T et al (1991) Mater Trans Jpn Inst Met 32:1005
64. Inoue A, Shinohara Y, Gook JS (1995) Mater Trans JIM 36:1180
65. Kim YJ et al (1994) Appl Phys Lett 65:2136
66. He Y et al (1996) J Non-Cryst Solids 205–207:602
67. Kim YH, Inoue A, Masumoto T (1990) Met Trans JIM 31:747
68. Chen H, He Y, Shifet GJ, Poon RJ (1991) Scr Met 25:1421
69. Inoue A, Zhang T, Itoi T, Takeuchi A (1997) Mater Trans JIM 38:359
70. Inoue A (1998) Bulk amorphous alloys, preparation and fundamental characteristics, materials science foundations, vol 4. Trans. Tech. Publications, LTD, Zurich, p 194
71. Yoshizawa Y, Oguma S, Yamauchi K (1988) J Appl Phys 64:6044
72. Suzuki K, Kataoka N, Inoue A, Makino A, Masumoto T (1990) Mater Trans JIM 31:743

73. Croat JJ, Herbst JF, Lee RW, Pinkerton FE (1984) J Appl Phys 55:2078
74. Hirosawa S, Matsuura Y, Yamamoto H, Fujimura S et al (1986) J Appl Phys 69:873
75. Taranichev VE, Alenov MN, Nemova OY (1994) J Magn Magn Mater 131(3):229
76. Kubota T, Inoue A (2004) Mater Trans JIM 45:199
77. Deng JF, Zhang XP, Min EZ (1988) Appl Catal A 37:339
78. Baiker A, Maciejewski M, Tagliaferri S (1993) Berichte der Bunsengesellschaft für physikalische Chemie 97(3):286
79. Wang WJ, Qiao MH, Yang J, Song HX, Deng JF (1997) Appl Catal A 163:101
80. Chen H (1974) Acta Met 22:1505
81. Inoue A (1995) Mater Trans JIM 36(7):866
82. Inoue A, Nishiyama N (1997) Mater Sci Eng A226–228:401
83. Kui HW, Greer AL, Turnbull D (1984) Appl Phys Lett 45:615
84. Inoue A, Ohtera K, Kita K, Masumoto T (1998) Jap J Appl Phys 27:L2248
85. Inoue A, Kohinata NM, Tsai AP, Masumoto T (1989) Mater Trans JIM 30:378
86. Inoue A, Zhang T (1989) Mater Trans JIM 30:965
87. Inoue A, Yamaguchi H, Zhang T (1990) Mater Trans JIM 31:10
88. Inoue A, Zhang T, Masumoto T (1990) Mater Trans JIM 31:177
89. Inoue A, Zhang T, Masumoto T (1993) J Non-Cryst Solids 156–158:473
90. Inoue A, Nishiyama N, Amiya K, Zhang T, Masumoto T (1994) Mater Lett 19:131
91. Inoue A, Shibata T, Zhang T (1995) Mater Trans JIM 36:1420
92. Peker AL, Johnson WL (1993) Apl Phys Lett 63:2342
93. Inoue A, Zhang T, Zhang W, Takeuchi A (1996) Mater Trans JIM 37:99
94. Inoue A, Nishiyama N, Matsuda T (1996) Mater Trans JIM 37:181
95. Inoue A, Gook JS (1995) Mater Trans JIM 36:1180
96. Inoue A, Shinuhara Y, Gook JS (1995) Mater Trans JIM 36:1427
97. Inoue A, Zhang T, Takeuchi A (1996) Mater Trans JIM 37:1731
98. Inoue A, Koshiba M, Zhang T, Makino A (1997) Mater Trans JIM 38:577
99. Inoue A, Zhang T, Takeuchi A (1997) Appl Phys Lett 71:464
100. Russew K, Stojanova L, Sommer F (2000) Preparation of bulk and ribbon-like amorphous metallic alloys and study of their structure and rheological properties. Part I: preparation and structural study. J Mater Sci Technol 8(1):25
101. Russew K, Stojanova L, Sommer F (2000) Preparation of bulk and ribbon-like amorphous metallic alloys and study of their structure and rheological properties. Part II: study of rheological properties. J Mater Sci Technol 8(1):34

3 Experimental Methods for Determination of the Magnitude and Temperature Dependence of the Viscosity of Amorphous Metallic Alloys

Abstract

The importance of knowing the magnitude and the temperature dependence of glassy metals viscosity is presented. The possibility for estimation of the critical cooling rate for obtaining the metallic alloys in amorphous state via rapid cooling from the melt is analyzed. An overview about the bibliographic sources devoted to the determination of glassy metals viscosity under isothermal and nonisothermal experimental conditions is proposed. The experimental techniques are taken under consideration. A thorough analysis of the method for precise determination of glassy metals viscosity under nonisothermal conditions is presented. A real estimation of the critical cooling rate with Fe–B glassy alloy as an example is demonstrated.

Keywords

Amorphous metallic alloys • Glassy metals • Glass transition • Viscosity determination • Relaxation

In studying the nature of glass formation, the kinetic approach [1] is often used, which assumes that all materials, including metals and alloys, can be vitrified, provided their melts are cooled down with sufficiently high cooling rate which has to be at least equal or somewhat greater than the critical cooling rate T_{cr}, specific for the material under consideration. Uhlmann and Onorato [2, 3] have proposed a method of estimating the critical cooling rate of glass-forming systems using accepted theories of nucleation, crystal growth, and volume transformation kinetics [4]. They have shown that the time, t, for a small fraction, ζ (usually accepted as 10^{-6}), crystallized at temperature T, equals

K. Russew, L. Stojanova, *Glassy Metals*,
DOI 10.1007/978-3-662-47882-0_3

$$t = \left[\frac{9.3\eta(T)}{kT}\right]\left\{\left(\frac{a_o\xi}{f^3N_v}\right)\frac{\exp\left(\frac{1.024}{T_r^3\Delta T_r^3}\right)}{\left[1-\exp\left(\frac{-\Delta H_m\Delta T_r}{RT}\right)\right]^3}\right\}^{0.25} \tag{3.1}$$

where $\Delta T_r = (T_m - T)$, $T_r = T/T_m$, a_o is the average atomic diameter, N_v is the number of atoms per unit volume, f is the fraction of sites at the melt/crystalline interface where atoms are preferentially added or removed, and ΔH_m is the molar enthalpy of fusion. The critical cooling rate is taken as

$$\dot{T}_{cr} = \frac{(T_m - T_n)}{t_n}, \tag{3.2}$$

where T_n and t_n are the temperature and the time at the nose of the temperature–time–transformation diagram showing the temperature dependence of the time, needed for reaching 10^{-6} volume fraction of crystallized melt. Such a TTT diagram is schematically shown in Fig. 3.1. Computation of the TT′I′ curve requires the temperature dependence of the melt viscosity $\eta(T)$ within the temperature range from melting temperature T_m to the glass transition temperature T_g to be known.

Typical temperature dependence of the viscosity η of a glass-forming metallic alloy is schematically shown in Fig. 3.2 in a plot of $\log\eta$ against $1/T$. At temperatures higher than T_m, the melt viscosity η is very low, approximately 10^{-3} Pa s, with a weak temperature dependence of Arrhenius-type [5, 6]:

$$\eta = \eta_o \exp\left(\frac{Q_a}{RT}\right), \tag{3.3}$$

where η_o is the melt viscosity at temperatures much higher than T_m (usually $\eta_o = 10^{-4}$ Pa s) [5, 6], Q_a is the activation energy for viscous flow, which is usually in the range of 10–40 kJ/mol, and R is the gas constant. When low and moderate cooling rates are used, small undercooling below T_m is sufficient for rapid crystallization of the melt to occur, resulting in a large, discontinuous increase in viscosity up to values of 10^{14}–10^{15} Pa s (Fig. 3.2).

By cooling the melt with $\dot{T}_{cr}$ or with a higher quenching rate, nucleation and crystal growth are avoided and η increases continuously following the temperature changes along the metastable equilibrium viscosity curve in Fig. 3.2 until the glass transition temperature T_g is reached and the material becomes solid at viscosity values of approximately 10^{12} Pa s. The temperature dependence of η in the temperature range T_m–T_g is much stronger than that predicted by Eq. (3.3) and is fairly well described by the empirical equation of Vogel–Fulcher–Tammann (VFT)

$$\eta = \eta_o^{VFT} \exp\left(\frac{B}{T - T_o}\right) \tag{3.4}$$

where η_o^{VFT} and B are empirical constants and T_o is known as ideal glass transition temperature. The VFT equation applies strictly only to the temperature dependence

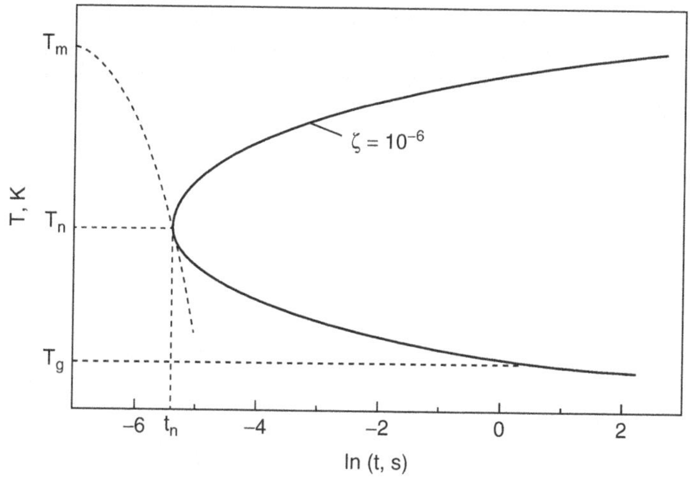

Fig. 3.1 Scheme of the temperature–time–transformation (TTT) curve for estimation of the critical cooling rate $\dot{T}_{cr}$ for obtaining of amorphous metallic alloy

Fig. 3.2 Schematic presentation of viscosity temperature dependence of glass-forming metallic melt. The *solid* parts of the curves present the temperature ranges where the viscosity could be experimentally determined

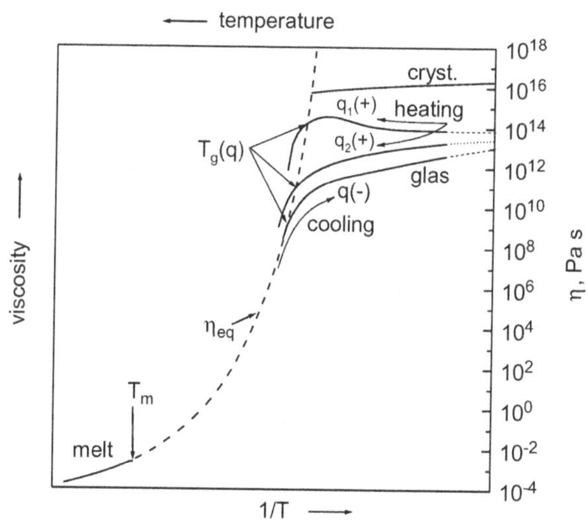

of the metastable equilibrium viscosity [5]. At temperatures lower than T_g, atomic mobility becomes so low that iso-configurationally frozen amorphous structure arises, which remains practically unchanged for very long periods of time near the ambient temperature T_{amb}. The transition to glassy state is connected with sudden deviation of η temperature dependence from the metastable equilibrium viscosity curve, and for the temperature range T_g–T_{amb}, it can be described again with an empirical equation closer to a relation of Arrhenius-type.

Typical temperature dependence of the viscosity of glass-forming metallic alloy, combined with the viscosity temperature dependence of the viscosity of vitrified metallic alloy by its reverse heating from glassy state back to molten state, is schematically shown in Fig. 3.2.

At temperatures lower than T_g, the atomic mobility decreases dramatically to a state, where the structural changes become practically infinite low in the timescale of experiment (very high cooling rate). This dramatic decrease of atomic mobility causes sudden deviation of the viscosity temperature dependence from the temperature dependence of the (quasi-)equilibrium viscosity. The structure of vitrified material becomes highly nonequilibrium from thermodynamic point of view. Consequently, the viscosity of vitrified metallic alloys depends not only on temperature but also on the duration of thermal treatment (by isothermal heating) and on the temperature reached with definite rate of heating or cooling (by nonisothermal treating) of the alloy. A very thorough consideration of the thermal stability of metallic glasses in the framework of their viscous behavior and viscosity temperature dependence is given in the synopsis articles of Spaepen and Taub [7], Cahn [8], and Davies [9].

It becomes quite clear from the above discussion, how important it is, to know not only the viscosity temperature dependence of undercooled metallic melts but also the viscosity temperature dependence of metallic glasses obtained by rapid solidification from the melt.

Experimental Methods for Viscosity Determination of Amorphous Metallic Alloys Under Isothermal and Nonisothermal Conditions

Viscosity measurements under isothermal conditions have been performed for different glassy metallic alloys by many authors [10–19]. Description of tensile creep experimental equipments used is given by Taub and Spaepen [10], Mulder et al. [12], Anderson et al. [13], Bauali et al. [14], etc. In these creep tests the specimen is usually fixed with a stationary pin at the top of the test rig by a special attachment. The bottom grip is connected to the core of a linear variable differential transformer (LVDT). The length changes of the specimen are measured by monitoring the displacement of the LVDT core to which the loading weight is connected. The experiments are performed under vacuum or in an inert atmosphere with argon as purging gas. The viscosity of the alloys studied is determined according to the Newtonian relation of homogeneous viscous flow

$$\eta = \frac{\tau}{\dot{\varepsilon}}, \tag{3.5}$$

where $\dot{\varepsilon}$ is the strain rate of the specimen under applied shear stress τ.

As the metallic glasses are in configurationally frozen states, which are less stable than other more relaxed or crystallized states of the material, they crystallize upon annealing at a sufficiently high temperature for a given time. Various changes of properties, however, take place before the crystallization occurs. These changes can be attributed to subtle changes in the atomic structure toward the metastable equilibrium liquid structure, known as structural relaxation. This structural relaxation is reflected in a most sensitive way by the viscosity of glassy alloys. In isothermal annealing practically all glassy metals exhibit linear increase of viscosity with annealing time [10–19]. It should be expected that the viscosity will reach a constant value when the glass has reached unstable equilibrium at its annealing temperature. In most cases, however, before the viscosity reaches a constant value, crystallization occurs with sudden contraction of the specimens. This is why there are few isothermal viscosity measurements of metallic glasses near the glass transition temperature due to the onset of rapid crystallization [13]. The only studies of metallic glasses in which equilibrium has been clearly reached prior to crystallization were the viscosity measurements of AuGeSi [20, 21], $Pd_{40}Ni_{40}P_{19}Si_1$ glasses [17, 18], and $Pd_{40}Ni_{40}P_{20}$ [22]. Therefore, isothermal viscosity measurements of glassy metallic alloys are more suitable to study the relaxation phenomena in these thermodynamically unstable systems than to gain useful information about their viscous flow behavior in the temperature range T_m–T_g. The authors of [23–36] have shown that the experimental difficulties, imposed by crystallization, can be somewhat overcome by measuring η under continuous heating rather than under isothermal conditions. At temperatures lower than T_g, information could be obtained about the viscous behavior of the glassy alloy in the frozen "isoconfigurational" state. This is so because the time needed for the occurrence of substantial structural relaxation under continuous heating conditions is much longer than the time needed to perform the creep experiment.

Several authors before us have described experimental equipments and methods for viscosity determination of amorphous metallic alloys under continuous heating conditions. Anderson and Lord [23] have developed an experimental equipment for viscous flow measurements under rapid heating conditions with heating rates up to 200 K/s. They used a vertical load train with stainless steel grippers of the specimen. A force has been exerted on the glassy alloy ribbon by a calibrated spring. The ribbon has been rapidly heated by passing an electric current through it and as the ribbon flowed due to the stress applied by a calibrated spring; its length change was monitored by a linear voltage displacement transformer (LVDT). A bridge has been used by Anderson and Lord [13] to measure resistivity of the ribbon in order to determine its temperature and heating rate at the beginning of the flow. The output voltage from the LVDT (proportional to the ribbon elongation) and the sample electric resistivity have been recorded on a storage oscilloscope. The whole scheme of the experiment is quite sophisticated, what leads to not sufficiently high accuracy of measurements. Nevertheless, Anderson et al. reported about measured values of viscosity for METGLAS 2826 amorphous alloy up to temperatures 40 K higher than the glass transition temperature T_g and determined the approximate critical cooling rate of the alloy studied. In later studies Anderson and Lord [27] disowned this experimental scheme and performed the sample heating in an electro-resistivity furnace. With the aid of program-controlled furnace voltage supply, they achieved constant heating rates from 1 to 25 K/min.

Vlasak et al. [28] have described an apparatus for thermal dilatation measurements including induced volume changes, i.e., viscosity and creep measurements. For thermal dilatation, viscosity, and creep measurements, the sample is placed in a heater. The heater is wound bifilarly from a Kanthal ribbon on a fused silica glass tube surrounded by a ceramic shield in such a manner as to obtain a homogeneous zone longer than 100 mm. The temperature within the oven is controlled by a thyristor regulator. A platinum resistor, placed directly on the heater, is used as a temperature control probe. The temperature of the sample is measured by a chromel–alumel thermocouple touching the sample in the middle. The elongation of the sample placed in the oven is transformed into a capacity change of the capacitor used in the capacity displacement transducer. The resolution of the transducer is reported to be 2×10^{-7} m and 2×10^{-8} m for linear heating and isothermal regimes, respectively. Heating rates from 5×10^{-3} to 5 K/s in the temperature range 300–1300 K are available. No information is given about the existence of loading train, about the maximal displacement possible to be measured, as well as about any viscosity measurements performed with the aid of this apparatus.

Taub [29] has measured the viscous flow of $Fe_{40}Ni_{40}P_{14}B_6$ and $Fe_{78}B_{13}Si_9$ glassy alloys near their glass transition temperatures using continuous healing. The flow tests have been performed by means of an Instron static tensile-testing machine set in the load-maintaining mode. In this mode, limit switches automatically drive the crosshead up and down to maintain the load within the desired range. A three-zone split oven centered along the length of the specimen has been used to heat the ribbon. Each zone of the furnace has been independently controlled in order to maintain a flat temperature profile over the entire hot zone length of 25 cm. Heating with constant heating rates up to 140 K/min has been maintained with a Model 5600 data track (Research Inc.) and Love temperature controllers. In this study Taub [29] has observed a decrease of viscosity with temperature at a rate that depends on the heating rate. Viscosities as low as 10^{10} Pa s have been obtained. No details are given about the way of data analyzing and viscosity calculating.

Bhatti and Cantor [30] have performed viscous flow experiments on $Fe_{78}B_{13}Si_9$ glassy alloy, using continuous heating conditions to obtain viscosity data in the vicinity of the glass transition temperature as a function of applied stress, heating rate, and prior annealing (thermal history) of the glassy alloy. A DuPont 1090/943 thermomechanical analyzer has been used. Specimens of 1.5–2-mm initial length have been clamped between a fixed book and a mobile hook attached to a tension probe, with a feedback system, designed to maintain a contact stress in the range 13–26 MPa during testing. Changes in the sample length have been monitored by a LVDT as a function of temperature, during continuous heating under argon at a heating rate in the range 15–75 K/min. No details have been given about data analysis and viscosity calculations.

Jingtang et al. [31] have studied the viscous flow behavior of $Ni_{89.3}P_{10.7}$ and $Fe_{78}Si_9B_{13}$ glassy alloys under continuous heating conditions using Perkin-Elmer thermomechanical analyzer TMS-2. The flow tests have been performed under tensile and compressive stresses using home-made silica glass assemblies especially designed for the tests. The authors claim that the viscosity values obtained and their temperature dependence may be extended to the liquid state of the glassy alloys studied according to the formula

$$\eta = \eta_o A\ \exp\left\{\frac{B + bT^{(1-1/T)}}{T - T^o_{vs}}\right\}. \tag{3.6}$$

No comments are given about the meaning and the magnitude of the coefficients η_0, A, B, b, and T^o_{vs} in Eq. (3.6) as well as about the data analysis and viscosity calculations.

Fu-Qian [32] has studied the viscous flow of the metallic glass $Ni_{30}Zr_{70}$ under continuous heating using a self-made small tensile-testing machine. No details are given about the experimental scheme except the information that the tensile stresses applied have been in the range 1–20 MPa and that the linear heating has been maintained for four heating rates in the range 0.8–30 K/min. The elongation of the specimens has been measured by means of LVDT. No details are given about viscosity calculations.

Myung et al. [33, 34] have studied the glass transition and viscous flow behavior of $Fe_{83\text{-}x}Me_xP_{17}$ (Me = Cr, V or Mo) and of metallic compositions of WC/amorphous $Ni_{73}Si_{10}B_{17}$ using thermomechanical analyzer (TMA) to clarify the effect of M-substitution and of the volume fraction of WC dispersed in the amorphous NiSiB matrix on the viscous flow features of the glassy alloys studied. The stresses, applied to induce viscous flow, have been in the range 2.5–20 MPa. No details are given about the experimental equipment used and no analysis has been performed in order to separate the contribution of viscous flow to the overall strain of the specimens studied from the strain contributions of other phenomena such as thermal expansion, anelasticity, relaxation, etc.

Russew et al. have recently reported a large number of experimental studies on the viscous flow behavior of $Cu_{56}Zr_{44}$ [25, 35], $Fe_{82}B_{18}$ [36], $Pd_{82}Si_{18}$ [37], $Fe_{82\text{-}x}V_xB_{18}$ [38], $Fe_{82\text{-}x}Cr_xB_{18}$ [39], $Fe_{40}Ni_{40}Si_6B_{14}$ [40], Ni_xZr_{100} [41], $Ni_{100\text{-}x}P_x$ [42], $Fe_{100\text{-}x}P_x$ [43], etc., glassy metallic alloys using Perkin-Elmer TMS-2 or Heraeus TMA 500 thermomechanical analyzers under continuous heating conditions. In these studies a comprehensive discussion is carried out about the experimental details and data evaluation in order to obtain reliable viscosity values.

From the foregoing survey, it can be recognized that the scope of activity in studying the viscous flow behavior of glassy alloys under continuous heating conditions is so large at present that no single source can call itself definitive. In most cases, however, there is a lack of sufficient information about the experimental equipment used and about the way in which the experimental data are treated and the viscosity evaluated. In the next book section, the reader will find a discussion of the most suitable experimental techniques for studying the high temperature creep of glassy alloys under continuous beating conditions. He/she will also find an analysis of the different contributions to the overall strain observed through such experiments in order to obtain clear understanding about the viscous flow contribution, itself to the overall strain and to calculate the viscosity values in a proper way.

This analysis will be represented in such a way that the unexperienced researcher can obtain the information he/she needs to carry out his/her experiments when studying the viscous flow behavior of a specific glassy alloy with the degree of precision and care pertinent to his/her objective.

Experimental Techniques for Viscosity Study of Metallic Glasses Under Nonisothermal Conditions

It becomes clear from the foregoing section that in studying the high temperature creep of glassy alloys, standard equipments for thermomechanical analysis manufactured by different firms, i.e., Perkin-Elmer, for studies discussed in [31, 35–41], Heraeus, for studies discussed in [41], DuPont, for studies discussed in [30], and Netzsch, are most frequently used.

All experimental systems for studying the viscous flow behavior of glassy alloys under isothermal and/or continuous heating conditions, regardless of their complexity, consist of some basic components, such as:

- System for analyzing unit consisting of a rigid frame, grippers, specimen, loading system, probe, and displacement-signal convertor.
- Heating unit including furnace with homogeneous zone longer than the specimen itself with constant temperature throughout the whole temperature interval, and temperature regulator, which ensures a constant temperature or a constant heating rate.
- Control temperature programming and monitoring of the specimen elongation under constant applied load, data acquisition, and processing unit.

The thermomechanical analyzing systems measure dimensional changes of a sample as a function of temperature and applied load. The research experience of the authors in the field of viscous flow behavior of glassy metals is mainly connected with Perkin-Elmer TMS-2 and Heraeus TMA 500 thermomechanical analyzers. This is why these equipments will be used as an example when discussing the most important features and measurement principles of TMA devices.

Figure 3.3 shows the schematic of a Perkin-Elmer TMS-2 analyzing unit.

All changes of sample dimensions are tracked by the probe rod, the movement of which is converted into a signal by a linear variable differential transformer (LVDT). As LVDT is most frequently used by displacement measurements, a short description of its measurement principle will be given here. The LVDT transducer is an electromechanical device which provides an output voltage proportional to the displacement of a movable Fe core. The transformer consists of a primary and two secondary wound to a common cylindrical form and a core which is axially oriented within the hollow body of the transformer. The construction of the transformer and the connection of the secondary windings are such that if the iron core is placed in a position which induces equal electromotive forces in the two

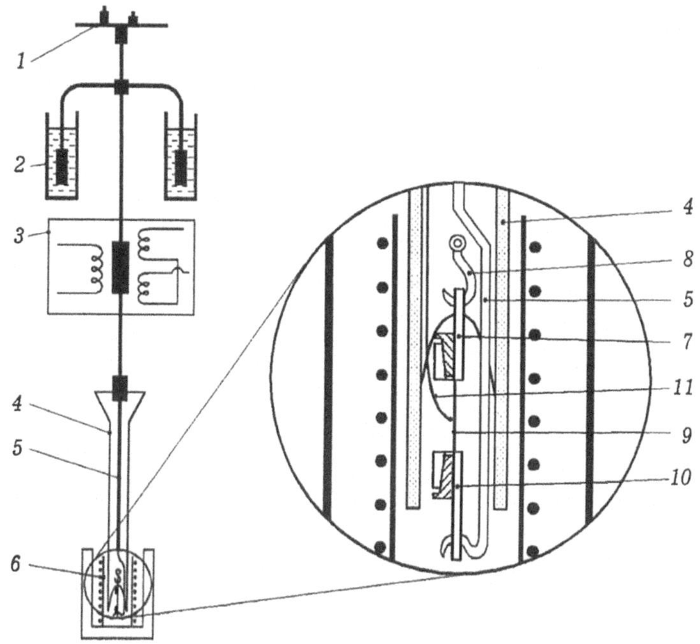

Fig. 3.3 A schematic of the Perkin-Elmer TMS-2 analyzing unit with a home-made silica glass assembly for high temperature creep measurements: (*1*) weight tray, (*2*) lifting float, (*3*) LVDT (linear voltage displacement transformer), (*4*) supporting silica glass tube (rigid frame), (*5*) silica glass probe with a hook at the end applying load and connecting moving grip to LVDT, (*6*) furnace, (*7*) unmovable wedge-shaped grip of Invar alloy, (*8*) stationary silica glass hook, (*9*) specimen, (*10*) movable wedge-shaped grip of Inver alloy, and (*11*) thermocouple

secondary windings, the net output voltage will be zero [44]. If the core is displaced toward one end of the unit, mutual inductance increases between the primary and one secondary and decreases between the primary and the other secondary – the output electromotive forces of the two secondary windings are no longer equal and a net output voltage results. The lower portion of the fixed silica glass sample tube together with the gripping assembly and the sample is positioned inside a cylindrical furnace with heating element which functions both as a heater and a resistance thermometer. The temperature of the sample space is monitored by a thermocouple. Constant heating rates up to 40 K min^{-1} in the temperature range 300–1000 K can be used.

The three-point bending assembly consists of a small piece of silica glass tube of 8 mm outer diameter with a groove, close to the edge of its upper horizontal cut.

The ribbon tested is stretched along the diameter of the upper cut of the silica glass tube. Both endpoints of the ribbon are fixed by a twisted platinum wire aligned in the groove. The wedge-shaped silica glass probe is positioned in contact with the middle of the ribbon, the initial length $2l_o$ of which is 6.5 mm. Preliminary experiments [36] have shown that irrespective of the good stretching capability of

Fig. 3.4 Magnified scheme of the three-point bending assembly

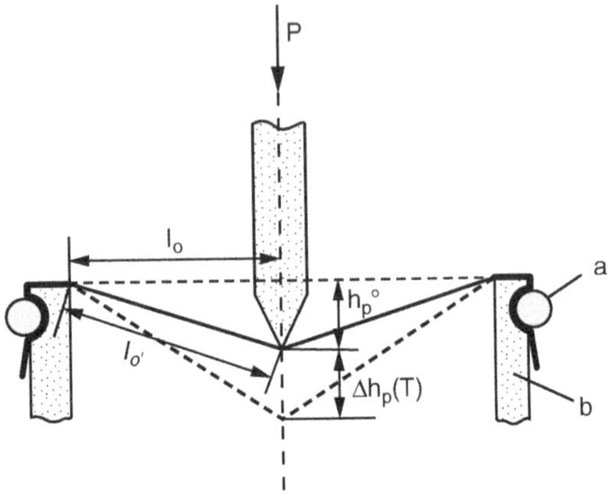

the ribbon between both end grippers, minimal spring effects existed which could only be eliminated at room temperature after applying a load of 0.02 kg at the middle of the ribbon. When using a zero load or loads lower than 0.02 kg, these effects disappeared in an uncontrollable, continuous way with increasing temperature, which led to unreliable results. That is why it has not been possible to measure the length changes of the ribbon by heating it under zero load. Such changes are expected to be due only to thermal expansion and possible structural relaxation. An understanding of these changes is needed in order to correct the temperature dependence of the ribbon elongation under applied load P and to obtain this part of the elongation which is due only to the viscous flow. These shortcomings can be overcome by measuring the deflection $h_p{}^\circ$ of the middle point of the ribbon under applied load P at room temperature (the initial length l_0 changes to l_0' (see Fig. 3.4)) and the subsequent change $\Delta h_p(T)$ of the deflection by a constant heating rate from room temperature to temperatures higher than the onset temperature of crystallization of the glassy alloy studied. Another problem which must be taken into account when using the three-point bending assembly for viscosity measurements is the circumstance that during the measurements, the shear stress, inducing viscous flow of the glassy ribbon tested, depends on the temperature via the deflection $h_p(T)$ reached at a definite temperature. It was shown [36] that under the experimental conditions of continuous heating using three-point bending assembly, the absolute value of the shear stress $\tau(T)$, developed under an applied load P at the middle of the ribbon, is given by the expression

$$\tau(T) = \frac{\sigma(T)}{3} = \left(\frac{Pg}{6S}\right)\left\{\frac{\left[l_o^2 + h_p^2\right]^{0.5}}{h_p(T)}\right\}, \tag{3.7}$$

where $\sigma(T)$ is the tensile stress, g is the gravity constant, S is the cross-sectional area of the ribbon tested, and $h_p(T) = h_p{}^\circ + \Delta h_p(T)$. This is illustrated in Fig. 3.5 with $Pd_{82}Si_{18}$ glassy alloy as an example [37]. The advantage of the above-described

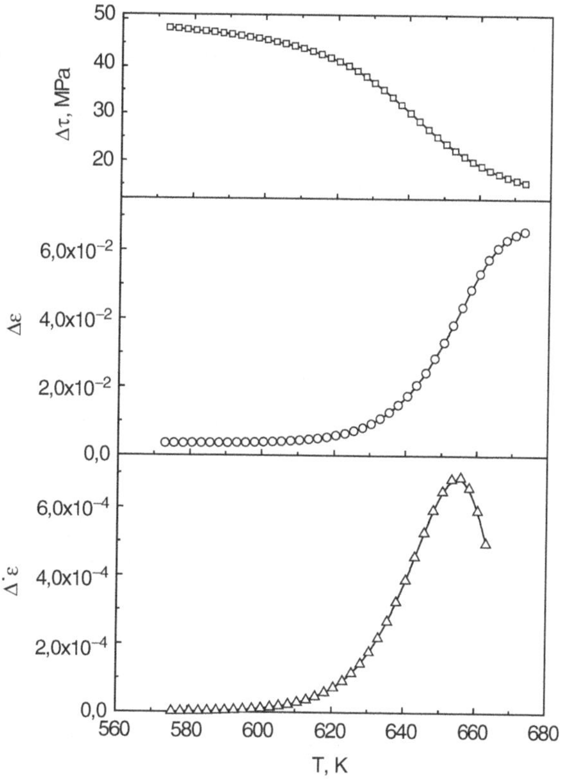

Fig. 3.5 Temperature dependence of the deformation $\Delta\varepsilon$, deformation rate $\Delta\dot{\varepsilon}$ and the tangential stress $\Delta\tau$, causing the viscous flow into the amorphous metallic alloy $Pd_{82}Si_{18}$ studied with the aid of a three-point bending assembly

three-point bending assembly consists in its relative simplicity and the possibility to make it of easily available means. Its disadvantage consists in the dependence of the applied shear stress τ on the instantaneous value of the ribbon deflection $h_p(T)$ reached during the experiment.

Another possibility to perform viscous flow measurements of glassy metallic alloys using silica glass dilatometer Heraeus TMA 500 with standard silica glass assembly for tensile creep measurements has been reported by Russew et al. [41]. Similar assembly for tensile creep measurements is offered by the firm Netzsch as a standard accessory of the silica glass dilatometers Netzsch TMA 402. A home-made assembly for tensile creep measurements was adapted by the authors to be used in combination with a Perkin-Elmer TMS-2. The schematic of the assembly itself is shown in Fig. 3.3. The ribbon sample is stretched between one stationary and one movable grip of Invar alloy. The load is applied to the sample via a specially shaped silica glass rod with a hook at the end which acts simultaneously as a probe, connecting the movable grip to the core of LVDT. The temperature calibration of the equipment has been carried out by using strips of pure metals Sn, Pb, and Zn (see Fig. 3.6).

The simple scheme of loading and the constant value of shear stress throughout the time of the creep experiment is the obvious advantage of the silica glass assembly for tensile creep measurements. The necessity to use sample grippers of

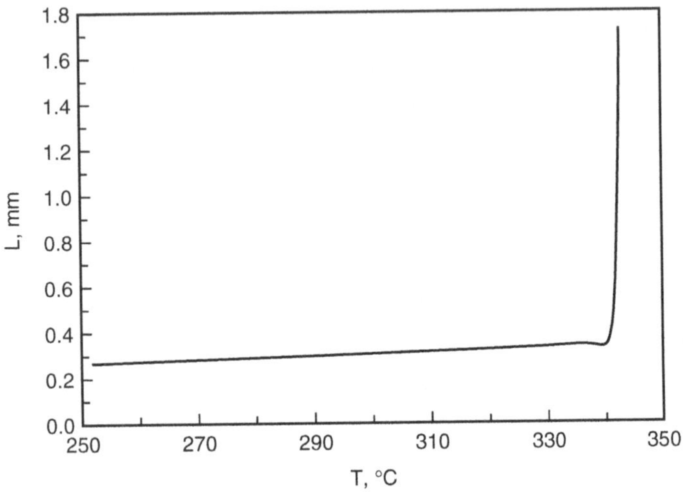

Fig. 3.6 Experimentally registered *curve* for temperature calibrating the device by using a sample strip of pure Pb. The heating velocity is 20 K/min. The difference between $T_m^{measured}$ and T_m^{real} is 338.0–327.5=11.5°C, e.g., the correction needed is −11.5 K

Invar alloy represents the main disadvantage of this assembly type as the thermal expansion of these grippers, although relatively low, must be amounted to and taken into account together with the thermal expansion of glassy alloy ribbon when considering the overall elongation and strain of the specimen. This has to be done in order to separate for further evaluation that part of the overall strain which is due to viscous flow contribution only. The main details of both assemblies described above are shown in Figs. 3.3 and 3.4.

Analysis of the Overall Deformation Curves Obtained Under Continuous Heating Conditions and Viscosity Calculations

Regardless of the loading scheme, the overall strain of a glassy alloy ribbon reached at temperature *T* under applied tensile stress under continuous heating conditions can be represented as

$$\varepsilon(T) = \frac{l(T) - l_o}{l_o} = \varepsilon_\sigma^{el}(T) + \varepsilon_\sigma^{an} + \varepsilon_\sigma^{rel}(T) + \varepsilon_\sigma^{te}(T) + \varepsilon_\sigma^{vf}, \tag{3.8}$$

where l_o and $l(T)$ are the initial length and the length of the specimen at temperature T, $\varepsilon_\sigma^{el}(T) = \sigma/E(T)$ represents the elastic strain of the ribbon with $E(T)$ the Young's modulus of the material, and $\varepsilon_\sigma^{an}(T)$ represents the possible anelastic contributions to the overall strain, $\varepsilon_\sigma^{rel}(T)$ takes into account the contribution of any relaxation effects to the overall strain, $\varepsilon_\sigma^{te}(T)$ represents the contribution of the thermal expansion to the overall strain, and $\varepsilon_\sigma^{vf}(T)$ takes into account the contribution of viscous flow to the

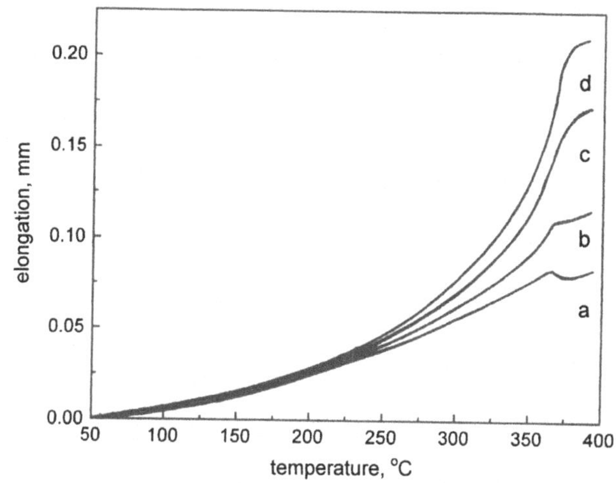

Fig. 3.7 Temperature (time) dependences of the elongation of $Ni_{81.3}P_{18.7}$ glassy alloy by applying four different loads: *a* 0.005 kg, *b* 0.050 kg, *c* 0.100 kg, and *d* 0.150 kg

overall strain. Typical experimental elongation $(l(T) - l_0)$ temperature (time) curves are shown in Fig. 3.7 with $Ni_{81.3}P_{18.7}$ glassy alloy as an example [42].

As is seen, in the low temperature range (up to approximately 523 K), the curves registered under the different loads coincide practically with each other. Quite visible influence of the load is demonstrated in the high temperature range of measurements, where the contribution of viscous flow to the overall elongation (deformation) is particularly substantial. The reason becomes clear from the following considerations, with the $Fe_{82}B_{18}$ glassy alloy as an example [36].

(A) According to Hausch and Török [45], Young's modulus *E(T)* of as-quenched $Fe_{82}B_{18}$ amorphous alloy is approximately 139 GPa at 610 K and 137 GPa at 713 K, i.e., the temperature dependence of E is very weak. This is a common feature for Young's modulus of metals and alloys at temperatures much lower than their melting point [46]. As the maximal absolute values of the stresses by our creep experiments do not usually exceed 12 MPa, the magnitude of the elastic strain contribution to the overall strain is estimated to be lower than 1×10^{-4} in most cases and thus can be neglected.

(B) The anelastic contribution $\varepsilon_{\sigma}^{an}(T)$ to the overall strain almost certainly cannot be neglected at lower temperatures. One might expect, however, that the anelastic response near T_g (at temperatures around (T_g -50)K and higher) does not significantly affect the viscosity measurements. Volkert and Spaepen [18] have shown that the pre-annealing of a $Pd_{40}Ni_{40}P_{19}Si_1$ amorphous ribbon under the test stress at temperature of 527 K, which is 50 K lower than T_g of this amorphous alloy, has no observable effect on the subsequent viscosity measurements. This procedure was performed by Volkert and Spaepen (1988) in order to anneal out the transients resulting from anelastic processes.

Later on, when studying the relaxation of bend stresses in metallic glasses (see ▶ Chap. 7, section "Relaxation of Bend Stresses in Ribbonlike Amorphous Metallic Alloys: Model Considerations," and author's article [1]), we have doubtless shown that the anelastic deformation fully disappears at temperatures much lower than the temperature at which viscous flow becomes experimentally detectable. Hence, $\varepsilon_{\sigma}^{an}(T)$ can be neglected. Since the coefficient of thermal expansion [7] and the rate of relaxation [15] are independent of the stress, the strain contributions $\varepsilon_{\sigma}^{te}(T)$ and $\varepsilon_{\sigma}^{rel}(T)$ to the overall strains $\varepsilon_{\sigma I}(T)$ and of ribbons tested under conditions differing only in the applied tensile stresses, $\sigma_I(T)$ and $\sigma_2(T)$, should be equal. Taking into account these considerations, one obtains

$$\Delta\varepsilon_{1,2}(T) = \varepsilon_{\sigma 1}(T) - \varepsilon_{\sigma 2}(T) \approx \varepsilon_{\sigma 1}^{vf}(T) - \varepsilon_{\sigma 2}^{vf}(T), \tag{3.9}$$

where $\Delta\varepsilon_{I,2}(T)$ is caused by an effective stress $\sigma_{I,2} = \sigma_I - \sigma_2$. Applying the Newtonian relation for viscous flow in Eq. (3.9) and taking into account that the shear stress

$\tau = \sigma/3$, one obtains

$$\Delta\dot{\varepsilon}_{1,2} = \frac{\Delta\tau_{1,2}}{\eta(T)}, \tag{3.10}$$

where $\Delta\dot{\varepsilon}_{1,2}(T) = \dot{\varepsilon}_1(T) - \dot{\varepsilon}_2(T)$ is the difference of strain rates caused by the applied tangential stresses τ_I and τ_2, and $\Delta\tau_{I,2} = \tau_I$-$\tau_2$. The validity of Eq. (3.10), which supposes a linear dependence of $\Delta\varepsilon(T)$ upon $\Delta\tau$, can be proven via a graphical presentation of $\Delta\varepsilon(T)/\Delta\tau$ and is shown in Fig. 3.8, again with the amorphous $Ni_{81.3}P_{18.7}$ as an example [42].

This confirms the expectations that under continuous heating conditions, using the experimental equipment described, it is actually possible to separate the viscous strain contribution from the overall strain of glassy alloys in order to gain useful information about the magnitude and the temperature dependence of the viscosity in the vicinity of the glass transition temperature T_g, including the temperature interval $T_g \pm 30$ K. In the next section, we shall try to clarify this possibility with the aid of several examples of viscous flow studies performed by the authors on different glassy alloy systems in order to gain information about the influence of different factors on the viscous flow behavior of glassy alloys.

Estimation of the Critical Melt Cooling Rate for Obtaining the Alloy in Glassy State

As it was discussed earlier, one of most important characteristics of metallic glasses is their so-called critical cooling rate, which determined the possibility itself for their laboratory and industrial production. For the determination of the critical cooling rate of a definite glass-forming melt, the method of Uhlmann [2, 3] is

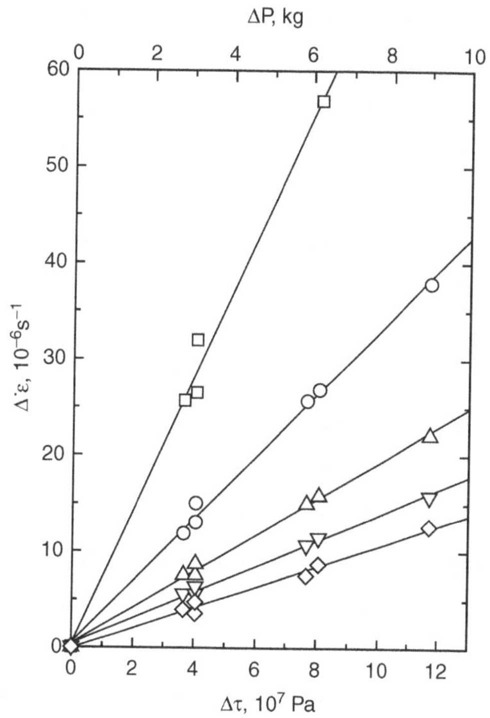

Fig. 3.8 Linear dependence of the strain rate $\Delta\dot{\varepsilon}$ on the tangential stress $\Delta\tau$ at different temperatures: ◇-563 K, ∇-583 K, Δ-603 K, O-623 K, and □-643 K. The glassy alloy $Ni_{81.3}P_{18.7}$ is used as an example

usually used (see Eqs. (3.1) and (3.2)). For this purpose one needs to introduce into Eq. (3.1) the proper viscosity temperature dependence of the undercooled melt, obtained from direct or extrapolated experimental data, which should be as close as possible to the real temperature dependence of the quasi-equilibrium viscosity in the temperature range T_m–T_g. The authors will try to demonstrate this approach by using as an example the study upon the amorphous $Fe_{82}B_{18}$ metallic alloy [36]. The viscosity temperature dependence determined at a heating rate of 20 K/min is shown in Figs. 3.9 and 3.10. The viscosity curve shows a typical form with two almost linear parts crossing each other at the glass transition temperature T_g, and a higher temperature segment of the temperature dependence after the onset temperature of crystallization T_{on}. In this section the viscosity grows up, due to the appearance of crystalline regions, imbedded into the amorphous parent phase. This can be taken into account by using the equation of Einstein for viscous flow of suspension in a viscous medium:

$$\eta_{eff} = \eta(1 + 2.5\varsigma). \tag{3.11}$$

In the temperature range T_g – (T_g + 30 K), the temperature dependence of the viscosity was corrected to take into account the volume fraction of the imbedded

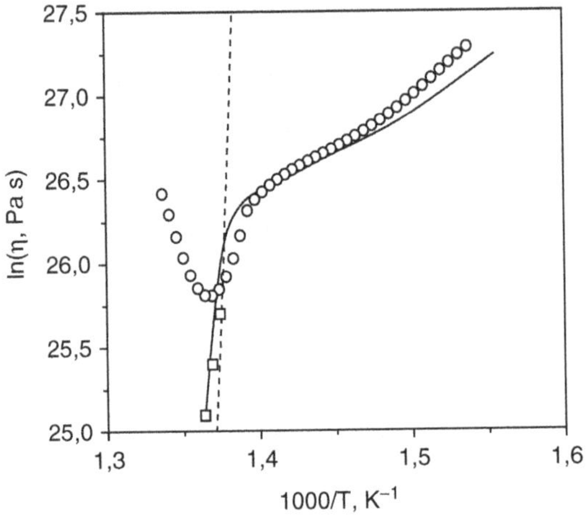

Fig. 3.9 Temperature dependence of the viscosity of the amorphous alloy $Fe_{82}B_{18}$ at heating rate 20 K/min. O – without a correction for crystallization influence, □ – with a correction by using the equation of Einstein for viscous flow of suspension in a viscous medium, – – – temperature dependence of the quasi-equilibrium viscosity η_{eq}, – temperature dependence of the nonequilibrium viscosity η. The curves are calculated with the aid of (▶ Eqs. 5.8), (▶ 5.17), and (▶ 5.9)

into the amorphous parent-phase growing crystalline particles. After this correction the temperature dependence of the viscosity can be presented with an empirical Arrhenius-type equation as follows:

$$\ln\,(\eta, \text{Pa s}) = -11.762 + 226719/RT, \tag{3.12}$$

with a correlation coefficient of 0.991. The viscosity temperature dependence of $Fe_{82}B_{18}$ metallic alloy in the temperature interval T_m–T_g in the coordinates of the Vogel–Fulcher–Tammann was calculated on the assumption that in the melting point T_m (1440 K), the melt viscosity of $Fe_{82}B_{18}$ metallic alloy is almost equal to viscosity of pure Fe in its melting point. It varies between 0.005 and 0.12 Pa s [47, 48]:

$$\ln\,(\eta, \text{Pa s}) = -8.9 + 2864/(T - 645), \tag{3.13}$$

$$\ln\,(\eta, \text{Pa s}) = -8.9 + 3663/(T - 622). \tag{3.14}$$

For comparison purposes, the VFT equation proposed by Davies [1] for the amorphous $Fe_{83}B_{17}$ is given:

$$\ln\,(\eta, \text{Pa s}) = -8.2 + 4630/(T - 638). \tag{3.15}$$

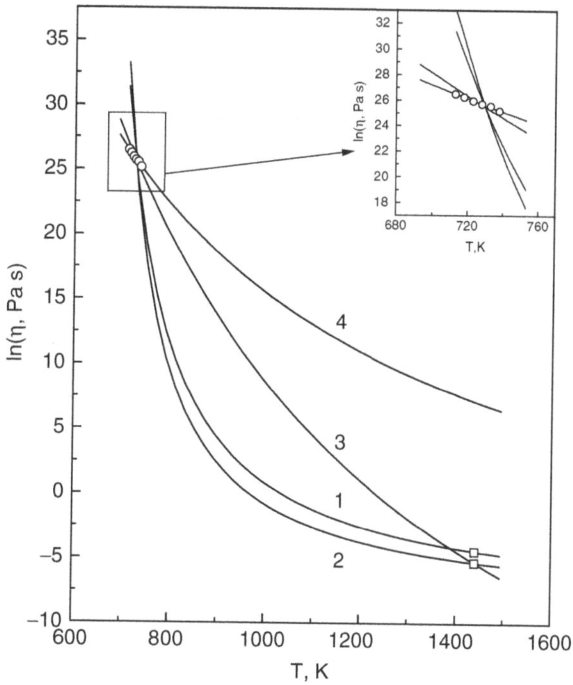

Fig. 3.10 Viscosity temperature dependence *η(T) of* $Fe_{82}B_{18}$ metallic glass in coordinates *ln(η)/T*: *1* according VFT – Eq. (3.13), *2* according VFT – Eq. (3.14), *3* according Arrhenius – Eq. (3.16), *4* according Arrhenius – Eq. (3.12). O – experimentally determined viscosity, □ – melt viscosity at 1440 [47]

The equation of Arrhenius-type, which takes into account a probable value of 0.005 Pa s for the viscosity in the melting point (1440 K) of the alloy studied, is

$$\ln(\eta, \text{Pa s}) = -36.804 + 377719/\text{R}T, \tag{3.16}$$

with a correlation coefficient of 0.998. The presumable temperature dependences of the $Fe_{82}B_{18}$ metallic glass in the temperature range T_m–T_g, according to Eqs. (3.12), (3.13), (3.14), and (3.16), are shown in Fig. 3.10.

The above-presented viscosity temperature dependences were used for estimation of the melt critical quenching rate $\dot{T}_{cr}$ according to the method of Uhlman [2, 3] – Eqs. (3.1) and (3.2).

By the calculation of the time t needed for the crystallization of a tiny volume fraction $\xi = 10^{-6}$ at a temperature T, it was accepted that the parameter f in Eq. (3.1) equals 1 (rough crystalline surface, each atom in contact with the crystalline nucleus, is incorporated into the crystalline structure). By analogy to Anderson and Lord [27], the term 1.024 in the first exponent of Eq. (3.1), which corresponds to the free energy change for creation of a critical crystalline nucleus $\Delta G^* = 50\ kT$ at $\Delta T_r = 0.2$ in the original equation of Uhlmann [2], was replaced with value of 1.331. This corresponds to nucleation free energy change $\Delta G^* = 65\ kT$. According to Onorato and Uhlmann [3], this value of ΔG^* leads to better

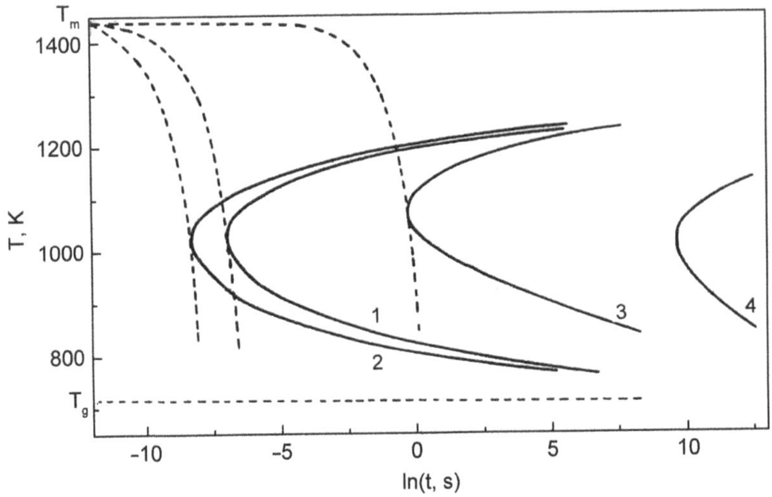

Fig. 3.11 Temperature–time–transformation (TTT) curves for estimation of the critical quenching rate for $Fe_{82}B_{18}$ glass-forming melt, calculated according to Eqs. (3.12), (3.13), (3.14), and (3.16), respectively. The curves corresponding to sequence of equations are *4*, *1*, *2*, and *3*, respectively

agreement between the estimated critical melt-quenching rates and the laboratory experimental tests with glass-forming metallic melts.

Figure 3.11 shows the TTT curves, calculated on the basis of viscosity values, determined with the aid of Eqs. (3.12), (3.13), (3.14), and (3.16).

As is seen, the derived empirical VFT equations lead to estimated values for the critical melt-quenching rate $\dot{T}_{cr} = 2.13 \times 10^6 \mathrm{K/s}$ and $\dot{T}_{cr} = 4.5 \times 10^5 \mathrm{K/s}$, respectively. Such critical melt-quenching rates are in accordance with the experimentally determined critical quenching rate [1] for the glassy alloy studied. On the contrary, the empirical Eqs. (3.12) and (3.16) of Arrhenius-type lead to unrealistically low critical quenching rates of $\dot{T}_{cr} = 2.6 \times 10^{-2} \mathrm{K/s}$ and $\dot{T}_{cr} = 4.8 \times 10^2 \mathrm{K/s}$, respectively. Irrespective of the circumstance that the viscosity data in the temperature range T_g–T_m are best fitted by the equations of Arrhenius-type (curves 3 and 4 in Fig. 3.11), it is quite clear, that the viscosity temperature dependence in the very broad temperature range T_g–T_m cannot be described by a simple extrapolation of Eqs. (3.12) and (3.16). Most probably, it must be assumed that close to the glass transition temperature under nonisothermal conditions of the experiment, the glass-forming system "tries" to reach its quasi-equilibrium structure with the corresponding quasi-equilibrium viscosity η_{eq}. This viscosity possesses temperature dependence of VFT type. Unfortunately, the quite low thermal stability of the alloy studied makes it impossible to determine experimentally the viscosity values at temperatures higher than the temperature of maximal crystallization rate T_{max}. It could be concluded that the empirical VFT equations derived by the authors present a good approximation to the real viscosity temperature dependence of the amorphous system $Fe_{82}B_{18}$ in the temperature range T_g–T_m. The values of the constant

Table 3.1 Characteristic parameters of viscous flow of the metallic glass $Fe_{82}B_{18}$ at a heating rate 20 K/min – the data are interpreted in the framework of FVM

η_o, Pa s/K	ν_r, 1/s	Q_r, kJ/mol	Q_η, kJ/mol	B, K	T_o, K	$c_{f,o}$, -	T_g, K	$\eta(T_g)$, Pa s
4.6×10^{-21}	4×10^{23}	181.2	230	3017	615	4.9×10^{-11}	720	2.7×10^{11}

B^{VFT} in VFT – Eqs. (3.13) and (3.14) – 2864 ± 30 K and 3663 ± 37 K, respectively, are in good agreement with the value of the FVM model parameter $B = 3017$ K (see Table 3.1), derived by the aid of FVM (▶ Eqs. 5.8), (▶ 5.9) and (▶ 5.17) considered in ▶ Chap. 5 of the present book (see also the book Appendix).

The following conclusions can be made:

- The viscosity temperature dependence of the glassy $Fe_{82}B_{18}$ alloy (and generally for all possible metallic glasses) in the temperature range T_g–T_m is most realistically described by an empirical equation of Vogel–Fulcher–Tammann type.
- The viscosity determination of metallic glasses under continuous heating conditions with constant heating rate allows to obtain an information about their viscosity temperature dependence, not attainable with other means, in the high temperature range T_g–T_{on}.
- The method of Uhlmann, combined with reliable experimental viscosity data of a definite amorphous metallic alloy, allows to obtain good estimation of the critical melt-quenching rate needed for its vitrification

References

1. Davies HA (1980) In: Duhaj P, Mrafko P (eds) Proceedings of the conference on amorphous metallic materials'78. Veda Publ House, Bratislava, p 107
2. Uhlmann DR (1972) J Non-Cryst Solids 7:337
3. Onorato PIK, Uhlmann DR (1976) J Non-Cryst Solids 22:367
4. Köster U, Heroldt U (1981) In: Guentherodt HJ, Beck H (eds) Glassy metals I. Springer Verlag, Berlin, p 225
5. Battezzati L, Greer AL (1989) Acta Metall 37:1791
6. Chhabra RP, Sheth DK (1990) Zs Metallkde 81:264
7. Spaepen F, Taub AI (1983) In: Luborsky FE (ed) Amorphous metallic alloys. Butterworths, London, p 281
8. Cahn RW (1983) In: Cahn RW, Haasen P (eds) Physical metallurgy II. North-Holland Physics Publishing, Amsterdam, p 406
9. Davies HA (1983) In: Luborsky FE (ed) Amorphous metallic alloys. Butterworths, London, p 13
10. Taub AI, Spaepen F (1979) Scr Metall 13:195
11. Taub AI, Spaepen F (1980) Acta Metall 28:1781
12. Mulder AL, van der Zwaag S, Huizer E, van den Beukel A (1984) Scr Metall 18:515
13. Anderson PM III, Lord AE Jr (1980) Mater Sci Eng 44:279
14. Bouali A, Tete C (1988) Mater Sci Eng 97:473
15. Taub AI (1982) Acta Metall 30:2129
16. Huizer E, Mulder A, van den Beukel A (1986) Acta Metall 34:493
17. Volkert CA, Spaepen F (1989) Acta Metall 37:1355
18. Volkert CA, Spaepen F (1988) Mater Sci Eng 97:449
19. Volkert CA, Spaepen F (1990) Scr Metall 24:463
20. Chen HS, Turnbull D (1968) J Chem Phys 48:2560
21. Tsao SS, Spaepen F (1985) Acta Metall 33:881
22. Duine PA, Sietsma J, van den Beukel A (1992) Acta Metall Mater 40:743
23. Anderson PM III, Lord AE Jr (1980) J Non-Cryst Solids 37:217
24. Jingtang W, Dexing P, Bingzhe D, Maoshu B, Shuling L (1988) Acta Metall Sin 24:B374

25. Russew K, Stojanova L, Marchev K (1990) Viscosity measurements at a constant heating rate of amorphous $Cu_{56}Zr_{44}$ alloy by means of TM analyzer. Neue Huette 35(3):111
26. Jen SU, Chien MC, Lee CY (1990) Phys Stat Sol (a) 119:35
27. Anderson M III, Lord AE Jr (1980) Mat Sci Eng 43:267
28. Vlasak G et al (1983) J Phys E: Sci Instrum 16:1203
29. Taub AI (1985) In: Steeb S, Warlimont H (eds) Proceedings of the conference on rapidly quenched metals RQ5, vol 2. North Holland, Amsterdam, p 1365
30. Bhatti AR, Cantor B (1988) Mater Sci Eng 97:479
31. Jingtang W et al (1988) Mater Sci Eng 97:483
32. Fuqian Z et al (1988) Mater Sci Eng 97:487
33. Myung WN et al (1991) Mater Sci Eng A133:418
34. Myung WN et al (1991) Mater Sci Eng A133:513
35. Russew K, Marchev K (1987) Constant heating rate viscosity measurements of $Cu_{56}Zr_{44}$ amorphous alloy. In: Proceedings of the 1st international conference on rapidly quenched metals, Varna, 1987, Akademie der Wissenschaften der DDR, VEB Kongress- und Werbedruck, Oberlungwitz, p 114
36. Russew K, Stojanova L (1990) Viscous flow of $Fe_{82}B_{18}$ amorphous alloy under continuous heating conditions. Mater Sci Eng A123:59
37. Stojanova L, Russew K, Illekova E (1991) Study of the structural relaxation of $Pd_{82}Si_{18}$ metallic glass by thermal expansion and viscous flow measurements. Mater Sci Eng A133:529
38. Russew K, Stojanova L, Anestiev L et al (1989) Influence of vanadium alloying additions on the viscous flow and critical cooling rate of amorphous Fe-B alloys. In: Proceedings of the international conference on days of metallurgy 6, vol 1. Balatonaliga, Hungary, p 118
39. Russew K, Sommer F, Stojanova L (1993) Influence of Cr-alloying additions on the viscous flow behaviour of $Fe_{82-x}Cr_xB_{18}$ amorphous alloys. In: Duhaj P, Mrafko P, Svec P (eds) Proceedings of the conference on amorphous metallic materials AMM III. Trans Tech Publ, Slovakia, p 625
40. Russew K, Stojanova L, Lovas A (1993) Effect of processing conditions on the ribbon geometry and viscous flow behaviour of $Fe_{40}Ni_{40}Si_6B_{14}$ amorphous alloy. Int J Rapid Solidif 8:147
41. Russew K, Sommer F, Duhaj P, Bakonyi I (1992) Viscous flow behaviour of Ni_xZr_{100-x} metallic glasses from $Ni_{30}Zr_{70}$ to $Ni_{64}Zr_{36}$. J Mater Sci 27:3565
42. Russew K, Stojanova L (1993) Viscous flow behaviour and thermal stability of $Ni_{100-x}P_x$ metallic glasses from $Ni_{84}P_{16}$ to $Ni_{79}P_{21}$. Mater Lett 17:199
43. Russew K, Anestiev L, Stojanova L et al (1995) Thermal stability and viscous flow behaviour of $Fe_{100-x}P_x$ metallic glasses. J Mater Sci Technol 3(2):3
44. Swanson SR (1974) Handbook of fatigue testing, vol 566. ASTM Special Technical Publication, Philadelphia, p 28
45. Housh G, Török E (1985) In: Steeb S, Warlimont H (eds) Proceedings of the conference on rapidly quenched metals RQ5, vol 2. North Holland, Amsterdam, p 1341
46. Feltham P (1966) Deformation and strength of materials. Butterworths, Guildford, p 21
47. Barfield RN, Kitchener JA (1955) J Iron Steel Inst 180:324
48. Wertmann AA, Samarin AM (1969) Properties of liquid iron alloys. Nauka, Moscow, p 208 (In Russian)

Viscous Flow Behavior of Amorphous Ribbonlike Metallic Alloys Depending on Different Factors

4

Abstract

An attempt has been made to provide an information about the experimental evidence, as obtained by the authors, for the existence of a definite interrelation between the viscous flow features and the chemical composition of binary metal-metal, metal-metalloid, ternary metal-metal-metalloid, and quaternary glassy alloys. The thermal prehistory of metallic glasses and the influence of technological conditions by rapid quenching from the melt are also considered. An estimation of the critical quenching rate and its reliability for predicting the glass transition will also be given.

Keywords

Glassy metals • Amorphous • Viscosity • Glass transition • Alloy chemical composition • Critical quenching rate

Keeping in mind that this section has to stay within reasonable limits, one should not expect that it would be as general as the chapter title indicates. Therefore, rather than trying to cover all possible factors affecting the structure and, hence, the viscous flow of glassy metals, an attempt will be made to provide an information about the experimental evidence, as obtained by the authors, for the existence of a definite interrelation between the viscous flow features and the chemical composition of binary metal-metal (Ni-Zr) [1], metal-metalloid (Ni-P and Fe-P) [2, 3], ternary metal-metal-metalloid (Fe-TM-B, TM = V, Cr) [4, 5], and quaternary Al-based with RE additions (Al-Ni-Co-RE, Re = Ce, Gd, U) [6] glassy alloys. The pre-annealing (thermal history) of metallic glasses (Pd-Si and Fe-Ni-B) [7, 8] and the influence of technological conditions by rapid quenching from the melt (Fe-Ni-Si-B) [9] will be also considered. The estimation of the critical quenching rate [4, 5] and its reliability for predicting the glass transition will be given as well.

Beforehand, the stipulation should be introduced that in most cases the experimental results will be treated on an empirical level as the results have been initially

K. Russew, L. Stojanova, *Glassy Metals*,
DOI 10.1007/978-3-662-47882-0_4

published. Some of them are later on theoretically reconsidered in the framework of free volume model (FVM), further extended by the authors for description of the viscous flow behavior and other relaxation phenomena under nonisothermal (continuous heating) conditions.

Influence of Chemical Composition Upon the Viscous Flow Behavior and Thermal Stability of Some Binary Glassy Metals

Metallic glasses fall conventionally into two categories: metal-metalloid systems such as Fe-B, Co-P, Pd-Si, etc. and purely metallic systems such as Ni-Zr, Fe-Zr, Cu-Ti, Cu-Zr, etc. Of these two types of amorphous alloy, the former have received more attention because of their potential technical applications. However, the latter type is more attractive for studying of the fundamental properties of glassy metals, as the purely metallic amorphous alloys can usually be prepared over a wide range of compositions, making it possible to study systematic changes of properties as a function of composition within one alloy system.

Binary $Ni_{100-x}Zr_x$ Glassy Metals

In the case of the Ni-Zr amorphous system, such systematic studies have been carried out with respect to glass formation, stability [10], crystallization [11], hydrogen absorption [12], superconducting transition temperature and magnetic susceptibility [13, 14], etc. To our knowledge, no systematic studies of the viscous flow behavior as function of composition for Ni-Zr metallic glasses have been carried out, the only exception being the study of viscous flow of $Ni_{30}Zr_{70}$ metallic glass [15]. Here we describe a study of the viscous flow behavior of $Ni_xZr_{100\ x}$($x = 30, 40, 50, 64$ at.%) melt-spun amorphous alloys with the aid of nonisothermal dilatometry (Heraeus TMA 500) [1]. The glassy alloys used were melt-spun $Ni_{30}Zr_{70}$, $Ni_{40}Zr_{60}$, $Ni_{50}Zr_{50}$, and $Ni_{fi4}Zr_{36}$ narrow ribbons.

The temperature (time) dependence of viscous strains $\Delta\varepsilon_{0\ 05}$ obtained from the elongation-temperature (time) curves of the $Ni_xZr_{100\ x}$ samples, under applied loads of 0.1 and 0.5 kg, are shown in Fig. 4.1.

The amount of strain depends on both the applied stress and the alloy composition, but the composition dependence is stronger. For the case of $Ni_{40}Zr_{60}$ and $Ni_{64}Zr_{36}$ amorphous alloys, the strains are up to an order of magnitude higher than those observed for $Ni_{30}Zr_{?0}$ and $Ni_{50}Zr_{50}$ amorphous alloys. This can be explained by means of the Ni-Zr equilibrium phase diagram schematically shown in Fig. 4.2 and with the nature and number of crystallization products of glassy Ni-Zr. The $Ni_{40}Zr_{60}$ and $Ni_{64}Zr_{36}$ amorphous alloys possess nearly eutectic compositions. This correlates with better glass-forming ability and thermal stability. Our DSC results show that $Ni_{40}Zr_{60}$ and $Ni_{64}Zr_{36}$ exhibit complex DSC traces with three and two crystallization stages, respectively. The final crystallization products are $NiZr_2$ + NiZr and $Ni_{10}Zr_7 + Ni_2Zr$, respectively [16]. The complex structures and the

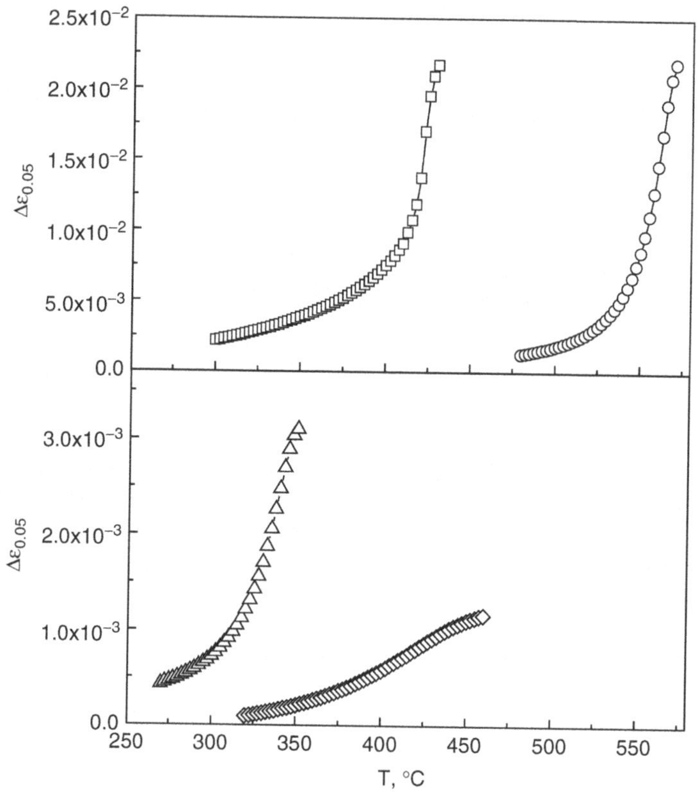

Fig. 4.1 Temperature (time) dependence of viscous strains $\Delta\varepsilon_{0\ 05}$ obtained from the elongation-temperature (time) curves of the $Ni_xZr_{100\ x}$ samples, under applied loads of 0.1 and 0.5 kg: Δ, $Ni_{30}Zr_{70}$; □, $Ni_{40}Zr_{60}$; ◇, $Ni_{50}Zr_{50}$; O, $Ni_{64}Zr_{36}$

number of crystallization products of the abovementioned glassy alloys require long-range atomic rearrangements, which are possible only when sufficiently high temperatures are reached (high thermal stability). The higher the temperature, the lower is the viscosity of the glassy structure. As a consequence of the low viscosity values, reached before the onset of crystallization, high strain levels are also reached before the crystallization starts. In the case of $Ni_{30}Zr_{70}$ and $Ni_{50}Zr_{50}$ glassy alloys, however, the compositions of which are very near to or practically coincide with the compositions of the peritectoid phases ($NiZr_2$ and NiZr, respectively; see Fig. 4.2); no long-range atomic diffusion of the large Zr atoms is needed for the formation of these crystalline compounds.

As a result, crystallization becomes possible at lower temperatures (low thermal stability) and at higher viscosity values. The amount of strain, reached before the onset of crystallization, is consequently much smaller than that reached before the onset of crystallization of nearly eutectic $Ni_{40}Zr_{60}$ and $Ni_{64}Zr_{36}$ glassy alloys.

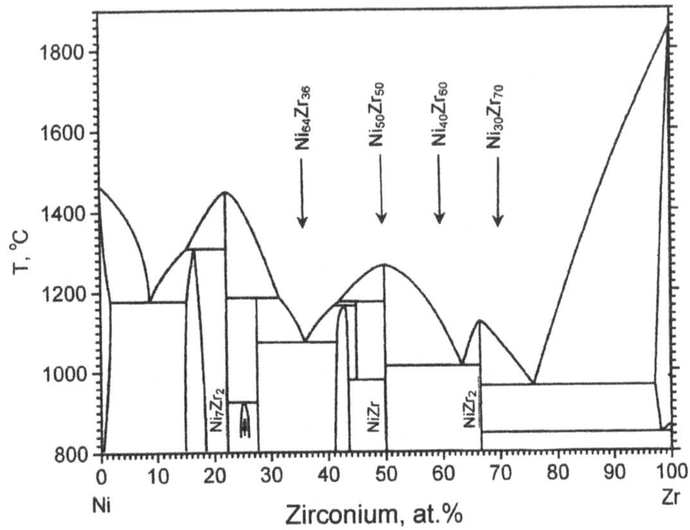

Fig. 4.2 Equilibrium phase diagram of Ni-Zr system (limited to the most important details)

Figure 4.3 shows the temperature dependence of the flow rates $\Delta\dot{\varepsilon}_{005}$ for the Ni-Zr glassy alloys studied.

All strain rate-temperature curves show an initial weak increase with temperature up to a specific temperature, which generally increases with increasing Ni content, after which a rapid increase in the strain rate is observed with an inflection point corresponding approximately to the onset temperature of crystallization. At temperatures which correspond approximately to the temperatures, T_{max}, of maximum heat evolution due to the first crystallization stage, the strain rates reach a maximum after which shrinkage due to crystallization prevails. The compositional dependence of strain rates is very well defined. The strain rate of $Ni_{40}Zr_{60}$ is almost 10^2 times higher than the strain rate of $Ni_{50}Zr_{50}$ glassy alloy. An explanation for this phenomenon is given above when considering the composition dependence of the strain (Fig. 4.1).

The temperature dependence of the apparent viscosity, η, of the glassy Ni-Zr alloys studied [1] is shown in Fig. 4.4 as a plot of ln η vs. $1000/T$. A common feature of all viscosity temperature dependences is the existence of two linear parts with different slopes. The crossover temperature is considered as the glass transition temperature T_g. At temperatures higher than the onset temperature of crystallization T_x, the values of ln η are influenced by the increasing volume fraction ζ of crystalline regions along with increasing temperature. As a result, viscosity values pass through a minimum and then increase rapidly. For the calculation of the true viscosity in this case, the Einstein equation for the flow of mixtures should be used [17]:

$$\eta_{eff} = \eta(1 + 2.5\xi), \tag{4.1}$$

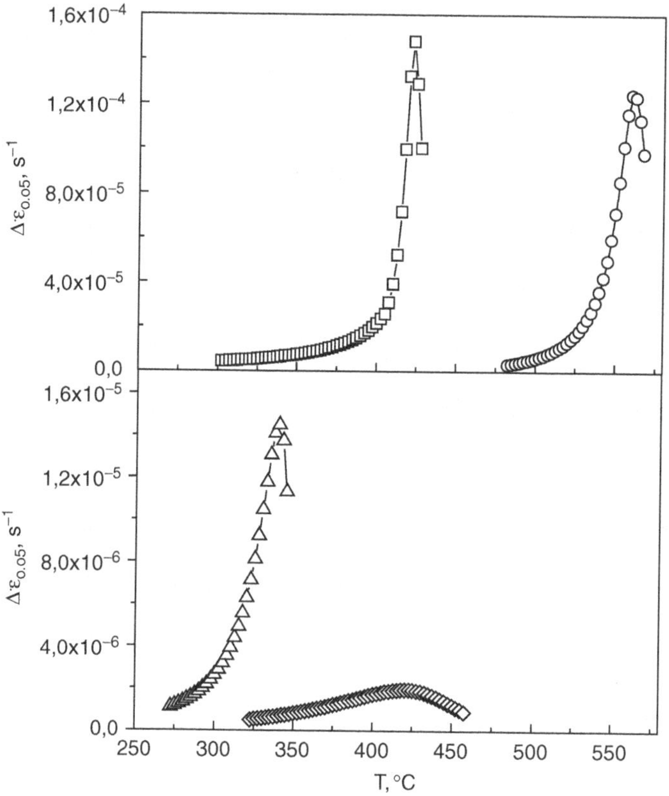

Fig. 4.3 Temperature dependence of strain rate (viscous flow rate) $\Delta\dot{\varepsilon}$ of amorphous alloys Ni_xZr_{100-x}: Δ, $Ni_{30}Zr_{70}$; □, $Ni_{40}Zr_{60}$; ◇, $Ni_{50}Zr_{50}$; O, $Ni_{64}Zr_{36}$

where ξ is the volume fraction of the suspended particles, η_o is the viscosity of the viscous medium, and η_{eff} is the viscosity of a mixture consisting of a small volume fraction of spherical particles suspended in the viscous medium. No corrections of the apparent viscosity values with Eq. (4.1) were made in this study, as the second linear parts of the viscosity temperature dependences (with the exception of $Ni_{50}Zr_{50}$ glassy alloy) were considered as well defined enough to calculate the activation energy for viscous flow from their slopes at temperatures higher than T_g, e.g., in the temperature range where the glassy alloys studied reach the quasi-equilibrium state of undercooled liquids.

The most characteristic parameters of crystallization and viscous flow processes in the Ni-Zr glassy alloys studied are the onset temperature of crystallization T_{on}, the temperature T_{max} at which maximum heat evolution due to the first crystallization stage is observed, the enthalpy ΔH_x of crystallization, the activation energy for crystallization Q_x determined according to the method of Kissinger [18], the glass transition temperature T_g, the viscosity values reached in the course of crystallization,

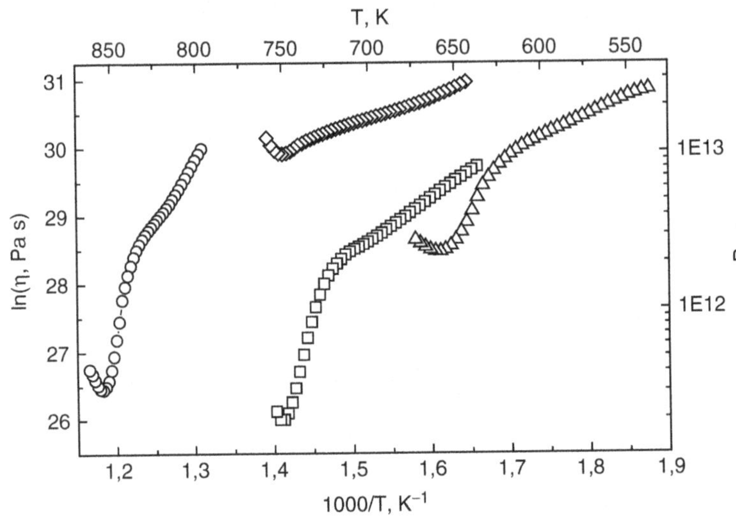

Fig. 4.4 Viscosity temperature dependence of amorphous metallic alloys Ni_xZr_{100-x}: Δ, $Ni_{30}Zr_{70}$; □, $Ni_{40}Zr_{60}$; ◇, $Ni_{50}Zr_{50}$; O, $Ni_{64}Zr_{36}$

Table 4.1 Dependence of the basic characteristic parameters of crystallization and viscous flow on the chemical composition of metallic glasses of Ni_xZr_{100-x} system [1]

C_{Ni}, at.%	Number of DSC peaks	T_{on}, K	T_{max}, K	ΔH_x, kJ/mol	Q_x, kJ/mol
30	1	612	617	−4.45	202
40	3	684	694	−3.47	371
50	1	660	674	−6.75	170
64	2	845	851	−6.48	478
C_{Ni}, at.%	T_g, K	Q_η, $(T < T_g)$ kJ/mol	Q_η, $(T > T_g)$ kJ/mol	$\eta(T_g)$, Pa s	η_{min}, Pa s
30	588	54	190	7.3×10^{12}	2.3×10^{12}
40	668	62	398	1.8×10^{12}	2.1×10^{11}
50	683	32	66	1.2×10^{12}	1.0×10^{13}
64	808	141	508	2.1×10^{12}	3.0×10^{11}

and the activation energies $Q_\eta(T < T_g)$ and $Q_\eta(T > T_g)$ of viscous flow at temperatures lower and higher than the glass transition temperature, respectively. These characteristic parameters are given in Table 4.1 as a function of composition.

Comparing the effective activation energies of crystallization Q_x with the activation energies for viscous flow Q_η, it could be concluded that with the exception of the metallic glass $Ni_{50}Zr_{50}$, both activation energies are almost equal, i.e., the crystallization is governed by viscous flow. Hence, it appears clear that:

- The viscous flow behavior of Ni_xZr_{100-x} glassy alloys over a range of compositions from 30 to 64 at.% of Ni shows very well-defined compositional

Table 4.2 Dependence of the basic characteristic parameters of crystallization and viscous flow for $Ni_{100-x}P_x$ glassy alloys upon the phosphorus content [2]

C_P, at.%	Number of DSC peaks	T_{on}, K	T_{max}, K	ΔH_x, kJ/mol	Q_x, kJ/mol
16.8	2	584.4	598.1	−0.93	187
17.7	2	619.6	622.4	−1.11	212
18.7	1	634.0	637.0	−1.06	226
19.1	1	640.0	643.6	−1.03	240
20.1	2	631.9	643.0	−0.85	200
20.3	2	634.5	649.4	−0.92	212
20.8	2	637.0	645.7	−0.88	214
C_P, at.%	T_g, K	Q_η ($T < T_g$), kJ/mol	Q_η ($T > T_g$), kJ/mol	$\eta(T_g)$, Pa s	η_{min}, Pa s
16.8	588	37	90	5.5×10^{12}	2.5×10^{12}
17.7	604	42	106	4.0×10^{12}	1.6×10^{12}
18.7	623	42	184/218*	3.6×10^{12}	7.3×10^{12}
19.1	637	44	183/225	3.2×10^{12}	8.3×10^{12}
20.1	639	41	70	3.8×10^{12}	2.6×10^{12}
20.3	641	32	107	4.5×10^{12}	2.3×10^{12}
20.8	648	40	130	5.0×10^{12}	2.8×10^{12}

dependence, which could be explained with the aid of the equilibrium phase diagram of this alloy system.

- The thermal stability of the glassy phase is greatest for the range of eutectic compositions ($x = 40$ and $x = 64$ at.%, respectively). This is due to the long-range atomic diffusion needed for the crystallization of at least two crystallization products.
- Glassy Ni_xZr_{100-x} alloys with compositions which correspond to the peritectoid phases $NiZr_2$ and NiZr ($x = 30$ and 50 at.%, respectively) show low thermal stability. This is due to the fact that no long-range diffusion of the large Zr atoms is needed for the formation of the stable crystallization end products $NiZr_2$ and NiZr.
- In all cases, the crystallization process is governed by viscous flow in the glassy alloys.
- Viscous flow measurements can be used as a sensitive tool for study of property changes against composition in amorphous metallic alloys.

The System $Ni_{100-x}P_x$

Here the results of a systematic study upon the viscous behavior and crystallization of binary $Ni_{100-x}P_x$ [2] amorphous metallic alloys depending on the *P* content are presented. This is done in order to find out the interrelation between these two processes.

Seven different compositions of the system $Ni_{100-x}P_x$ have been studied (see Table 4.2). The viscosity measurements were carried out with the aid of

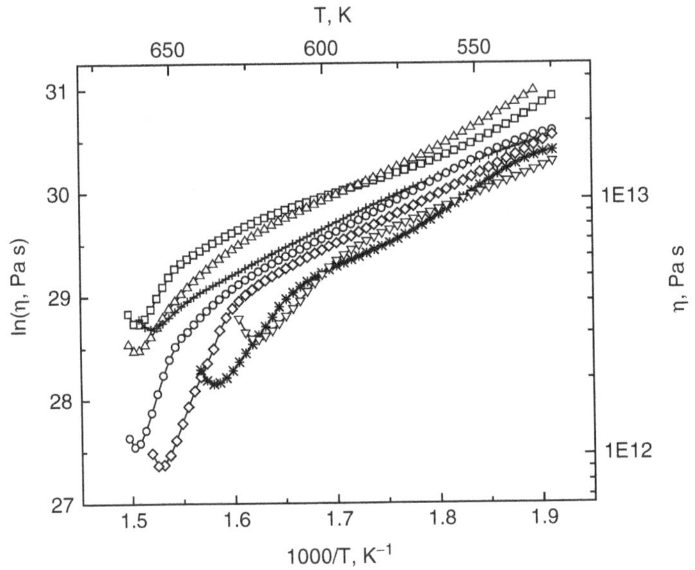

Fig. 4.5 Viscosity temperature dependence of the studied $Ni_{100-x}P_x$ amorphous metallic alloys: ∇, $Ni_{83.2}P_{16.8}$; *, $Ni_{82.3}P_{17.7}$; ◇, $Ni_{81.3}P_{18.7}$; O, $Ni_{80.9}P_{19.1}$; +, $Ni_{79.9}P_{20.1}$; Δ, $Ni_{79.7}P_{20.3}$; □, $Ni_{79.2}P_{20.8}$

thermomechanical analyzer Perkin-Elmer TMS-2 silica glass thermomechanical analyzer with a homemade silica glass assembly for creep measurements as described in ▶ Chap. 3, section "Experimental Techniques for Viscosity Study of Metallic Glasses Under Nonisothermal Conditions". The elongation-temperature (time) curves of these specimens under applied loads of 0.005 to 0.150 kg, at a heating rate of 20 K/min, were monitored up to temperatures higher than the temperature of fully completed crystallization (see Fig. 4.5).

A Perkin-Elmer DSC 2C differential scanning calorimeter has been used to study the crystallization behavior of $Ni_{100-x}P_x$ amorphous alloys at heating rates of 5, 10, 20, and 40 K/min, respectively. In order to test the extent to which the crystallization process in the alloys studied is governed by viscous flow, the activation energies for crystallization Q_x and viscous flow Q_η, respectively, have been compared. For this purpose the temperatures, T_{max}, at which maximum rate of heat evolution due to the crystallization at the different scanning rates is observed, have been determined, and the activation energy for crystallization, Q_x, has been calculated by the method proposed by Kissinger [18].

The temperature dependence of the apparent viscosity, η, of the glassy $Ni_{100-x}P_x$ alloys studied, shown in Fig. 4.5 as a plot of ln η against 1000/T, possesses the common feature of an existence of two nearly linear parts with different slopes. The crossover temperature is assumed as the glass transition temperature T_g. At temperatures higher than the onset temperature of crystallization T_{on}, the values of η are influenced by the increasing volume fraction ζ of crystalline regions along with increasing temperature. As a result, viscosity values go through a minimum at

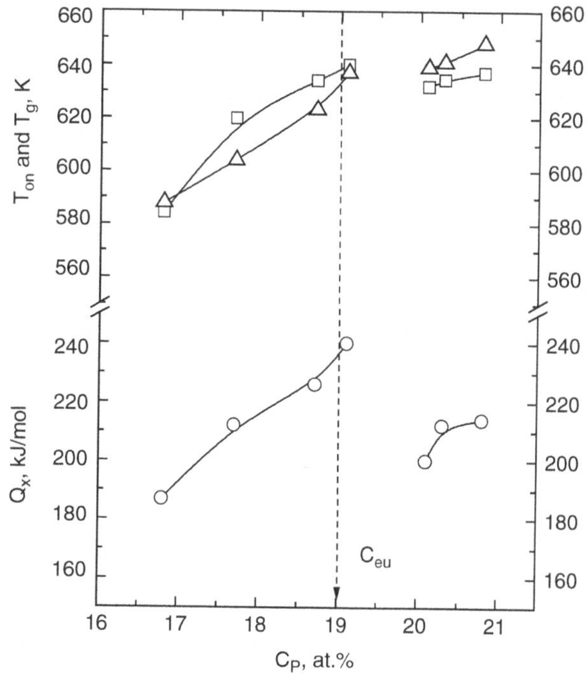

Fig. 4.6 Dependence of the characteristic parameters of crystallization and viscous flow on the phosphorus content in the system Ni_xP_{100-x}: □, onset temperature of crystallization T_{on}; Δ, temperature of glass transition T_g; O, activation energy of crystallization Q_x

approximately the temperature T_{max} of maximal rate of heat release due to the crystallization and then increase rapidly. For the calculation of the true viscosity in this case, the Einstein equation for flow of mixtures should be used.

The most characteristic parameters of crystallization and viscous flow processes in the $Ni_{100-x}P_x$ glassy alloys studied are the onset temperature of crystallization T_{on}, the temperature T_{max} at which maximum rate of heat evolution due to the crystallization is observed, the enthalpy ΔH_x of crystallization, the activation energy for crystallization Q_x, the glass transition temperature T_g, the viscosity values $\eta(T_g)$ at glass transition temperatures, the minimum viscosity values η_{min} reached during the course of crystallization, and the activation energies $Q_\eta(T < T)$ and $Q_\eta(T > T)$ of viscous flow at temperatures lower and higher than the glass transition temperature, respectively. These characteristic parameters are given in Table 4.2 as a function of composition.

Figure 4.6 shows the composition dependence of the onset temperature of crystallization T_{on} measured at heating rate of 20 K/min and the glass transition temperature T_g, respectively, together with the composition dependence of the activation energy of crystallization Q_x. A critical concentration in the vicinity of the eutectic composition at P ~ 19 at.% is observed, which separates the abovementioned parameters in two composition ranges with different crystallization and viscous flow behavior. This is also confirmed by Fig. 4.7, where the composition dependences of the enthalpy ΔH_X together with the concentration dependences of $\eta(T_g)$ and η_{min}, respectively, are shown.

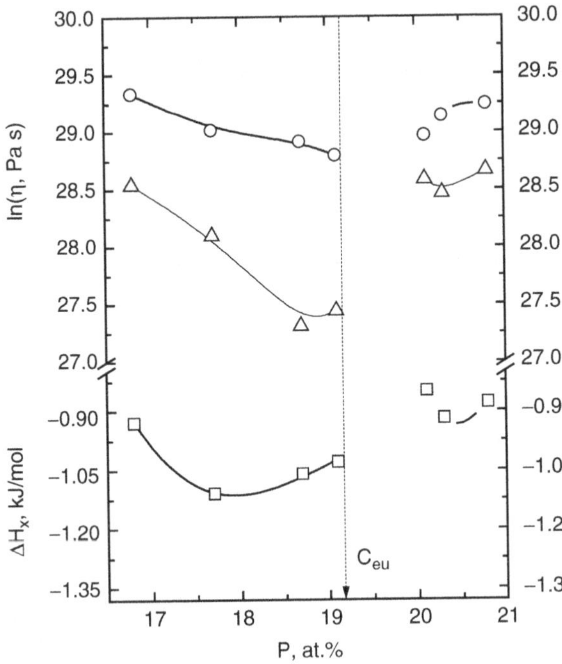

Fig. 4.7 Dependence of the crystallization enthalpy ΔH_x (□), together with the dependences of $\eta(T_g)$ (O) and η_{min} (Δ), upon the phosphorus content in $Ni_{100-x}P_x$ amorphous metallic alloys studied

The experimental results obtained have shown that thermal stability (related to T_{on}) increases rapidly along with increasing the P content up to the eutectic composition of P ~ 19 at.%, goes over small maximum, and then remains nearly constant. This is a common feature for the thermal stability of metal-metalloid amorphous alloys such as Fe-B, Fe-P, and Ni-P [19, 20]. It is evident that the higher the thermal stability, the higher the activation energy for crystallization (see Fig. 4.6).

The viscous behavior of melt-quenched $Ni_{100-x}P_x$ alloys studied (see Table 4.2, Fig. 4.7) demonstrates well-defined concentration dependence of the viscosity at T_g. The viscosity at glass transition temperatures is very near to the theoretically expected value of 10^{12} Pas and goes through a minimum at eutectic concentration of P ~ 19 at.%. The glass transition temperatures T_g vs. P content show the same trend as T_{on} vs. P content and are very near to the onset of crystallization. The temperature dependences of viscosities are shifted as a whole to higher viscosity values with increasing of P content. This could be explained with the role of chemical bonding in metal-metalloid amorphous alloys which exhibit stable intermetallic phases (such as Ni_3P in the case of $N_{100-x}P_x$) near their deep eutectics [20].

The composition dependence of viscosity values η_{min} reached in the course of crystallization (see Fig. 4.7) is complicated and strikingly similar to the similarly complicated composition dependence of the enthalpy of crystallization ΔH_x. This similarity could be explained with the fact that the temperature of the specimens at the moment when the maximal rate of crystallization heat release is reached is substantially higher than the preset temperature of the furnace monitored by the thermocouple. The higher the value of ΔH_x, the higher is the deviation of real specimen temperature from the preset temperature and the lower the values of η_{min}.

The existence of two almost linear parts in the viscosity temperature dependencies $\ln(\eta)/(1000/T)$ before and after the glass transition temperature is tempting to be treated as temperature dependences of Arrhenius-type and that their slopes can be used to determine two different effective activation energies for viscous flow before and after the glass transition temperature T_g, $Q_\eta(T < T_g)$, and $Q_\eta(T > T_g)$, respectively. $Q_\eta(T > T_g)$ should coincide with the activation energy for crystallization Q_x. Later on, when considering the free volume model (FVM) presentation of viscosity temperature dependence, it will be shown that this empirical approach is too naive and that the real activation energy for viscous flow is actually one and the same before and after the glass transition temperature. Irrespective of this, however, the expectation that Q_x should be very near to the effective activation energy for viscous flow $Q_\eta(T > T_g)$ is physically meaningful, provided that the crystallization is governed by the viscous flow. In the case of the studied-by-us N-P metallic glasses, this is true. There is a real good agreement between activation energy for crystallizations Q_x and $Q_\eta(T > T_g)$. The activation energies are in the range of 220–240 kJ/mol.

It can be concluded that:

- The viscous flow behavior of melt-quenched $Ni_{100-x}P_x$ ($16 < x < 21$ at.%) amorphous alloys is of Newtonian type with viscosity values $\eta(T_g)$ at glass transition temperatures T_g very near to the theoretically expected value of 10^{12} Pa s. The composition dependence of $\eta(T_g)$ goes through a minimum at the eutectic composition of P ~ 19 at.%. The glass transition temperature T_g increases rapidly along with increasing the P content up to the eutectic composition and then remains almost constant. The activation energy of viscous flow at $T > T_g$ is approximately 220 kJ/mol. A systematic shift of the viscosity temperature dependences to higher viscosity values with increasing the P content is observed, which is most probably due to the role of chemical bonding in Ni-P amorphous system.
- The thermal stability (T_{on}) of melt-quenched $Ni_{100-x}P_x$ amorphous alloys increases rapidly along with increasing of P content up to the eutectic composition of P ~ 19 at.%. The value of T_{on} is very near to T_g and shows similar composition dependence as T_g. The eutectic composition of P ~ 19 at.% is a critical one, separating two composition ranges with different crystallization behavior. The activation energy of crystallization follows the trend of the concentration dependences of T_{on} and T_g and practically coincides with the activation energy $Q_\eta(T > T_g)$ for the glassy alloys with $T_g < T_{on}$ (alloys with nearly eutectic composition). The crystallization processes in $Ni_{100-x}P_x$ amorphous system are governed by viscous flow.
- Viscous flow measurements under continuous heating conditions can be used as a sensitive tool for studying property changes against composition in amorphous metallic alloys.

The System $Fe_{100-x}P_x$

A systematic study upon the viscous flow and crystallization behavior of three rapidly quenched from the melt $Fe_{100-x}P_x$ (x = 13.7, 16.2 and 19.0 at.%) alloys has been carried out, depending on the phosphorus concentration around the eutectic composition – 16.8 at.% P [3]. The dimensions of the ribbonlike samples varied from $0.01 \times 0.048\ mm^2$ to $0.015 \times 0.8\ mm^2$.

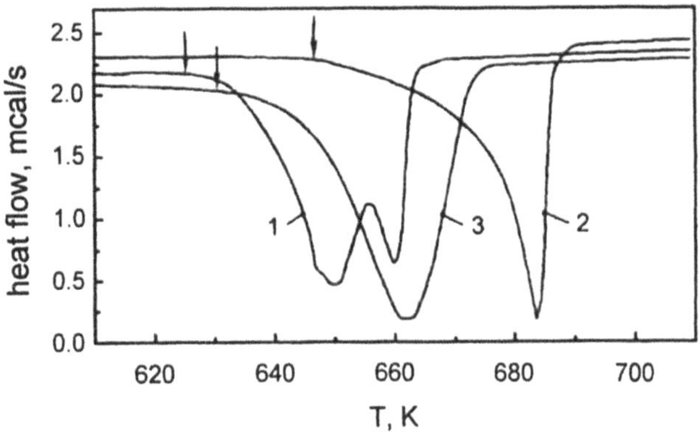

Fig. 4.8 DSC-crystallization peaks of $Fe_{100-x}P_x$ amorphous metallic alloys studied at heating rate 20 K/min: 1, $Fe_{86.5}P_{13.7}$; 2, $Fe_{83.8}P_{16.2}$; 3, $Fe_{81}P_{19}$

Table 4.3 Dependence of main crystallization and viscous flow parameter of $Fe_{100-x}P_x$ amorphous system upon the phosphorus concentration [3]

DSC					
C_P, at.%	Number and kind of DSC peaks	T_{on}, K	T_{max}, K	Q_x, kJ/mol	ΔH_x, kJ/mol
13.7	2	628	650	213	−6.5
16.2	1, asymmetric	648	684	235	−6.3
19.0	1, almost symmetric	633	663	182	−6.7
TMS					
C_P, at.%	T_g, K	$\eta(T_g)$, Pa s	Q_η, kJ/mol	Q_r, kJ/mol	$c_{f,o}$, –
13,7	628	2.92×10^{11}	200.5	144.4	2.15×10^{-11}
16,2	660	1.86×10^{11}	240.5	175.6	$2,15 \times 10^{-11}$
19,0	645	4.14×10^{11}	205.5	145.6	$2,15 \times 10^{-11}$
TMS					
C_P, at.%	B, K	T_o, K	η_o, Pa s/K	ν_r, 1/s	
13.7	6052	455	5.83×10^{-24}	3.4×10^{25}	
16.2	5146	492	1.28×10^{-24}	3.4×10^{25}	
19.0	6029	476	4.80×10^{-24}	3.4×10^{25}	

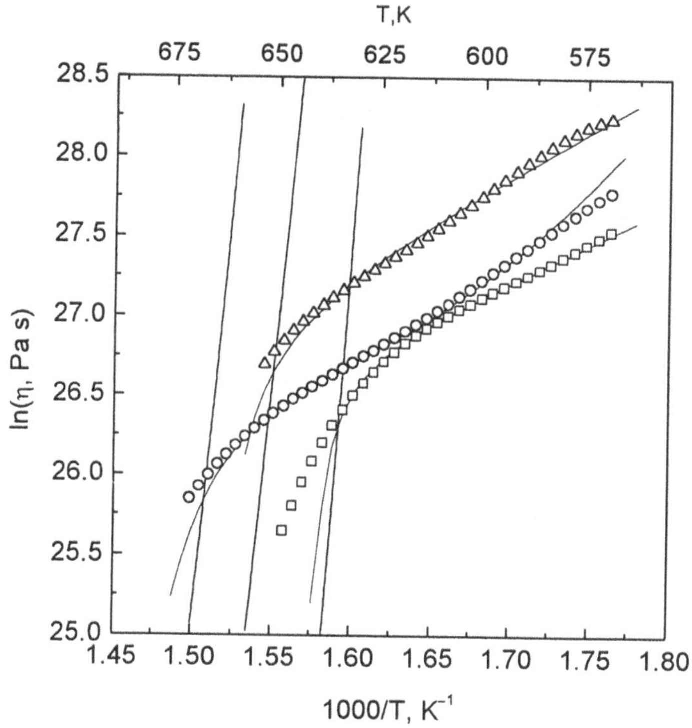

Fig. 4.9 Viscosity temperature dependence of $Fe_{100-x}P_x$ amorphous metallic alloys presented as *ln(η)* vs. 1000/*T*. Heating rate 20 K/min: □, $Fe_{86.5}P_{13.7}$; O, $Fe_{83.8}P_{16.2}$; Δ, $Fe_{81}P_{19.0}$

The crystallization behavior has been studied with the aid of Perkin-Elmer DSC 2C at heating rates of 5, 10, 20, and 40 K/min, respectively. The effective activation energy of crystallization Q_x was determined according to the method of Kissinger [18]. The viscosity determinations were carried out with the aid of Perkin-Elmer TMS-2 by using the already described earlier silica glass homemade tensile test equipment. The heating rate used was 20 K/min.

The DSC peaks of crystallization of $Fe_{100-x}P_x$ metallic glasses at a heating rate of 20 K/min are shown in Fig. 4.8. The characteristic crystallization parameters T_{on}, T_{max}, and ΔH_x and the effective activation energy of crystallization Q_x are given in Table 4.3.

Figure 4.9 presents the viscosity temperature dependencies of the three $Fe_{100-x}P_x$ amorphous metallic alloys studied in the accustomed coordinates ln(*η*)/(1000/*T*).

Chronologically the author's study upon the Fe-P glass-forming system was carried out 3 years later (1995) than the Ni-P studies (1992). At that time the author's free volume model studies upon the viscosity temperature dependence of glassy alloys under continuous heating conditions were already finished to a great extent. The FVM theoretical considerations are presented in detail in ▶ Chap. 4 of

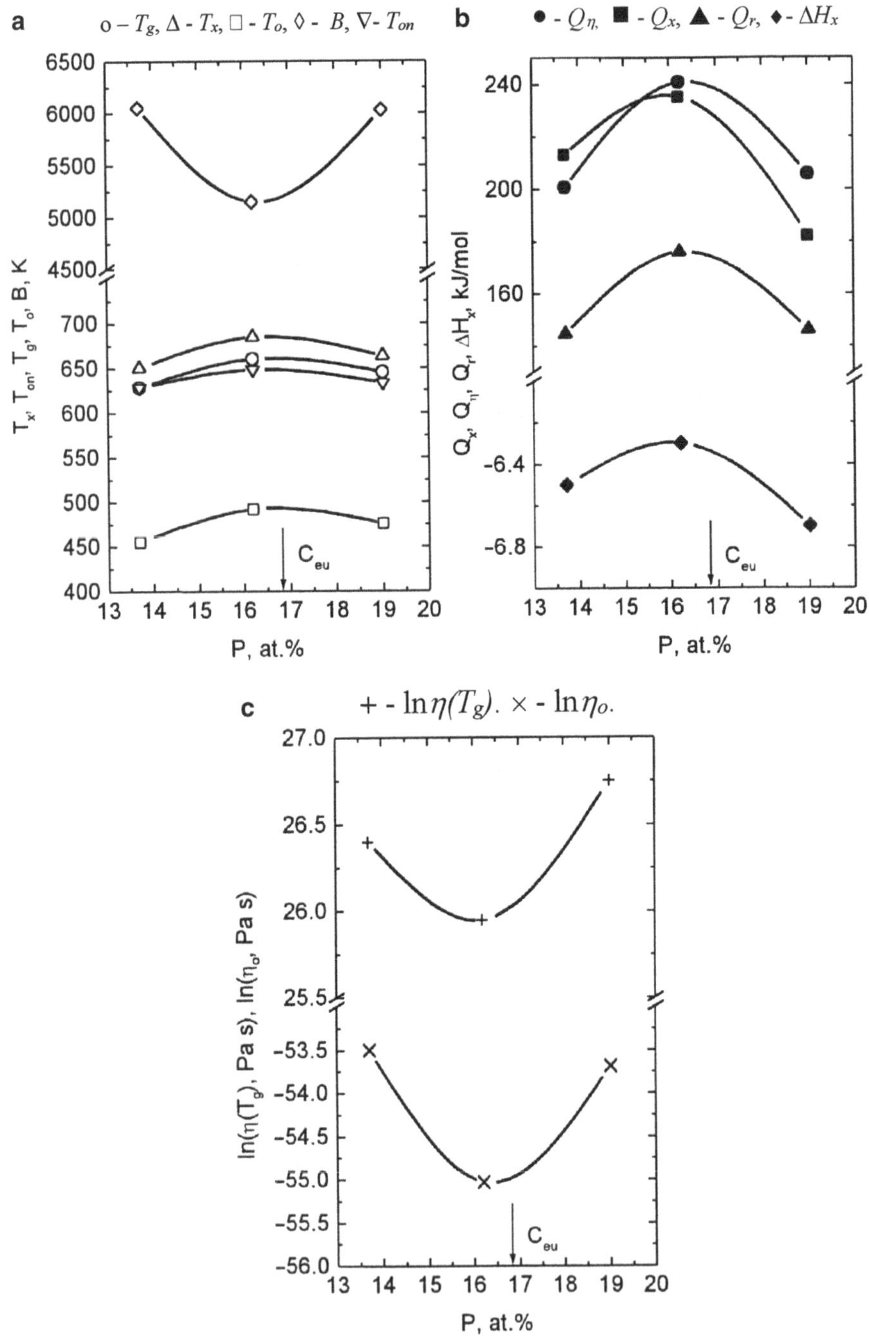

Fig. 4.10 Dependence of the main crystallization and viscous flow parameters of the glassy $Fe_{100-x}P_x$ system upon the phosphorus concentration: (**a**) o, T_g; Δ, T_x; □, T_o; ◇, B; ∇, T_{on}. (**b**) •, Q_η; ▪, Q_x; ▲, Q_r; ◆, ΔH_x. (**c**) +, $\ln\eta(T_g)$. × – $\ln\eta_o$

this book. At this locality of the book, without any motivation, it will be declared that the viscosity temperature dependence of glassy metals can be described with the aid of an integral equation with parameters: the heating rate q, the atomic attempt frequency ν_r, the pre-exponential viscosity factor η_o, the activation energy of structural relaxation Q_r, the initial concentration of flow defects $c_{f,o}$, and two model parameters B and T_o are incorporated as parameters, which are nearly similar to the empirical coefficients of the Vogel-Fulcher-Tammann equation.

Along with the increasing temperature (constant heating rate q), the structure of the metallic glass relaxes quickly and the viscosity tries to reach its quasi-equilibrium value η_{eq}. The quasi-equilibrium viscosity depends only on the momentous value of the temperature reached, following an equation, which incorporates the parameters B, T_o, and η_o and the activation energy of viscous flow Q_η. The abovementioned characteristic FVM parameters of viscous flow for the Fe-P system were determined with the aid of multiparameter nonlinear regression analysis of the viscosity experimental data and are given in Table 4.3.

Figure 4.10 shows the dependence of the main crystallization and viscous flow parameter of $Fe_{100-x}P_x$ amorphous system upon the phosphorus concentration.

As is seen from Fig. 4.10a–c, the eutectic composition C_{eu} of 16.8 at.% P, similar to the Ni-P glassy system, occupies a particular position in the concentration dependencies of the characteristic parameters of crystallization and viscous flow. It becomes obvious that the metallic Fe-P glass of eutectic composition possesses the highest thermal stability (the highest T_x and T_g values) and the most stable amorphous structure (the lowest crystallization enthalpy ΔH_x, the lowest $\eta(T_g)$, and the highest T_o values, respectively). The model parameters B, T_o, and η_o also show clearly expressed minima and maxima around the eutectic composition of 16.8 at.% P [3].

The conclusion can be drawn, that the thermal stability of $Fe_{100-x}P_x$ amorphous metallic alloys rapidly solidified from the melt goes over a maximum by the eutectic phosphorus concentration of 16.8 at/% P. The onset temperature of crystallization coincides or is lower from the glass transition temperature, which makes difficult the estimation of the glass transition via viscosity measurements. The effective activation energy for crystallization Q_x and the activation energy for viscous flow Q_η coincide practically.

Viscous Flow Behavior of Ternary $Fe_{82-x}TM_xB_{18}$ (TM = V, Cr) Amorphous Metallic Alloys

It is well known that the glass-forming ability and the thermal stability of glassy metals to a great extent depend on the chemical nature and number of metal components forming their composition [20, 21]. Nevertheless, it could be stated that very limited number of experimental data exists [22], concerning the influence of the type and concentration of metallic additions upon the viscous behavior, the thermal stability, and the critical cooling rate of thermally metal-metal-metalloid amorphous metallic alloys. At this place of the book, the author's results on the systematic studies upon the characteristic peculiarities of viscous flow in the ternary

$Fe_{82-x}V_xB_{18}$ [4] and $Fe_{82-x}Cr_xB_{18}$ [5] amorphous metallic alloys, depending upon the type and concentration of alloying additions, will be presented.

Influence of Vanadium Alloying Additions

The ternary metallic glasses $Fe_{82-x}V_xB_{18}$ ($x = 0 \div 9$ at.%) in the form narrow metallic ribbons were produced via the CBMS method. Their amorphousness was checked via X-ray and DSC analysis. The melting point T_m and the molar heat of fusion ΔH_m of the crystalline precursor alloys were determined with Perkin-Elmer DTA 1700 equipment.

The onset temperature and the molar heat of crystallization ΔH_x, respectively, were determined with the aid of differential scanning calorimeter Perkin-Elmer DSC 2C at a heating rate of 20 K/min. The obtained results as a function of the vanadium concentration are given in Table 4.4.

Viscosity measurements have been carried out in a Perkin-Elmer TMS-2 silica glass dilatometer at a heating rate of 20 K/min. The three-point bending assembly was used.

Figure 4.11 shows the temperature dependence of the shear viscosity of the glassy alloys studied. In the temperature range (T_g- 60 K) $\div$ T_g, the temperature dependences of the viscosities can be well expressed by Arrhenius-type empirical equations. The characteristic parameters of crystallization and viscous flow of the ternary metallic glasses studied are given in Table 4.4 as function of V concentration. The effective activation energies for viscous flow $Q_\eta(T > T_g)$ are approximately equal to 220–240 kJ/mol. They coincide practically with activation energy for nucleation and crystal growth in the amorphous system Fe-B [22].

The viscosity temperature dependence of the glassy alloys studied in the broad temperature range T_g–T_m was calculated in the coordinates of the Vogel-Fulcher-Tammann (VFT) empirical equation, assuming that viscosity of the ternary amorphous metallic alloys in their melting point T_m equals the viscosity of pure Fe in its melting point:

Table 4.4 Main characteristic parameters of crystallization and viscous flow of ternary $Fe_{82-x}V_xB_{18}$ metallic glasses, depending on the V concentration

C_V, at.%	T_{on}, K	T_m, K	ΔH_m, kJ/mol	ΔH_x, kJ/mol	T_g, K	$\eta(T_g)$, Pa s
0	716	1440	10.0	−7.55	713	0.32×10^{12}
2.3	743	1444	10.3	−7.04	729	0.26×10^{12}
9.0	792	1448	15.7	−6.40	758	0.14×10^{12}
C_V, at.%	Q_η $(T < T_g)$, kJ/mol	Q_η $(T > T_g)$, kJ/mol	η_o^{VFT}, Pa s	B, K	T_o, K	$\dot{T}_{cr}$, K/s
0	55	234	1.36×10^{-3}	2864	645	2.1×10^6
2.3	60	217	1.36×10^{-3}	2942	649	1.7×10^6
9.0	58	220	1.36×10^{-3}	2790	685	1.2×10^6

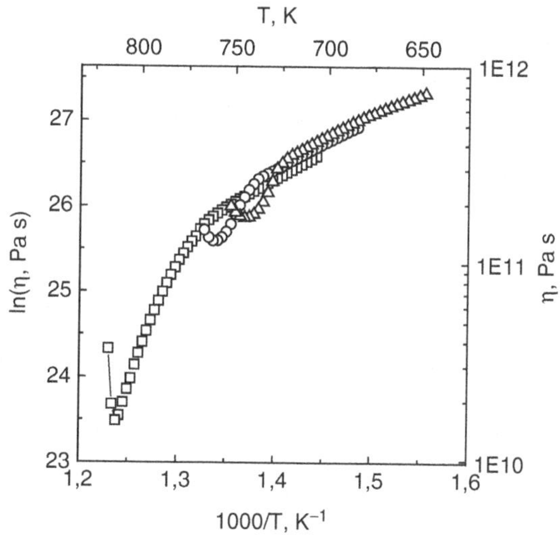

Fig. 4.11 Viscosity temperature dependence of ternary $Fe_{82-x}V_xB_{18}$ amorphous metallic alloys at a heating rate of 20 K/min: □, $Fe_{82}B_{18}$; Δ, $Fe_{79.7}$ $V_{2.3}B_{18}$; O, $Fe_{73.0}$ $V_{9.0}B_{18}$

$$\eta = \eta_o^{VFT} \exp\left(\frac{B^{VFT}}{T - T_o^{VFT}}\right) \tag{4.2}$$

According to Barfield et al. [23] and Wertmann et al. [24], this value varies between 0.005 and 0.012 Pa s. The calculated values of the empirical constants η_o^{VFT}, B, and T_o are given in Table 4.4.

Calculating according to Vogel-Fulcher-Tammann, viscosity temperature dependences were used for estimating critical cooling rate $\dot{T}_{cr}$ of the ternary glassy metallic alloys studied. This is performed by using the method of Uhlmann (see ▶ Chap. 3, Eqs. (3.1) and (▶ 3.2)) for the critical cooling rate needed to obtain the amorphous alloy $Fe_{82}B_{18}$ [25, 26]. The estimated critical cooling rates $\dot{T}_{cr}$ are given in Table 4.4. More details for the estimation of the critical cooling rate are presented in ▶ Chap. 3. The alloying additions of 9 at.% V also increase the thermal stability (T_{on}) of the alloys studied with about 80 K (see Table 4.4).

Influence of Chromium Alloying Additions

Three ternary $Fe_{82-x}Cr_xB_{18}$ ($x = 0$, 1.4, 4.2, 8.2 at.%) amorphous metallic alloys have been studied. Their thermal stability and viscous flow behavior under continuous heating conditions were investigated with the aid of differential scanning calorimeter Perkin-Elmer DSC 2C (by heating rate of 20 K/min) and silica glass dilatometer Heraeus TMA 500 tensile test equipment (by heating rate of 10 K/min) [5], respectively. The research scheme was analogous to the scheme used by the study of $Fe_{82-x}V_xB_{18}$ metallic glasses. The viscosities of the three Cr-containing metallic glasses, monitored at a heating rate of 10 K/min, are shown in Fig. 4.12. In the same figure, the viscosity temperature dependence of the binary $Fe_{82}B_{18}$ metallic glass is also presented, monitored by 20 K/min with the aid of

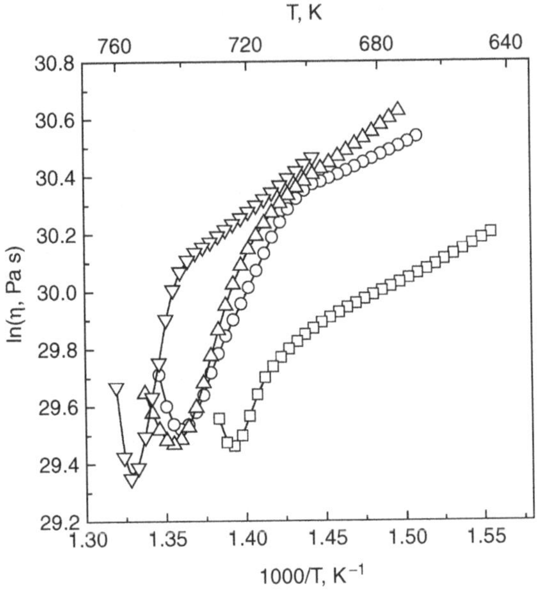

Fig. 4.12 Viscosity temperature dependence of the $Fe_{82-x}Cr_xB_{18}$ amorphous metallic alloys at a heating rate 10 K/min together with the viscosity temperature dependence of $Fe_{82}B_{18}$ metallic glass at a heating rate of 20 K/min: □, $Fe_{82}B_{18}$; O, $Fe_{80.6}Cr_{1.4}B_{18}$; Δ, $Fe_{77.8}Cr_{4.2}B_{18}$; ∇, $Fe_{73.8}Cr_{8.2}B_{18}$

Table 4.5 The main characteristic parameters of crystallization and viscous flow of the $Fe_{82-x}Cr_xB_{18}$ metallic glasses, depending on the Cr content [5]

C_{Cr}, at.%	T_{on}, K	T_m, K	ΔH_m, kJ/mol	ΔH_x, kJ/mol	T_g, K	$\eta(T_g)$, Pa s
0	716	1440	10.0	−7.6	713	0.32×10^{12}
1.4	725	1428	11.5	−7.7	709	11.5×10^{12}
4.2	732	1436	10.4	−7.4	725	14.0×10^{12}
8.2	745	1439	9.10	−7.0	744	12.0×10^{12}
C_{Cr}, at.%	$Q_\eta\ (T < T_g)$, kJ/mol	$Q_\eta\ (T > T_g)$, kJ/mol	η_o^{VFT} Pa s	B, K	T_o, K	$\dot{T}_{cr}$, K/s
0	55	234	1.36×10^{-3}	2929	645	2.1×10^6
1.4	53	209	1.36×10^{-3}	2905	634	1.8×10^6
4.2	87	272	1.36×10^{-3}	2883	644	1.5×10^6
8.2	98	309	1.36×10^{-3}	2788	666	0.9×10^6

Perkin-Elmer TMS-2 equipment for three-point bending tests. Due to the different equipments and experimental conditions used, the data for the three ternary $Fe_{82-x}Cr_xB_{18}$ metallic glasses and the data for the binary $Fe_{82}B_{18}$ metallic glass can be compared and considered only in a semiquantitative way.

The characteristic parameters of crystallization of metallic glasses studied as a function of the Cr content; the effective activation energies for viscous flow; the empirical VFT-constants η_o^{VFT}, B, and T_o; and the estimated critical cooling rates for vitrification are given in Table 4.5.

Considering the data presented in sections "Influence of Vanadium Alloying Additions and "Influence of Chromium Alloying Additions," it could be generally concluded that the alloying of binary metallic glass $Fe_{82}B_{18}$ with V and Cr additions increases both the glass-forming ability and the thermal stability of $Fe_{82-x}TM_xB_{18}$ metallic glasses [4, 5]. The additions of vanadium are more effective in this respect. It is interesting to point out also that to the same conclusion come Myung et al. [27] by studying the system Fe-TM-P.

Viscous Flow Behavior of Quaternary $Al_{85}Ni_5Co_2RE_8$ (RE = Ce, Gd, U) Amorphous Metallic Alloys

Some independent publications of Inoue and He [28–30] are reported at first in 1988 for ductile Al-based glasses. These thin ribbons of 15–25 μm thickness show excellent mechanical properties: the ultimate tensile stress is about three times larger than that for the best polycrystalline Al-based alloy. The Al-rich glasses are difficult to manufacture due to the large cooling rate and protective atmosphere necessary for their production. The largest critical thickness obtained so far is about 0.5 mm for $Al_{85}Gd_8Ni_7$ alloy [31]. For applications one needs bulk samples obtainable in this case only by consolidation of the amorphous flakes or powders prepared by rapid solidification techniques. This consolidation needs hot isostatic pressure or hot extrusion which should be carried out much below the crystallization temperature in order to prevent the devitrification during the compaction. Thus, it was of interest to search compositions rich in Al, with relative high crystallization temperature, T_x, and large $(T_x–T_g)$ values, where T_g is the glass transition temperature. The most frequent composition of the Al-based glasses is the ternary alloy: Al-LTM-RE, where LTM is a late transition element (Ni, Co, Fe) and RE stands for rare earth elements [32–34].

The aim of this work was to study the thermal stability and viscous flow behavior of quaternary $Al_{85}Ni_5Co_2RE_8$ (RE = Ce, Gd, U) alloys. In the case of uranium alloying, an actinide element (with extremely weak radioactivity) for the first time was checked as a possible new glass-forming element [33, 34]. Alloy ingots of the alloys were prepared by inductive melting starting from pure elements in water-cooled copper boat under purified argon protecting atmosphere. About 10 g of alloy quantities was rapidly solidified using the melt spinning method under protecting Ar atmosphere. The alloy compositions were calculated by the given ratio of the amounts of pure elements used. Additional chemical analysis of the composition was not performed as the alloys do not contain volatile components which may cause measurable deviations from the nominal composition. The peripheral speed of the casting wheel was 40 m/s. The amorphousness of the resulting ribbons was checked by X-ray diffraction (XRD, using Co-Kα radiation on a Bruker AXS 200 diffractometer) and by differential scanning calorimetry. The ribbons studied had a thickness of 0.02 ÷ 0.025 mm and width of 1.52 ÷ 1.90 mm.

Both the crystallization and the nonisothermal viscosity measurements were carried out at a heating rate of 20 K/min [6]. No appearance of primary crystallization during the heating has been detected from the DSC signal. The nonisothermal crystallization was followed using a Perkin-Elmer differential scanning calorimeter DSC 2C. A temperature accuracy (±1 K) was ensured according

Table 4.6 Onset temperature of crystallization, T_{on}, temperature of maximum of the crystallization peaks, T_{max}, enthalpy of crystallization, ΔH_c, and glass transition temperature from DSC measurements, T_g^{DSC}, of the rapidly solidified $Al_{85}Ni_5Co_2RE_8$ alloys

Alloy	Number of peaks	T_{on}, (K)	T_{max}, (K)	ΔH_{cr}, (kJ/mol)	T_g^{DSC}, (K)
$Al_{85}Ni_5Co_2Ce_8$	I	576	582.3	2.90	543
	II	596	601.6	1.62	
$Al_{85}Ni_5Co_2Gd_8$	I	568	569.9	1.97	550
	II	595	606.1	0.32	
	III	658	666.6	1.71	
$Al_{85}Ni_5Co_2U_8$	I	580	589.7	1.83	560

to the standard calibration procedure of the apparatus. The baseline of the DSC traces was determined by subtracting the signal of the "dead" crystallized from the signal of the "as-quenched" amorphous sample after its crystallization (a standard procedure of the software of the computer-aided DSC device).

Table 4.6 summarizes all DSC data for the rapidly solidified alloys studied – the glass transition temperatures (T_g^{DSC}), the onset temperatures of crystallization (T_{on}), the temperatures of maximum of the crystallization process (T_{max}), and the enthalpy of the crystallization (ΔH_{cr}). The highest thermal stability possesses the amorphous alloy $Al_{85}Ni_5Co_2U_8$ with $T_{on} = 580$ K.

The shear viscosity $\eta(T)$ was determined with the aid of a Perkin-Elmer thermomechanical device TMS-2, as described in ▶ Chap. 3, under applied loads between 20 and 100 g. The initial length of the samples was kept at 4 mm. A schematic of the Perkin-Elmer TMS-2 measuring unit with the homemade silica glass assembly is presented in ▶ Fig. 3.3. The temperature accuracy (±1 K) of the TMS-2 was calibrated by using 4 mm long strips of pure Sn, Pb, Zn, and Al with exactly known melting points.

The experimental data (points) for all the three alloys, $Al_{85}Ni_5Co_2RE_8$ (RE = Ce, Gd, U), shown in Fig. 4.13 are fitted using the free volume model equations (solid curves). The basic concepts of this model are discussed in ▶ Chap. 5. A combination of (▶ Eqs. 5.8) and (▶ 5.17) is used for obtaining the nonequilibrium viscosity curves and Eq. (▶ 5.9) for obtaining the quasi-equilibrium viscosity curves. The intersection of the nonequilibrium and quasi-equilibrium curves for each alloy studied determines the glass transition temperature T_g^{Visc}. The glass transition temperature T_g^{Visc} and the value of the viscosity $\eta(T_g^{Visc})$ are given in Table 4.7. The values of the model parameters in (▶ Eqs. 5.8), (▶ 5.9), and (▶ 5.17) (ν_r, Q_r, $c_{f,o}$, T_o, Q_η, B, and η_o) obtained by regression analysis of the experimental data are given in Table 4.7 also.

The temperatures of the glass transition are 545 K, 548 K, and 564 K, for RE = Ce, Gd, U, respectively, and indicate a rather narrow supercooled liquid region (~15 ÷ 30 K) for these Al-based amorphous alloys. The good thermal properties make these alloys promising for preparing bulk amorphous samples below the T_{on} by hot consolidation of flakes, obtained by cryomilling of amorphous ribbons.

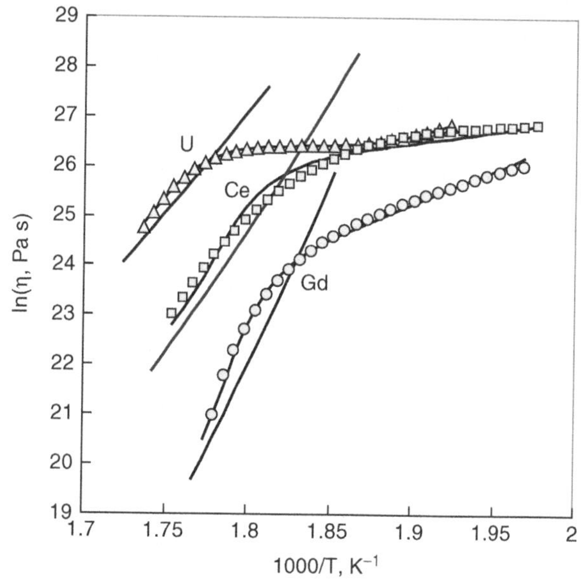

Fig. 4.13 Experimental temperature dependencies of the viscosity of the amorphous alloys $Al_{85}Ni_5Co_2RE_8$ (RE = Ce, Gd, U) at heating rate 20 K/min (points). *Solid curves* – calculated according to the FVM equation viscosity dependencies. *Steep solid lines* – temperature dependence of the quasi-equilibrium viscosity η_{eq}, according to FVM concepts

Table 4.7 Free volume model parameters, ν_r, Q_r, $c_{f,o}$, T_o, Q_η, B, *and* η_o, for the $Al_{85}Ni_5Co_2RE_8$ (RE = Gd, Ce, U) alloys studied. The glass transition temperature T_g^{Visc} and viscosity $\eta(T_g^{Visc})$ are given also

Parameters	Dimension	Alloy $Al_{85}Ni_5Co_2Ce_8$	Alloy $Al_{85}Ni_5Co_2Gd_8$	Alloy $Al_{85}Ni_5Co_2U_8$
ν_r	1/s	$2.7 \times 10^{+17}$	$5.6 \times 10^{+16}$	$8.6 \times 10^{+21}$
Q_r	kJ/mol	135	105	204
$c_{f,o}$	–	9.7×10^{-7}	1.7×10^{-8}	5.9×10^{-5}
T_o	K	411	424	397
Q_η	kJ/mol	182	230	175
B	K	1990	2390	1710
η_o	Pa s/K	9.9×10^{-16}	1.8×10^{-23}	$7.7 \times {}^{-13}$
T_g	K	545	548	564
$\eta(T_g)$	Pa s	1.5×10^{11}	2.6×10^{10}	2.1×10^{11}

Influence of Thermal Prehistory on the Viscous Flow Behavior of Amorphous Metallic Alloys

The Amorphous $Pd_{82}Si_{18}$ Alloy

A systematic study of the viscosity temperature dependence of amorphous $Pd_{82}Si_{18}$ [7] has been carried out under continuous heating conditions at constant heating rate of 20 K/min. Silica glass dilatometer Perkin-Elmer TMS-2 with three-point bending test assembly was used. The creep experiments included as-quenched amorphous

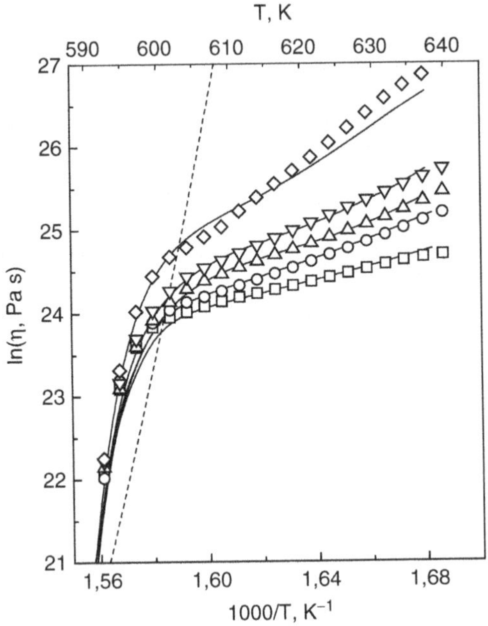

Fig. 4.14 The viscosity temperature dependences of the examined amorphous $Pd_{82}Si_{18}$ samples depending on their thermal prehistory: □, as-quenched sample; heat treated at O, 423 K; Δ, 473 K; ∇, 523 K; ◇, 563 K; −−−, temperature dependence of the calculated quasiequilibrium viscosity η_{eq}; *solid lines*, fitting curves, calculated via regression analysis of the data on the basis of the FVM

samples as well as samples, preliminary isothermally heat treated for 15 min at 423, 473, 523, and 563 K, respectively. The heat treatment was interrupted by rapid cooling of the samples to room temperature. After completing this procedure of preliminary heat treatment, the samples were subjected to creep measurements from room temperature to their crystallization temperature T_{on} (648 K). The crystallization enthalpy ΔH_x(−3.05 kJ/mol), the melting point T_m (1093 K), and the enthalpy of melting ΔH_m (7.3 kJ/mol) of the master crystalline alloy, respectively, were also measured. For this purpose Perkin-Elmer DTA 1700 was used. These data, together with the viscosity temperature dependence in the temperature range T_g–T_{on}, were used for estimation of the critical cooling rate needed to obtain the amorphous samples via rapid quenching from the melt.

The viscosity temperature dependences of the examined amorphous samples depending on their thermal prehistory are shown in Fig. 4.14.

As is seen, along with increasing of the preliminary heating for 15 min, the almost linear low-temperature parts of the viscosity temperature dependences are shifted to higher viscosity values up to reaching the temperature of glass transition. These viscosity temperature dependences are also characterized with increased slope along with increasing the temperature of pre-annealing. The authors have earlier (before the further development of the FVM for nonisothermal conditions of the experiments) erroneously interpreted this phenomenon as an increase of the effective activation energy for viscous flow (Arrhenius interpretation). After reaching the glass transition temperature T_g, an overlapping of the viscosity temperature dependences occurs, as if the samples "forget" their thermal prehistory. The reason for this phenomenon is the circumstance that their viscosity begins to follow the unique temperature dependence

Table 4.8 Characteristic parameters of viscous flow of the metallic glass $Pd_{82}Si_{18}$ depending on the thermal prehistory – empirical approach

Temperature of preliminary isothermal annealing (15 min), K	T_g, K	$\eta(T_g)$, 10^{10} Pa s	$Q(T < T_g)$, kJ/mol
As-quenched	633	2.2	63.0
423	632	2.3	91.8
473	631	2.9	97.2
523	630	3.4	111.6
563	628	5.5	193.5

Table 4.9 Characteristic parameters of viscous flow for $Pd_{82}Si_{18}$ amorphous alloy depending on the thermal prehistory. The data are interpreted on the basis of the FVM

Temperature of isothermal pre-annealing (15 min), K	Pre-exponential viscosity factor η_o, Pa s/K	Attempt frequency ν_r, 1/s	Activation energy for relaxation Q_r, kJ/mol	Activation energy for viscous flow Q_η, kJ/mol
As-quenched	6.84×10^{-20}	6.83×10^{19}	120	187.6
423	6.84×10^{-20}	6.83×10^{19}	120	187.6
473	6.84×10^{-20}	6.83×10^{19}	120	187.6
523	6.84×10^{-20}	6.83×10^{19}	120	187.6
563	6.84×10^{-20}	6.83×10^{19}	120	187.6
Temperature of isothermal pre-annealing (15 min), K	B, K	T_o, K	$c_{f,o}$, -	
As-quenched	2023	557	2.5×10^{-11}	
423	2023	557	6.8×10^{-12}	
473	2023	557	4.4×10^{-12}	
523	2023	557	2.8×10^{-12}	
563	2023	557	9.4×10^{-13}	

of the quasi-equilibrium viscosity of the undercooled melt (see ▶ Fig. 3.2). By the first publication of these data [7], they were interpreted on an empirical level, according to the author's knowledge and imaginations about the mechanism of amorphous metals viscous flow at that time.

The glass transition temperatures T_g of amorphous samples studied also depend on their thermal prehistory. They, together with the effective activation energies for viscous flow in the temperature range 593 K – T_g, are presented in Table 4.8.

After the relaxation pre-annealing at 563 K, the effective activation energy of viscous flow is calculated to be 193 kJ/mol. It practically coincides with the viscosity activation energy, determined by Taub and Spaepen for the same $Pd_{82}Si_{18}$ amorphous alloy [35].

Table 4.9 presents the characteristic parameters for viscous flow of $Pd_{82}Si_{18}$ amorphous alloy as a function of sample thermal prehistory, determined on the basis of the new author's FVM concepts for experiments carried out under nonisothermal conditions. As is seen, all FVM parameters have one and the same value, irrespective of the temperature of pre-annealing, except the initial concentration of flow defects $c_{f,o}$. This initial concentration decreases with more than an order of magnitude along with increasing of the pre-annealing temperature from 433 to 563 K, respectively.

The viscosity temperature dependence in the temperature range T_g–T_{on} was used for determination of the VFT-constants B^{VFT} (931.7 K) and T_o^{VFT} (600.5) in the empirical Vogel-Fulcher-Tammann equation in order to describe the viscosity temperature dependence in the temperature range T_g–T_m, assuming that the viscosity value of the master $Pd_{82}Si_{18}$ alloy at the melting point is 1 Pa s [36]:

$$\eta = 1.52 \times 10^{-2} \exp\left(\frac{931.7}{T - 600.5}\right). \tag{4.3}$$

The critical cooling rate $\dot{T}_{cr}$ for the metallic glass $Pd_{82}Si_{18}$ was calculated with the aid of Eqs. (4.1), (4.2), and (4.3) to be 1.1×10^4 K/s. This is in very good agreement with the experimental evidence that this glassy alloy possesses very high glass-forming ability [36].

The Amorphous $Fe_{40}Ni_{40}B_{20}$ Alloy

The amorphous $Fe_{40}Ni_{40}B_{20}$ samples were with 0.02 mm thickness by 1.0 mm width. The samples were subjected to varying preliminary isothermal treatment for 0 (as-quenched) and 50 min at temperatures 548, 573, 598, and 633 K, respectively. The nonisothermal viscosity determinations were carried out at heating rates 10 and 20 K/min, respectively. The higher heating rate was used only in the case of the as-quenched samples. The equipment used was Perkin-Elmer TMS-2 with homemade silica glass accessory for high-temperature tensile tests. The results about the influence of the preliminary heat treatment upon the viscous flow behavior are published in [8]. The viscosity temperature dependences of the $Fe_{40}Ni_{40}B_{20}$ samples studied are shown in Fig. 4.15. The steeper line without symbols represents the temperature dependence of the (quasi)equilibrium viscosity η_{eq} (▶ Eq. 5.9). The solid curves represent the regression curves, calculated according to FVM temperature dependence of viscosity at a constant heating rate (▶ Eqs. 5.8) and (▶ 5.17).

As is seen, the viscosity temperature dependences are shifted upward along with increasing the temperature of preliminary isothermal heat treatment. The slope of the low temperature, almost linear part of the curves before reaching the glass transition temperature T_g, increases along with the temperature of preliminary heat treatment. At temperatures higher than the glass transition temperature, the samples as if "forget" their thermal prehistory. The viscosity temperature dependences of all

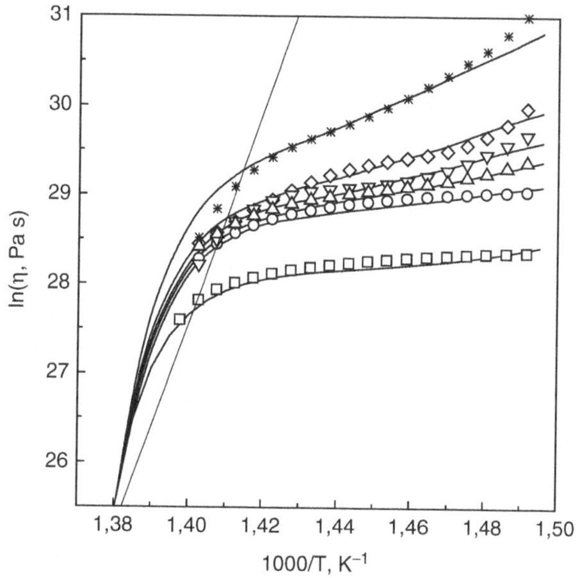

Fig. 4.15 Viscosity temperature dependence of the amorphous $Fe_{40}Ni_{40}B_{20}$ alloy depending on the preliminary heat treatment conditions: □, as-quenched sample measured at 20 K/min; O, as-quenched sample measured at 10 K/min. Samples, preliminary heat treated at Δ, 548°K; ∇, 573°K; ◇, 598°K; *, 633°K

samples overlap each other and begin to follow a unique course, i.e., the effect of preliminary heat treatment disappears. The reason for this circumstance is that at temperature higher than the glass transition temperature, the rate of structural relaxation becomes so high that all samples reach the temperature dependence of the quasi-equilibrium viscosity of the amorphous alloy studied. This viscosity temperature dependence is unique for each sample and depends only on the transitory value of the temperature reached.

The experimental data as mentioned above are considered in the framework of the free volume model (FVM) (see ▶ Chap. 5). The calculation of the FVM parameters is carried out on the basis of nonlinear regression analysis. All experimental data are subjected to the regression analysis *simultaneously*. The FVM parameters obtained (for heating rate 10 K/min) are shown in Table 4.10. The values of the glass transition temperature T_g and the viscosity $\eta(T_g)$ at heating rate 20 K/min are 713 K and 1.1×10^{12} Pa s, respectively.

As is seen, all FVM regression parameters of one and the same type coincide with each other, regardless of the sample thermal prehistory, except the initial concentration of flow defects $c_{f,o}$. The dependence of the initial concentration of flow defects upon the temperature of the preliminary isothermal heat treatment is shown in Fig. 4.16. The concentration $c_{f,o}$ decreases rapidly more than four times along with increasing the temperature of isothermal pre-annealing from 548 to 643 K.

This effect is called annealing-out of flow defects, due to structural relaxation.

Table 4.10 Characteristic parameters of viscous flow of $Fe_{40}Ni_{40}B_{20}$ metallic glass depending on the sample thermal prehistory. The experimental data are considered in the framework of the FVM

Parameters, dimensions	Parameter values and temperatures of the preliminary heat treatment				
	As-quenched	3×10^3 s at 548 K	3×10^3 s at 573 K	3×10^3 s at 598 K	3×10^3 s at 633 K
ν_r, 1/s	7.3×10^{22}	7.3×10^{22}	7.3×10^{22}	7.3×10^{22}	7.3×10^{22}
Q_r, kJ/mol	150.6	150.6	150.6	150.6	150.6
$c_{f,o}$	5.9×10^{-10}	1.77×10^{-13}	8.6×10^{-14}	4.25×10^{-14}	1.24×10^{-14}
T_o, K	543	543	543	543	543
B, K	5355	5355	5355	5355	5355
Q_η, kJ/mol	210.4	210.4	210.4	210.4	210.4
η_o, Pa.s/K	1.42×10^{-20}	1.42×10^{-20}	1.42×10^{-20}	1.42×10^{-20}	1.42×10^{-20}
T_g, K	710	710	709	709	707
$\eta(T_g)$, Pa s	2.3×10^{12}	2.5×10^{12}	2.8×10^{12}	3.2×10^{12}	5.4×10^{12}

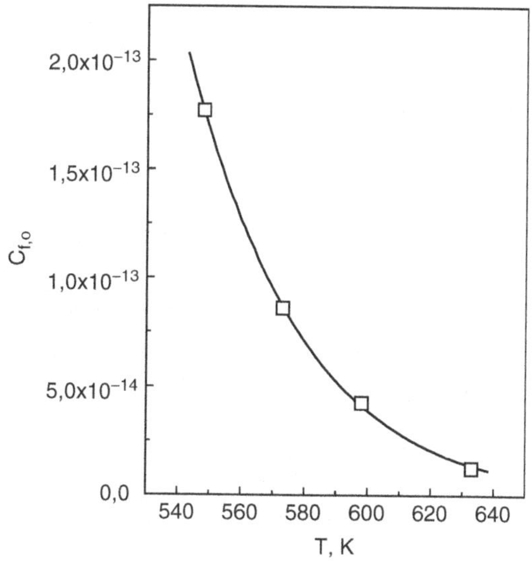

Fig. 4.16 Decrease of the initial concentration of flow defects $c_{f,0}$ in amorphous $Fe_{40}Ni_{40}B_{20}$ glassy alloy along with increasing the temperature of isothermal pre-annealing for 50 min

Influence of Technological Production Parameters Upon the Viscous Flow of the Amorphous Metallic Alloy $Fe_{40}Ni_{40}Si_6B_{14}$

This study was carried out by using samples of ribbonlike $Fe_{40}Ni_{40}Si_6B_{14}$ metallic glass. The amorphous ribbons were produced via rapid quenching from the melt according to the CBMS method. The technological parameters of rapid quenching were controlled very precisely, as described in ▶ Chap. 3 of this book. The viscosity determinations were carried out by using silica glass dilatometer Perkin-Elmer TMS-2 with silica glass assembly for high-temperature tensile tests. The melting point ($T_m = 1230$ K) of the crystalline master alloy was determined with DTA equipment Perkin-Elmer DTA 1700. The onset temperature of crystallization T_{on} and the crystallization enthalpy ΔH_x of the amorphous samples were measured with the aid of differential scanning calorimeter Perkin-Elmer DSC 2C at a heating rate of 20 K/min. The viscosity temperature dependences of the $Fe_{40}Ni_{40}Si_6B_{14}$ amorphous samples, depending upon the technological conditions of the ribbon production, are shown in Fig. 4.17 [9] in coordinates $\ln(\eta)$ vs. $1000/T$.

It is obvious that the experimental data are grouped in two well-differentiated zones, depending upon the peripheral velocity of the quenching disk used – 20 or 40 m/s. The amorphous ribbons that produced at $V_s = 20$ m/s demonstrate up to three times higher viscosity values (see Table 4.11) as compared to those that produced at $V_s = 40$ m/s. This circumstance could be explained with the relation of Ruhl [37], for the dependence of the melt-quenching rate upon the

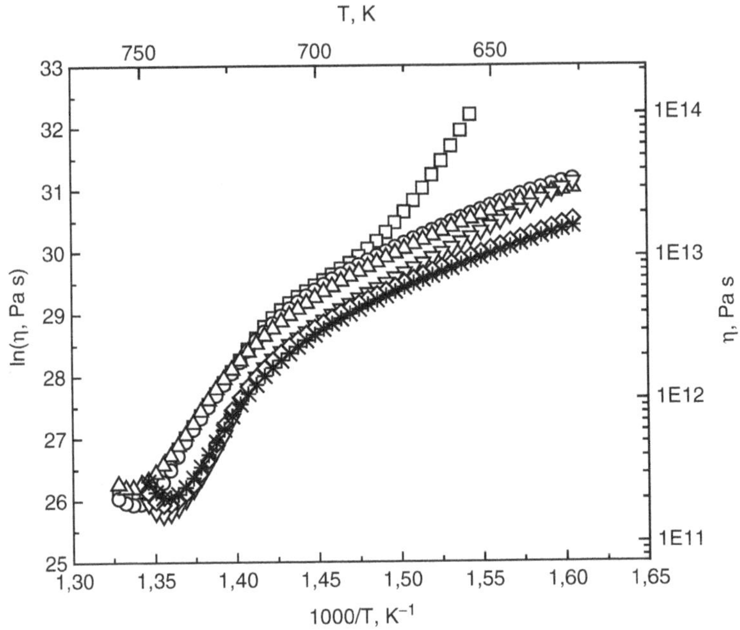

Fig. 4.17 Viscosity temperature dependences of $Fe_{40}Ni_{40}Si_6B_{14}$ amorphous ribbons, depending upon the melt overheating ΔT_m and the peripheral quenching disk velocity V. $V_s = 20$ m/s: Δ, $\Delta T_m = 40$ K; O, $\Delta T_m = 100$ K; □, $\Delta T_m = 400$ K; $V_s = 40$ m/s: ∗, $\Delta T_m = 40$ K; ◇, $\Delta T_m = 300$ K; ∇, $\Delta T_m = 400$ K

technological production parameters of amorphous ribbons upon the rate of rapid quenching $\dot{T}$:

$$\dot{T} = \left\{ \frac{h(T^m - T^s)}{\rho^l c_p^l} \right\} / t \approx \left\{ \frac{h(T^m - T^s)}{\rho^l c_p^l} \right\} (V_s)^{0.75}, \tag{4.4}$$

where h is the coefficient of heat transfer, c_p^l is the specific heat capacity of the supercooled melt, and T^m and T^s are the melt temperature and the surface temperature of the quenching disk, respectively.

Assuming that the surface temperature T^s of the quenching disk is constant and doesn't depend upon the variations of the melt overheat ΔT^m, the increase of ΔT^m from 40 to 400 K should cause an increase of $\dot{T}$ by a factor of 1.4, while the twofold increase of V_s from 20 to 40 m/s should increase the cooling rate $\dot{T}$ by a factor of 1.7. While the estimation about the melt cooling rate increase depending upon increase of peripheral disk velocity seems to be quite realistic, on the contrary, the increase of quenching rate depending upon the melt overheat endures serious criticism. Obviously, the quenching disk surface temperature, and the temperature of the amorphous ribbon in contact with it, should be considerably higher by melt overheat of 400 K as compared to the case by melt overheat of 40 K [38]. Taking

Table 4.11 Characteristic parameters of crystallization and viscous flow of $Fe_{40}Ni_{40}Si_6B_{14}$ metallic glass depending upon the technological production conditions [9]

V_s, m/s	ΔT_m, K	T_{on}, K	ΔH_x, kJ/mol	T_g, K	$\eta(T_g)$, Pa s
20	40	715	−3.87	707	2.9×10^{12}
20	100	–	–	707	3.1×10^{12}
20	400	726	−5.37	707	3.2×10^{12}
30	40	723	−3.77	–	–
30	400	714	−3.65	–	–
40	40	714	−3.69	710	1.1×10^{12}
40	300	–	–	710	1.1×10^{12}
40	400	719	−3.89	710	1.4×10^{12}
50	40	715	−3.77	–	–
50	400	708	−3.26	–	–
V_s, m/s	ΔT_m, K	Q_η $(T < T_g)$, kJ/mol	Q_η $(T > T_g)$, kJ/mol	B^{VFT}, K	T_o^{VFT}, K
20	40	150	280	3311	619
20	100	160	320	3311	619
20	400	170	310	3311	619
30	40	–	–	–	–
30	400	–	–	–	–
40	40	100	330	3306	620
40	300	100	370	3306	620
40	400	120	400	3306	620
50	40	–	–	–	–
50	400	–	–	–	–

into account this circumstance, one has to accept the idea that considerable additional annealing of the ribbons takes place still during the process of ribbon production, leading to higher degree of structural relaxation and, consequently, to higher viscosity values of the metallic glasses produced at melt overheats of 440 K as compared to the case at melt overheat of 40 K.

In the framework of a differentiated group of viscosity data, obtained by one and the same peripheral quenching disk velocity V_s, it is obvious that, except the data for ribbons obtained at melt overheat $\Delta T^m = 400$ K, all other viscosity temperature dependences overlap each other. The viscosity curves posses a typical form with two almost linear parts crossing each other at the glass transition temperature T_g and a higher temperature segment of the temperature dependence after the onset temperature of crystallization T_{on}. In this section the viscosity grows up, due to the appearance of crystalline regions, imbedded into the amorphous parent phase. The slopes of both almost linear parts of viscosity temperature dependences were used for calculation of the empirical effective activation energies $Q_\eta(T < T_g)$ and $Q_\eta(T > T_g)$, respectively.

Specific feature, only in the case of ribbons obtained by melt overheat $\Delta T^m = 400$ K, is the presence of a third low-temperature curved region in the viscosity temperature dependence at $T < T_g - 30$ K, of stronger viscosity temperature dependence, showing considerably higher viscosity values; see Fig. 4.17. This peculiarity is most probably due to the structural relaxation, annihilation of free volume, and densification of the ribbon amorphous structure still during the contact of the ribbon with the quenching disk, with consequent viscosity increase.

When discussing the viscous behavior of the $Fe_{40}Ni_{40}Si_6B_{14}$ amorphous samples produced under different production conditions, one has always to keep in mind that the different samples are produced by melt-quenching rates of 10^6 K/s order of magnitude, but which nevertheless differ from each other by a factor of 2–3, depending on the production parameters used. These differences in the melt-quenching rates are reflected in somewhat different structural states of the amorphous ribbons obtained. The samples in all cases remain very far from quasi-equilibrium amorphous state and differ from each other in the concentration of structural defects. During the procedure of viscosity determinations, the ribbons are subjected to additional heating from room temperature with a constant heating rate of 20 K/min. This rate of heating is several orders of magnitude lower than the rate of ultra-rapid cooling by their production. At high enough temperatures, at which the atomic mobility becomes measurable, the metallic glass slowly begins to relax, trying to approach the quasi-equilibrium structural state, following trajectory, which depends upon the thermal prehistory of melt solidification [39]. This is very weak reflected by the viscosity data, discussed above. According to Chen [40], the increase of glass transition temperature T_g from 707 K at $V_s = 20$ K/min to 710 K at $V_s = 40$ K/min can be related to the differences in the ultra-rapid melt quenching in the following way:

$$\frac{dT_g}{d\ln\dot{T}} \approx \frac{kT_g^2}{Q_\eta(T > T_g)}. \tag{4.5}$$

By $Q_\eta(T > T_g) = 360$ kJ/mol (3.8 eV), for the ratio $\dot{T}(V_s = 40\ \mathrm{m/s})/\dot{T}(V_s = 20\ \mathrm{m/s})$, one obtains approximately the value of 1.5. This is in very good agreement with the estimation, obtained with the aid of Eq. (4.5).

On the contrary to the expectations, that at temperatures higher than the glass transition temperature the two groups of viscosity temperature dependences will overlap each other building a common quasi-equilibrium viscosity temperature dependence $\eta_{eq}(T)$, they only demonstrate a tendency for approach to each other. This is somewhat similar to the behavior of $Pd_{82}Si_{18}$ metallic glass, depending upon the sample thermal prehistory, discussed in section "Influence of Chemical Composition Upon the Viscous Flow Behavior and Thermal Stability of Some Binary Glassy Metals" of this chapter, as well as with the observations of Volkert and Spaepen [17] for the viscous flow behavior of $Pd_{40}Ni_{40}P_{19}Si_1$ metallic glass. In the case of this amorphous alloy, the iso-configurational viscosities of two different structural states – of as-quenched sample and of preliminary annealed sample – demonstrate opposite directions of structural relaxation by their approach to reach their quasi-equilibrium structural state for a definite constant temperature. Due to the rapid sample crystallization, however, it is impossible to reach high enough temperatures, preserving the amorphous state of the samples, thus ensuring the quasi-equilibrium structural state.

It follows from the above described that:

- The twofold increase of the peripheral quenching disk rate exerts much greater influence upon the viscous flow behavior of $Fe_{40}Ni_{40}Si_6B_{14}$ metallic glass as compared to the influence exerted by one order of magnitude increase of melt overheat.
- The melt overheat increase from 40 to 400 K, respectively, causes additional relaxation annealing of the amorphous ribbon still during the production process.
- Even in the temperature range between the glass transition temperature T_g and the onset temperature of crystallization T_{on}, the sample amorphous structure doesn't reach the quasi-equilibrium state.

References

1. Russew K, Sommer F, Duhaj P, Bakonyi I (1992) Viscous flow behaviour of Ni_xZr_{100-x} metallic glasses from $Ni_{30}Zr_{70}$ to $Ni_{64}Zr_{36}$. J Mater Sci 27:3565
2. Russew K, Stojanova L (1993) Viscous flow behaviour and thermal stability of $Ni_{100-x}P_x$ metallic glasses from $Ni_{84}P_{16}$ to $Ni_{79}P_{21}$. Mater Lett 17:199
3. Russew K, Anestiev L, Stojanova L et al (1995) Thermal stability and viscous flow behaviour of $Fe_{100-x}P_x$ metallic glasses. J Mater Sci Technol 3(2):3
4. Russew K, Stojanova L, Anestiev L et al (1989) Influence of vanadium alloying additions on the viscous flow and critical cooling rate of amorphous Fe-B alloys. In: Proceedings of international conference on days of metallurgy'6 vol 1, Balatonaliga, 1989, p 118
5. Russew K, Sommer F, Stojanova L (1993) Influence of Cr-alloying additions on the viscous flow behaviour of $Fe_{82-x}Cr_xB_{18}$ amorphous alloy. In: Duhaj P, Mrafko P, Svec P (eds) Proceedings of conference on amorphous metallic materials AMM III. Trans Tech Publications, Slovakia, p 625
6. Stojanova L, Russew K, Fazakas E, Varga LK (2012) Thermo-mechanical study of rapidly solidified amorphous alloys $Al_{85}Ni_5Co_2RE_8$. J Alloys Compd 540:192
7. Stojanova L, Russew K, Illekova E (1991) Study of the structural relaxation of $Pd_{82}Si_{18}$ metallic glass by thermal expansion and viscous flow measurements. Mater Sci Eng A133:529
8. Russew K, Stojanova L, Koleva E (1989) Thermal pre-history and non-isothermal viscous flow of $Fe_{40}Ni_{40}B_{20}$ glassy alloy. In: National conference metal science and heat treatment of metals'98. Scientific communications on STU machine building vol V, June 1998, p 80
9. Russew K, Stojanova L, Lovas A (1993) Effect of processing conditions on the ribbon geometry and viscous flow behaviour of $Fe_{40}Ni_{40}Si_6B_{14}$ amorphous alloy. Int J Rapid Solidif 8:147
10. Dong YD, Gregan G, Scott MG (1981) J Non-Cryst Solids 43:403
11. Kolb-Teliens A, Shu-Song T (1988) J Non-Cryst Solids 107:122
12. Araki T, Abe T, Tanaka K (1989) Mater Trans JIM 30:748
13. Altounian Z, Strom-Olsen JO (1983) Phys Rev B 27:4149
14. Sharma SK, Mukhopadhyay P (1990) Acta Metall Mater 38:129
15. Fuqian Z et al (1988) Mater Sci Eng 97:487
16. Altounian Z, Guo-hua T, Strom-Olsen JO (1983) J Appl Phys 54:3111
17. Volkert CA, Spaepen F (1990) Scr Met Mater 24:463
18. Kissinger HE (1957) Anal Chem 29:1702

19. Kim C-O, Kim T-K, Takahashi M (1982) In: Masumoto T, Suzuki K (eds) Proceedings of conference on rapidly quenched metals RQ4 vol 1, Japan Institute of Metals, p 723
20. Sommer F (1981) Zs Metallkde 72:219
21. Takayama S (1976) J Mater Sci 11:164
22. Greer AL (1982) Acta Metall 30:171
23. Barfield RN, Kitchener JA (1955) J Iron Steel Inst 180:324
24. Wertmann AA, Samarin AM (1969) Properties of liquid iron alloys. Nauka, Moskow, p 208 (In Russian)
25. Uhlmann DR (1972) J Non-Cryst Solids 7:337
26. Onorato PIK, Uhlmann DR (1976) J Non-Cryst Solids 22:367
27. Myung WN et al (1991) Mater Sci Eng A133:513
28. Inoue A, Yamamoto M, Kimura HM, Masumoto T (1987) J Mat Sci Lett 6:194
29. Inoue A, Ohtera K, Tsai AP, Masumoto T (1988) Japn J Appl Phys 27:L479; L280; L1796
30. He Y, Poon SY, Shiflet GJ (1988) Scr Met Mater 22:1813
31. Sanders WS, Warner JS, Miracle DB (2006) Intermetal 14:348
32. Fazakas E, Varga LK (2007) J Mater Sci Technol 15(4):211
33. Fazakas E, Varga LK (2008) Rev Adv Mater Sci 18:494
34. Michalik S, Bednarcik J, Jóvári P et al (2010) J Phys Condens Matter 22:404209. doi:10.1088/0953-8984/22/40/404209
35. Taub AI, Spaepen F (1980) Acta Metall 28:1781
36. Davies HA (1980) In: Duhaj P, Mrafko P (eds) Proceedings of conference on amorphous metallic materials'78. Veda Publication House, Bratislava, p 107
37. Ruhl RC (1967) Mater Sci Eng 1:313
38. Patterson JP, Jones DRH (1980) Acta Metall 28:657
39. Volkert CA, Spaepen F (1988) Mater Sci Eng 97:449
40. Chen HS (1982) In: Masumoto T, Suzuki K (eds) Proceedings of conference on rapidly quenched metals RQ4 vol 1, Japan Institute of Metals, p 495

5 Basic Concepts of the Free Volume Model: Structural Relaxation and Viscosity in the Framework of the Model

Abstract

The basic concept and ideals of the free volume model (FVM) of Turnbull and Cohen are considered. The relation between structural relaxation and atomic free volume in amorphous condensed matter is pointed out. The viscosity temperature dependence of amorphous metallic alloys in the framework of the free volume model, as considered by F. Spaepen, A. van den Beukel, P. Duine, etc., is presented and discussed. The contribution of the authors to the extension of the FVM ideas for describing the viscosity temperature dependence of glassy metals under nonisothermal experimental conditions is presented. An experimental proof of the correctness of this FVM extension is given.

Keywords

Glassy metals • Amorphous metallic alloys • Free volume model • Basic concepts • Viscosity • Relaxation • Nonisothermal • Free volume annihilation/production

Basic Ideas of the Free Volume Model

The free volume model (FVM) describes the atomic mobility in liquids, undercooled melts, and amorphous solids via the free volume available for a single atom within the material structure. The understanding of the term *structural relaxation* requires a definition for the term *amorphous structure*. The lack of translational structural symmetry is the principal difference between the amorphous metallic alloys and their crystalline counterparts. This is due to the circumstance that each atom is surrounded of a free volume of different forms and sizes. In the free volume theory, initially developed by Cohen and Turnbull [1, 2] to describe the atomic mobility in liquids, to each single atom is ascribed a definite volume v'. This volume can be demonstrated with the constructions of Voronoi or Wigner–Seitz.

K. Russew, L. Stojanova, *Glassy Metals*,
DOI 10.1007/978-3-662-47882-0_5

The Wigner–Seitz cell, named after Eugene Wigner and Frederick Seitz, is a type of Voronoi cell used in the study of condensed matter in solid-state physics. The Wigner–Seitz cell around a structural point is defined as the site (locus) of points in space that are closer to that structural point than to any of the other structural points. The unique property of a crystal is that its atoms are arranged in a regular three-dimensional array called a lattice. All the properties attributed to crystalline materials originate from this highly ordered structure. Such a structure exhibits discrete translational symmetry. In order to model and study such a periodic system, one needs a mathematical "handle" to describe the symmetry and hence draw conclusions about the material properties consequent to this symmetry. The Wigner–Seitz cell is a means to achieve this. The cell may be chosen by first picking a single atom within the structure. Then, lines are drawn to all nearby (closest) atoms. At the midpoint of each line, another line is drawn normal to each of the first set of lines. In the case of a three-dimensional structure, a perpendicular plane is drawn at the midpoint of the lines between the structural single atom sites. By using this method, the smallest area (or volume) is enclosed in this way and is called the Wigner–Seitz primitive cell. All area (or space) within the structure will be filled by this type of primitive cell and will leave no gaps. It is made by drawing planes normal to the segments joining nearest structural atomic sites to a particular structural point, through the midpoints of such segments.

This is shown schematically in Fig. 5.1 for two-dimensional amorphous structure. The internal polyether enclosed by these planes possesses a volume υ'.

Further on:

- The free volume of a single atom v_f is defined as the difference between the mean value $<\upsilon>$ of space volume around a single atom and volume Ω, which represents the mean value belonging to a single atom in a dense random packed atomic structure (an ideal amorphous structure).
- The atomic transport in the environment of a definite atom is only possible if its free volume υ_f is greater than the so-called critical free volume υ^*. An atom, which possesses free volume equal or greater than υ^*, is called structural defect.
- The redistribution of free volume between the atoms does not require free energy change of the system.

It follows from the last assumption [1] that the probability for the appearance of a void of volume υ' is

$$P(\upsilon') = \frac{\gamma}{\upsilon_f}\exp\left(\frac{-\gamma\upsilon'}{\upsilon_f}\right), \tag{5.1}$$

where γ is geometric overlap factor, in which the value is between 0.5 and 1. The full probability $P(\upsilon^*)$ for the appearance of a void of a volume greater than υ^* is $P(\upsilon^*) = \int_{\nu*}^{\propto} P(\upsilon^*)d\upsilon = \exp(-\gamma\upsilon^*/\upsilon_f)$. This probability is called concentration of structural defects c_f:

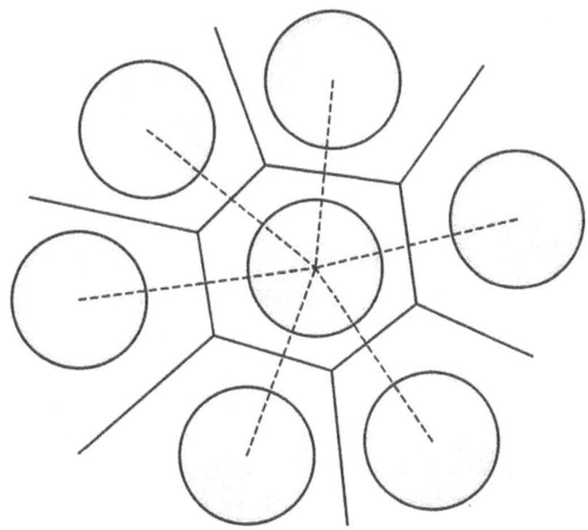

Fig. 5.1 Construction of a cell – atomic free volume around an atom in two-dimensional amorphous structure according to Wigner–Seitz

$$c_f = \exp\left(\frac{-\gamma v^*}{v_f}\right) = \exp\left(\frac{-1}{x}\right), \tag{5.2}$$

where $x = \frac{v_f}{\gamma v^*}$ is the so-called reduced free volume. It is a dimensionless quantity of about several hundredths.

Structural Relaxation and Atomic Free Volume

By rapid quenching of glass-forming melts, their structure is frozen into a metastable state. Should it be possible to track out the mean atomic free volume v_f as a function of temperature during the rapid quenching, it should decrease in a manner shown in Fig. 5.2.

At comparatively high temperatures in the temperature range between the alloy melting point T_m and $T_{m.eq.}$ (see Fig. 5.2), the melt is in quasi-equilibrium structural state. The corresponding quasi-equilibrium atomic free volume is given, according to Turnbull and Cohen [1, 2], as

$$v_{f,eq}(T) = x_{eq}(T)\gamma v^* = \frac{T - T_o}{B}\gamma v^*, \tag{5.3}$$

where B and T_o are two model parameters.

At lower temperatures, $T < T_{m.eq.}$, the atomic rearrangements become too slow and are no more able to follow the quasi-equilibrium temperature dependence structural arrangement. As a result, in the structure is frozen-in nonequilibrium excess free volume. By additional isothermal heat treatment at temperature T_1, the frozen-in excess free volume annihilates, trying to reach its quasi-equilibrium value $v_{f,eq.}(T_1)$; see Fig. 5.2. Provided that after a rapid increase and arrest of the temperature to T_2 occurs, the free volume should change to its new quasi-

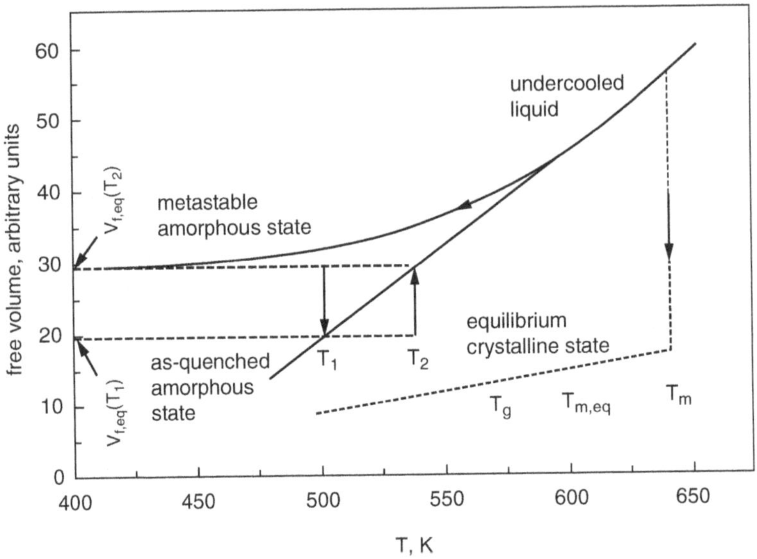

Fig. 5.2 Change of mean atomic free volume υ_f per a single atom as a function of temperature (schematic)

equilibrium value $\upsilon_{f,eq.}(T_2)$. These free volume changes along with the temperature/time changes are, of course, not instantaneous, but possess a definite mechanism and kinetics, about which the free volume theory of Turnbull and Cohen *does not propose any indications and which should be specified experimentally*.

The structural relaxation of a nonequilibrium amorphous structure with concentration of structural defects higher than the quasi-equilibrium concentration of structural defects $c_{f,eq.}(T)$ is identified as structure, striving to reach the quasi-equilibrium for a given temperature structural state with concentration of structural defects $c_{f,eq.}(T)$ via annihilation or production of free volume.

Viscosity Temperature Dependence of Amorphous Metallic Alloys in the Framework of the Free Volume Model

The viscosity of amorphous metallic alloys under moderate applied tangential stress τ is described by the well-known equation of Newton (▶ Eq. 3.5), i.e., the viscosity η is in a direct ratio to the applied tangential τ stress and in reverse ratio to the macroscopic deformation rate $\dot{\varepsilon}$. Spaepen [3] has further developed the free volume theory in order to describe the viscous flow behavior of metallic glasses, introducing the concept of a structural flow defect. The basic characteristics of the flow defect are that by applying a tangential stress τ, it undergoes local tangential deformation ε_o, which is elastically transmitted to the sample surface and causes macroscopic deformation. The rate of tangential deformation (viscous flow) should be consequently presented as

$$\dot{\varepsilon} = \frac{c_f}{\Omega} \upsilon_o \varepsilon_o k_r, \tag{5.4}$$

where c_f is the concentration of structural flow, υ_o is the volume of the flow defect, and k_r is a frequency factor of atomic rearrangements in the vicinity of flow defects.

Applying the theory of absolute reaction rate [4], Spaepen proposes the following expression for k_r:

$$k_r = \nu_r \exp\left(-\frac{Q_f}{kT}\right) \sinh\left(\frac{\tau \varepsilon_o \upsilon_o}{kT}\right), \tag{5.5a}$$

where ν_r is vibrational (attempt) frequency and Q_f is the activation energy for atomic jump over. At $\frac{\tau \varepsilon_o \upsilon_o}{kT} << 1$, Eq. (5.5a) reduces to

$$k_r = \nu_r \left(\frac{\tau \varepsilon_o \upsilon_o}{kT}\right) \exp\left(-\frac{Q_f}{kT}\right), \tag{5.5b}$$

where k_r denotes the rate constant for relaxation.

Combining Eq. (5.3) with Eqs. (5.4) and (5.5b), one obtains for the viscosity η

$$\eta = \exp\left(\frac{Q_f}{kT}\right) \frac{kT\Omega}{(\varepsilon_o \upsilon_o)^2 \nu_r} \frac{1}{c_f}. \tag{5.6}$$

A. van den Beukel et al. [5] have shown on the basis of empirical analysis of own experimental data, and data of other researchers, that in the temperature range of measurable viscosity data, the product $(\varepsilon_o \upsilon_o)$ depends exponentially on the temperature according to the expression:

$$\varepsilon_o \upsilon_o = A \exp\left(-\frac{Q_s}{kT}\right). \tag{5.7}$$

Taking into account this circumstance, one obtains for the viscosity:

$$\eta = \eta_o T \exp\left(\frac{Q_\eta}{RT}\right) \left(\frac{1}{c_f}\right), \tag{5.8}$$

where $Q_\eta = Q_f + 2Q_s$ is the molar activation energy for viscous flow in (J/mol), and $\eta_o = \frac{k\Omega}{A^2 \nu_r}$, in (Pa s/K).

Taking into account the dependence of the flow defect concentration on the free volume (Eq. (5.3)) and the temperature dependence of the quasi-equilibrium free volume, one obtains for the quasi-equilibrium viscosity η_{eq}

$$\eta_{eq} = \eta_o T \exp\left(\frac{Q_\eta}{RT}\right) \exp\left(\frac{B}{T - T_o}\right). \tag{5.9}$$

Equation (5.9) is the so-called hybrid equation [5] which is a combination of an Arrhenius temperature dependence and temperature dependence of the type of the empirical equation of Vogel–Fulcher–Tammann. Equations (5.8) and (5.9) deserve a special attention, as far as they provide a tool for the analysis of the experimental viscosity data for amorphous metallic alloys gained under isothermal and nonisothermal experimental conditions. These equations also give the possibility to determine the time–temperature dependence of structural defect concentration in glassy metals. In this way, it becomes possible to determine the mechanism and kinetics of glassy metal structural relaxation and to analyze quantitatively its determining role and influence upon a series of glassy metal properties, such as the thermal expansion anomalies, the specific heat, the relaxation of bend stresses, relaxation anomalies in glassy metals, density changes, etc.

The Influence of Structural Relaxation Upon Viscosity: Useful Tool for the Study of the Mechanism and Kinetics of Free Volume Relaxation Changes

Viscosity and Relaxation Under Isothermal Conditions

When an amorphous metallic alloy is not in internal (quasi-) equilibrium, the concentration of structural defects is not constant, but changes, due to the topological rearrangements of atoms. Taub and Spaepen [6] were the first who have found that the defect concentration changes cause linear increase of glassy metal viscosity along with increasing the time of isothermal annealing far from their metastable equilibrium. In that case, it follows from Eq. (5.8) that the concentration c_f of flow defects decreases inversely proportional to the time of annealing, e.g.,

$$\frac{dc_f}{dt} = -k_r c_f^2, \tag{5.10}$$

where $k_r = \nu_r \exp\left(-\frac{Q_r}{RT}\right)$ in that case is considered as the rate constant of the process of annihilation of flow defects and Q_r the activation energy of annihilation. Equation (5.10) supposes that the process of defect annihilation is a bimolecular reaction, i.e., the process of defect annihilation occurs only in the case if two defects encounter each other. Since Eq. (5.10) is valid only if $c_f >> c_{f,eq.}$, it describes only the relaxation annihilation of free volume at temperatures considerably lower than the glass transition temperature T_g. By relaxation annealing at temperatures, at which it is quite close to its quasi-equilibrium state (temperatures a little bit lower or a little bit higher than T_g), Eq. (5.10) should include also a term, taking into account the production of new defects. This problem is considered by Koebrugge et al. [8]. Equation (5.10) has to be expanded including term which takes into account the production of defects:

$$\frac{dc_f}{dt} = -k_r c_f^2 + k_p P, \tag{5.11}$$

where P is a constant and k_p is the rate constant of defect production process. Objectively, it should be equal to k_r. When the system comes to equilibrium, the total change of defect concentration should equal to 0 and c_f becomes equal to $c_{f,eq}$. In that case

$$\frac{dc_f}{dt} = -k_r \left(c_f^2 - c_{f,eq}^2 \right). \tag{5.12}$$

Another possibility is that the production of defects occurs at already existing defects, i.e., $P = c_f S$. In that case, the constant S becomes equal to $(k_r/k_p)c_{f,eq.}$, and the differential equation obtains the form

$$\frac{dc_f}{dt} = -k_r c_f \left(c_f - c_{f,eq} \right). \tag{5.13}$$

Tsao and Spaepen [8], on the other hand, supposed also a mechanism, in which the sites for annihilation and production of defects depend upon the number of defects in excess, i.e., on the difference $(c_f\_c_{f,eq.})$.

The differential equation transforms in that case to

$$\frac{dc_f}{dt} = -k_r \left| c_f - c_{f,eq} \right| \left(c_f - c_{f,eq} \right). \tag{5.14}$$

None of the proposed possibilities can be a priori accepted as the proper description of the mechanism of annihilation and production of structural defects. Equation (5.14), as proposed by Tsao and Spaepen, is unrealistic from the physical point of view because it supposes different mechanisms of relaxation at the opposite sites of the metastable structural equilibrium. This is why this possibility should be rejected. One has to choose between Eqs. (5.12) and (5.13). This problem is subject for a thorough analysis of Duine et al. [9], in an experimental study of the isothermal viscous flow behavior of $Pd_{40}Ni_{40}P_{20}$ amorphous metallic alloy at the opposite sites of the metastable structural equilibrium. The analysis of Duine et al. has shown that Eq. (5.13) describes better the time dependence of viscosity during the isothermal annealing at the opposite sites of the (quasi-) equilibrium, trying to reach its (quasi-) equilibrium value upward and downward at two different temperatures of isothermal annealing. On the basis of Eq. (5.13) and the law of Newton for viscous flow (▶ Eq. 3.5), Duine et al. have derived the following equation for the sample elongation as a function of the temperature and the time of isothermal annealing:

$$\varepsilon(t) = \frac{\tau \exp\left(\frac{-Q_\eta}{RT} \right)}{\eta_o T} \left(c_{f,eq} t + \frac{1}{k_r} \ln \left(1 - \frac{c_{f,o} - c_{f,eq}}{c_{f,o}} \right) \exp(-k_r c_{f,eq} t) - \frac{1}{k_r} \ln \frac{c_{f,eq}}{c_{f,o}} \right). \tag{5.15}$$

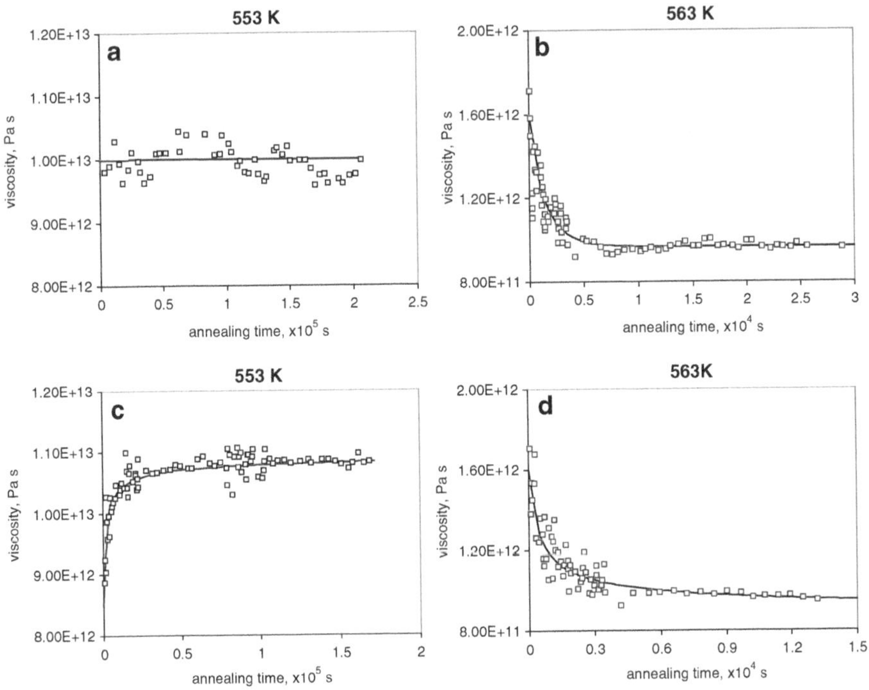

Fig. 5.3 Reversible cyclic viscosity changes of $Pd_{40}Ni_{40}P_{20}$ metallic glass during the time of annealing striving to reach its quasi-equilibrium value at two different temperatures of isothermal annealing (After Duine et al. [9]). (**a**) Quasi-equilibrium viscosity at 553 K; (**b**) decrease of viscosity along with the increasing time of annealing after sudden change of the temperature from 553 to 563 K, till reaching the new quasi-equilibrium value; (**c**) increase of viscosity along with the increasing time of annealing after sudden change of the temperature from 563 to 553 K; (**d**) repeated decrease of viscosity by the new rapid temperature increase from 553 to 563 K till reaching the new quasi-equilibrium viscosity value at 563 K

The data of Duine et al. as reproduced from the original article [9] are shown in Fig. 5.3.

Equation (5.15) has been used by Duine et al. to calculate the best fit of the experimentally determined $Pd_{40}Ni_{40}P_{20}$ sample elongation data with the aid of nonlinear multiparameter regression analysis. Equation (5.15) incorporates six regression coefficients – the viscosity pre-exponential factor η_o, the activation energy for viscous flow Q_η, the attempt (atomic vibrational) frequency ν_r, the activation energy for relaxation Q_r, the model parameters B and T_o, and the initial concentration of flow defects $c_{f,o}$. The successful application of Duine et al. [9] approach is possible only for a very limited number of amorphous metallic alloys with very high thermal stability. The most important for practical use glassy metals crystallize quickly in the vicinity of the glass transition temperature. This circumstance makes impossible the viscosity determination around T_g. However, as it was extensively described in ▶ Chap. 2, the viscosity determination around T_g becomes possible via nonisothermal sample elongation creep measurements with a constant heating rate. The necessity to find out an equation, describing the viscosity temperature dependence of amorphous metallic alloys under nonisothermal experimental conditions, becomes obvious. This requires further development of the theoretical concepts of Cohen et al. [1, 2], Spaepen et al. [3, 6], Koebrugge et al. [7], Duine et al. [9], and van den Beukel et al. [10]. The authors provide such a development in order to make possible the application of the FVM for the description of the viscosity temperature dependence of glassy metals under nonisothermal experimental conditions.

Viscosity and Relaxation Under Nonisothermal Conditions

Stemming from the research results of Duine et al. [9], Russew et al. [11] have carried out an experimental determination of the viscosity temperature dependence of ribbonlike $Pd_{40}Ni_{40}P_{20}$ glassy samples under continuous heating conditions by a constant heating rate of 10 K/min. The idea was to establish a link to the isothermal experimental viscosity data of Duine et al. The goal was to develop a universal experimental method for viscosity determinations of all kinds of amorphous metallic alloys, irrespective of their thermal stability, together with a method for FVM description of the viscosity temperature dependence.

Under continuous heating conditions with a constant heating rate q, the differential Eq. (5.13), describing the concentration change of structural flow defects as a function of the isothermal annealing time, should be transformed in order to describe the c_f changes depending upon the current temperature and the heating rate q. Taking into account that $\mathrm{d}c_f/\mathrm{d}t = (\mathrm{d}c_f/\mathrm{d}T)q$, Eq. (5.13) transforms to

$$\frac{dc_f}{dT} + P(T)c_f = c_f^2 Q(T), \tag{5.16}$$

where $P(T,q) = -\frac{\nu_r}{q}\exp\left(-\frac{Q_r}{RT} - \frac{B}{T-T_0}\right)$ and $Q(T,q) = -\frac{\nu_r}{q}\exp\left(-\frac{Q_r}{RT}\right)$. This is an equation of Bernoulli [12] of second order. Its solution is

$$c_{f,high}^{-1}(T,q) = \left(c_{f,0}^{-1} - \int_{T_0}^{T} Q(\theta,q)\exp\left(-\int_{T_0}^{\theta} P(\theta'',q)d\theta''\right)d\theta\right)\exp\left(\int_{T_0}^{T} P(\theta',q)d\theta'\right). \tag{5.17}$$

The parameter T_o (the physical meaning of which is the ideal glass transition temperature in the empirical equation of Vogel–Fulcher–Tammann) is considered as the starting temperature of heating. The combination of Eq. (5.8) with Eq. (5.17) represents the FVM description of glassy alloy viscosity temperature dependence under continuous heating conditions with a constant heating rate q. The FVM parameters can be determined with the aid of a nonlinear multiparameter regression analysis of the viscosity experimental data in a similar way as Duine et al. [9] have done it for isothermal experimental conditions of the experiment.

At temperatures, considerably lower than the glass transition temperature T_g, Eq. (5.13) can be simplified taking into account that in that case $c_f(t)$ and $c_{f,o}$ are considerably greater than $c_{f,eq.}$ at all experimental times:

$$\frac{dc_f}{dt} = -k_r c_f\left(c_f - c_{f,eq}\right) \cong -k_r c_f^2. \tag{5.18}$$

The combination of Eq. (5.8) with Eq. (5.18) represents the viscosity dependence upon the annealing time at a constant annealing temperature in the low range, considerably lower than the glass transition temperature. Equation (5.18) is very useful when considering the relaxation phenomena in metallic glasses in the low temperature range as far as it provides the possibility to avoid the necessity to introduce the model parameters B and T_o when performing the regression analysis of the experimental data. In this way, the solution of Eq. (5.18) for nonisothermal annealing conditions with a constant heating rate q obtains the form

$$c_{f,low}^{-1}(T,q) = c_{f,0}^{-1} + \frac{\nu_r}{q}\int_{T_B}^{T} \exp\left(-\frac{Q_r}{RT'}\right) dT', \tag{5.19}$$

where T_B is the starting temperature of heating, for example, the room temperature.

The combination of Eq. (5.19) with Eq. (5.8) describes the temperature dependence of η under nonisothermal experimental conditions at temperatures, considerably lower than the glass transition temperature T_g:

$$\eta(T,q) = \eta_0 T\exp\left(\frac{Q_\eta}{RT}\right)\left[c_{f,0}^{-1} + \frac{\nu_r}{q}\int_{T_B}^{T} \exp\left(-\frac{Q_r}{RT'}\right) dT'\right]. \tag{5.20}$$

In this temperature range, however, the viscosity of the amorphous metallic alloys is too high (usually higher than 10^{16} Pa.s) and cannot be determined directly from the creep deformation curves neither under isothermal nor under nonisothermal heating conditions. A suitable method for indirect determination of viscosity under such circumstances is the study of the relaxation of bend stresses of ribbonlike glassy alloys. This method will be considered in the incoming ▶ Chap. 7. In the concrete case, the validity of Eq. (5.17) was proven [11], with the aid of nonisothermal viscosity determinations of $Pd_{40}Ni_{40}P_{20}$ amorphous ribbons at constant heating rate of 10 K/min by using Heraeus TMA 500 silica glass dilatometer for tensile creep tests. The viscosity temperature dependence obtained was compared with the isothermal viscosity data of Duine et al. [9], for the same amorphous alloy. The fitting procedure of the nonisothermal viscosity data was carried out based on Eqs. (5.8) and (5.17) by using the model parameters, determined by Duine for the same glassy alloy.

The samples from the $Pd_{40}Ni_{40}P_{20}$ amorphous metallic alloy used were in ribbon form with dimensions of 1.45×0.049 mm^2. The temperature accuracy was ± 2 K.

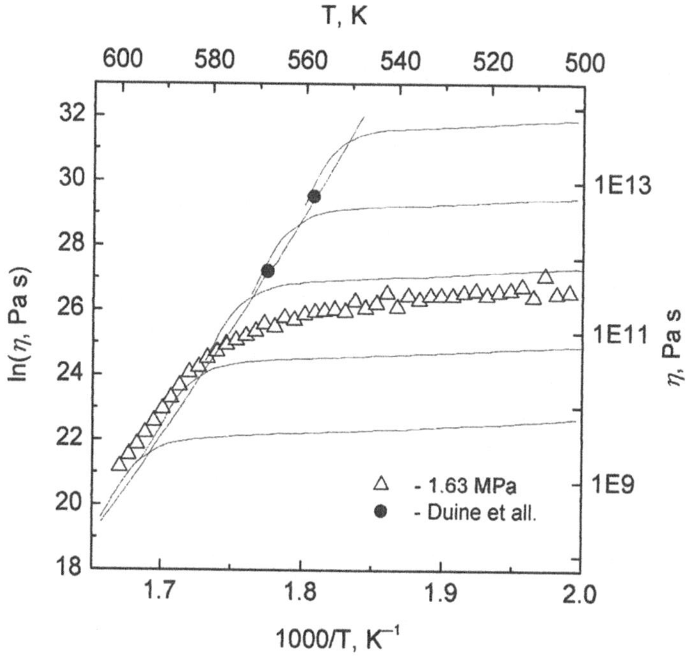

Fig. 5.4 Viscosity temperature dependences of the $Pd_{40}Ni_{40}P_{20}$ glassy metal for heating rates of 0.01, 0.1, 1, 10, and 100 K/min, as well as the temperature dependence of the (quasi-) equilibrium viscosity η_{eq} – (——); (quasi-) equilibrium viscosity, experimentally determined by Duine et al. [9], at 553 и 563 K, – (•); some representative values of the nonisothermally determined viscosity [11] at heating rate of 10 K/min – (Δ)

The sample elongation was monitored by accuracy, better than ±1 μm. The viscosity determination was carried out according to the authors' equation of Newton for nonisothermal experimental conditions (see ▶ Eq. 3.5, ▶ Chap. 3). Figure 5.4 shows the viscosity temperature dependence of the amorphous alloy studied at a heating rate of 10 K/min which is experimentally obtained by Russew et al. [11]. On the same figure, the calculated viscosity temperature dependences of $Pd_{40}Ni_{40}P_{20}$ amorphous metallic alloy at heating rates of 0.01, 0.1, 1, 10, and 100 K/min, as well as the temperature dependence of the quasi-equilibrium viscosity η_{eq} (Eq. (5.9)), according to Eqs. (5.8) and (5.17), are shown. By performing these calculations determined by Duine et al. [9], values of the FVM parameters, namely, $\eta_o = 3.26 \times 10^{-23}$ Pa s/K, $Q_\eta = 193$ kJ/mol, $\nu_r = 3.4 \times 110^{25}$ 1/s, $Q_r = 158$ kJ/mol, $B = 6600$ K, $T_o = 355$ K, and $c_{f,o} = 2.14 \times 110^{-11}$, were used. On the same figure, for comparison purpose, the two values of the quasi-equilibrium viscosity of $Pd_{40}Ni_{40}P_{20}$ glassy alloy are presented also, determined by Duine et al. [9], at 553 и 563 K, as well as several representative values of our experimentally determined viscosity temperature dependence at heating rate of 10 K/min. The almost excellent coincidence of isothermally and nonisothermally obtained viscosity data is obvious.

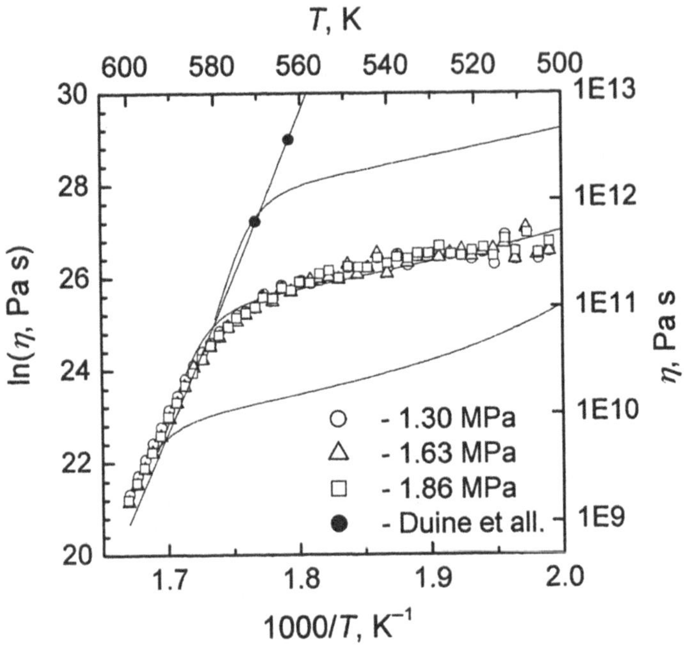

Fig. 5.5 Measured viscosity temperature dependence of $Pd_{40}Ni_{40}P_{20}$ at a heating rate of 10 K/min – (○, Δ, □); calculated viscosity temperature dependence at heating rates of 1, 10, and 100 K/min – (——); quasi-equilibrium viscosity at two constant temperatures – (•), experimentally determined by Duine et al.

There is a striking coincidence between the calculated by us quasi-equilibrium viscosity $\eta_{eq}(T)$ at a heating rate of 10 K/min and the isothermal quasi-equilibrium viscosity values, experimentally determined by Duine et al. [9]. There is a slight discrepancy between the author's nonisothermal experimental viscosity data and those calculated by using the FVM parameters of Duine et al in the temperature range under 575 K. That can be explained with the circumstance that the samples used by Duine and the samples used by the authors have been taken from different batches with somewhat different initial concentration of flow defects $c_{f,o}$.

Figure 5.5 presents the experimental data for the viscosity temperature dependence of $Pd_{40}Ni_{40}P_{20}$ glassy alloy at a heating rate of 10 K/min with all details, as received. As is seen, the viscosity does not depend upon the magnitude of the applied tangential stress, which confirms the statement that the measurements have been carried out in the moderate stress range of homogeneous Newton's viscous flow. The FVM parameter values for calculating the temperature viscosity dependences shown in the figure are as follows: $\eta_o = 3.03 \times 10^{-23}$ Pa s/K, $Q_\eta =$ 196 kJ/mol, $\nu_r = 1.77 \times 10^{23}$ 1/s, $Q_r = 1.26$ kJ/mol, $B = 6900$ K, $T_o = 355$ K, and $c_{f,o} = 5.11 \times 10^{-12}$. These values resulted from nonlinear multiparameter regression analysis of the author's nonisothermal viscosity data according to the FVM equations (5.8), (5.9) and (5.17).

In the incoming sections of the present book, it will be repeatedly shown that the experimental method for nonisothermal viscosity determinations and their FVM interpretation have been confirmed for numerous amorphous metallic alloys, such as $Pd_{82}Si_{18}$, $Fe_{82}B_{18}$, $Fe_{40}Ni_{40}Si_6B_{14}$, $Fe_{25}Zr_{75}$, $Al_{25}La_{50}Ni_{25}$, $Co_{(100-x)}Zr_x$, etc.

References

1. Cohen MH, Turnbull D (1959) J Chem Phys 31:1164
2. Turnbull D, Cohen MH (1970) J Chem Phys 52:3038
3. Spaepen F (1980) In: Balian R, Kléman M, Poirer JP (eds) Physics of defects, Les Houches 1980, session XXXV. North Holland, Amsterdam
4. Gladstone S, Laidler KJ, Eyring H (1941) The theory of rate processes. McGraw-Hill, New York, p 480
5. Van den Beukel A, Huizer E, Mulder AL, van der Zwaag S (1986) Acta Metall 34:483
6. Taub AI, Spaepen F (1980) Acta Metall 28:1781
7. Koebrugge GW, Sietsma J, van den Beukel A (1992) Acta Metall Mater 40(4):753
8. Tsao SS, Spaepen F (1985) Acta Metall 33:881
9. Duine PA, Sietsma J, van den Beukel A (1992) Acta Metall Mater 40(4):743
10. Van den Beukel A, Sietsma J (1990) Acta Metall Mater 38(3):383
11. Russew K, Zappel BJ, Sommer F (1995) Nonisothermal viscous flow behaviour of $Pd_{40}Ni_{40}P_{20}$ glassy alloy considered as a free volume related phenomenon. Scr Met Mater 32:271
12. Korn GA, Korn TM Mathematical Handbook for Scientists and Engineers: Definitions, Theorems and Formulas for Reference and Review. ISBN-13: 978-0486411477, ISBN-10: 0486411478

Glass Forming Ability, Crystallization, and Viscosity of Rapidly Solidified Amorphous $Co_{100-x}Zr_x$ and $Cu_{100-x}Zr_x$ Alloys

6

Abstract

The glass forming ability (GFA) of glassy alloys with Co-Zr and Cu-Zr as examples is studied, considered, and discussed in relation to their nonisothermal viscous flow and crystallization behavior. The glass forming ability (GFA) and melt fragility number of Angell and its Moynihan interpretation are presented. An FVM interpretation of the melt fragility numbers of Angell and Moynihan is proposed and experimentally proven. It is demonstrated that the exact viscosity measurement of glassy metals allows the determination of FVM parameters describing its temperature dependence. In this way it is possible to calculate the melt fragility number of Angell and to estimate the glass forming ability (GFA) of metallic alloys.

Keywords

Glassy metals • Co-Zr • Cu-Zr • Glass forming ability (GFA) • Angell's melt fragility number • Moynihan melt fragility number • FVM interpretation

Glass Forming Ability (GFA) and Melt Fragility Number. Equation of Angell and Moynihan Interpretation

The glass transition is the fundamental characteristic of amorphous metallic alloys obtained by rapid solidification from the melt. The viscosity of molten metallic alloys is very low – of 10^{-2}–10^{-3} Pa s order of magnitude. This is why extremely high cooling rates about 10^6 K/s or more are needed in order to avoid their crystallization and to obtain them in a vitreous state. When a glass forming melt solidifies in amorphous state, its viscosity becomes very high, causing rapid decrease of the atoms. This can happen in different temperature ranges. The best glass formers possess high viscosity still at high temperatures (higher than their melting point). Nevertheless, there exist some good glass forming substances, which possess relatively low viscosity at higher temperatures. Typical for these

K. Russew, L. Stojanova, *Glassy Metals*,
DOI 10.1007/978-3-662-47882-0_6

substances is the circumstance that the nucleation and crystal growth is suppressed during the rapid cooling. This suppression can be due to the fact that the crystal structure is very complicated and it is difficult to be realized [1]. Keeping in mind this consideration, it is not surprising that, although the good glass forming ability correlates with high melt viscosity values at high temperatures, it seems to be only necessary but not crucial condition for the vitrification. Metallic glasses obey this general behavior. Many molten metallic alloys can be vitrified by using very high quenching rates in the form of thin ribbons without formation of stable or metastable crystalline phases [2]. Recently, a series of metallic glasses have been discovered [3], which can be obtained by moderate or low melt cooling rates, comparable to the cooling rates needed for conventional silicate glasses. It is generally established that their melt viscosity is very high still in their melting point $T_{m.}$ At decreasing temperatures, the melt viscosity grows rapidly up with several orders of magnitude up to reaching the glass transition temperature about 0.6 T_m. There exist, however, metallic alloys, which vitrify at temperatures about 0.5 T_m [4]. The reasons for these differences in the behavior of glass forming metallic melts are still not very well understood.

Angell C.A. [5–7] has pointed out that not the absolute value of viscosity but the velocity of viscosity change after reaching the glass transition temperature is the determinant factor for the glass forming ability (GFA) of molten metallic alloys. He introduced the concept of strength and fragility of glass forming melts. It should be recalled to attention here that Angell bears in mind not the usual meaning of strength and fragility in the sense of mechanical strength and fragility of silicate glasses at room temperature. The classical meaning of these notions describes the catastrophic demolition of glassy objects subjected to critical mechanical overloading. Fragility in this case has no direct relationship with the colloquial meaning of the word "fragility," which more closely relates to the brittleness of a material. Angell has in mind the "architecture" of the glass forming (metallic) melt and its hardiness to the temperature changes under cooling. Angell classifies the glass forming melts as "strong" and "fragile" depending on the their viscosity temperature dependence in coordinates $\log_{10}(\eta)$ vs. T_g/T. The slope of this temperature dependence at $T_g/T = 1$ defines the so-called melt fragility number of Angell m_A. The higher is m_A, the more fragile is the liquid and vice versa. In the physics of amorphous bodies, *fragility* characterizes how rapidly the dynamics of a material slow down as it is cooled toward the glass transition temperature $T_{g.}$ Materials with a higher fragility have a relatively narrow glass transition temperature range, while those with low fragility have a relatively broad glass transition temperature range. Physically, fragility may be related to the breakdown of the usual Stokes-Einstein relationship between viscosity and diffusion. Liquids of Arrhenius-type are described as "strong," and these of Vogel-Fulcher-Tammann type as "fragile." The strong liquids possess a melt fragility number m_A lower than 10, while the "fragile" liquids have melt fragility numbers about 100 or more. Metallic glasses occupy an intermediate position with melt fragility numbers between 30 and 60. The higher the melt fragility number of Angell, the lower is the glass forming ability of metallic melts. The melt fragility number of Angell can be conveniently used as a measure for GFA. This is mathematically presented, as mentioned above, with aid of the temperature dependence of the viscosity in coordinates $\log_{10}(\eta)$ vs. T_g/T

$$m_A = \left[\frac{d(\log\eta)}{d\left(\frac{T_g}{T}\right)}\right]_{T=T_g}. \tag{6.1}$$

Equation (6.1) is based on the assumption that melt viscosity of glass forming substances follows the Vogel-Fulcher-Tammann (VFT) temperature dependence and at $T = T_g$ the viscosity value is 10^{12} Pa s, although this is only a very rough empirical estimate. In the case of FVM interpretation [8–16], the viscosity temperature dependence should be presented by the hybrid (▶ Eq. 5.9), ▶ Chap. 5, instead of a VFT type, and the glass transition temperature is presented by the cross point of the (non)equilibrium viscosity temperature dependence (▶ Eq. 5.8) and the (quasi) equilibrium viscosity temperature dependence (▶ Eq. 5.9), respectively. The index of Angell m_A in this case equals

$$m_A = 0.434\left[\frac{BT_g}{\left(T_g - T_o\right)^2} + \frac{Q_\eta}{RT_g} - 1\right]. \tag{6.2}$$

The m_A can be easily calculated by using the FVM parameters Q_η, B, and T_o, presented in all tables, containing the main characteristic viscous flow parameters of the metallic alloys studied by the authors of the present book.

A simple method for calculation of the melt fragility number is proposed by Moynihan [7, 17]

$$\frac{\Delta T_g}{T_g} \cong \frac{2}{m_M}, \tag{6.3}$$

where $\Delta T_g = T_g - T_{on}$, T_g is the glass transition temperature, and T_{on} is the onset temperature of crystallization of the glass forming material. The approach expressed by Eq. (6.3) is based on the assumption that the viscosity value at $T = T_{on}$ is two orders of magnitude lower than the viscosity at T_g, $\log_{10}[\eta(T_g)/\eta(T_{on})] = 2$, which is a very rough estimate, and that the equation of Vogel-Fulcher-Tammann describes the temperature dependence of the (quasi)equilibrium viscosity. The approach of Moynihan is, however, tempting for application as far as it provides the possibility to use DSC measurements only in order to gain the needed estimation about the melt fragility number. This approach can be easily improved, as far as one has the possibility to calculate the real value of $\log_{10}[\eta(T_g)/\eta(T_{on})]$ from the (quasi)equilibrium viscosity obtained by the nonisothermal viscosity determinations. Numerous studies of Russew et al. [18–20] have shown that the ratio $\mu = \ln\eta(T_g)\text{-}\ln\eta(T_{on})$ varies in a broad range. It is interesting to mention, for example, that for the glassy alloy $Ni_{80}P_{20}$, $\mu \approx 0.3$, while for the glassy alloy $Cu_{46}Zr_{54}$, $\mu \approx 5.3$.

Taking into account also that the hybrid (▶ Eq. 5.9) should be used instead of the classical Vogel-Fulcher-Tammann equation, one obtains

$$m'_M = 0.434\left[\frac{BT_{on}}{(T_g - T_o)(T_{on} - T_o)} + \frac{Q_\eta}{R}\left(\frac{2}{T_g} - \frac{1}{T_{on}}\right) - \mu - 1\right]. \tag{6.4}$$

Unfortunately, in such a case it becomes impossible to use the DSC crystallization peaks of the metallic glasses solely, without need of direct viscosity determinations, in order to perform the calculation of the Moynihan melt fragility index. The more correct approach of the authors of the present book will be demonstrated on the example with crystallization and viscosity studies upon the amorphous systems $Co_{100-x}Zr_x$ and $Cu_{100-x}Zr_x$.

Crystallization, Viscosity, and Glass Forming Ability of Rapidly Solidified Amorphous $Co_{100-x}Zr_x$ Alloys

Metallic alloys transition metal-zirconium (TM-Zr) are characterized with numerous interesting phase transformations. One of them is their capability to form metallic glasses in a broad concentration range of Zr. Some of them are interesting for application as ferromagnetic materials with very low magnetostriction coefficient [21–23] or as materials of high mechanical strength [24]. The knowledge of the viscosity temperature dependence of these alloys and the crystallization characteristics are very important. The characteristic viscous flow parameters allow to determine the glass forming ability of Co_{100-x} Zr_x glassy alloys and in this way to determine the possibility itself to obtain them in an amorphous state. The knowledge of crystallization parameters gives information about the temperature range for carrying out a heat treatment in order to improve their magnetic and/or mechanical properties [25, 26].

The crystallization and viscous flow behavior of nine rapidly solidified Co_{100-x} Zr_x ($7 \leq x \leq 78.5$ at. %) metallic alloys were studied under continuous heating conditions as a function of the chemical composition. The ribbon-like samples were produced by using the CBMS method under low vacuum and He atmosphere. The ribbon width varied within 0.93–1.55 mm, by thickness within 0.0018–0.045 mm, respectively. With the aid of DSC studies, the number of crystallization stages and the activation energy of crystallization have been determined, respectively. The activation energy for crystallization was calculated according to the method of Kissinger [27]. Seven of the metallic alloys studied were with fully amorphous structure, while two of them ($Co_{93}Zr_7$, and $Co_{50}Zr_{50}$) possessed mixed amorphous, nanocrystalline, and microcrystalline structures, respectively. A good correlation between the onset temperature of crystallization and the viscosity temperature dependence was established. On the contrary to the crystallization behavior, the viscosity features (glass transition temperature and the viscosity value at this temperature) are substantially influenced by the structural state of the alloys studied and demonstrate smooth dependence upon the Zr content.

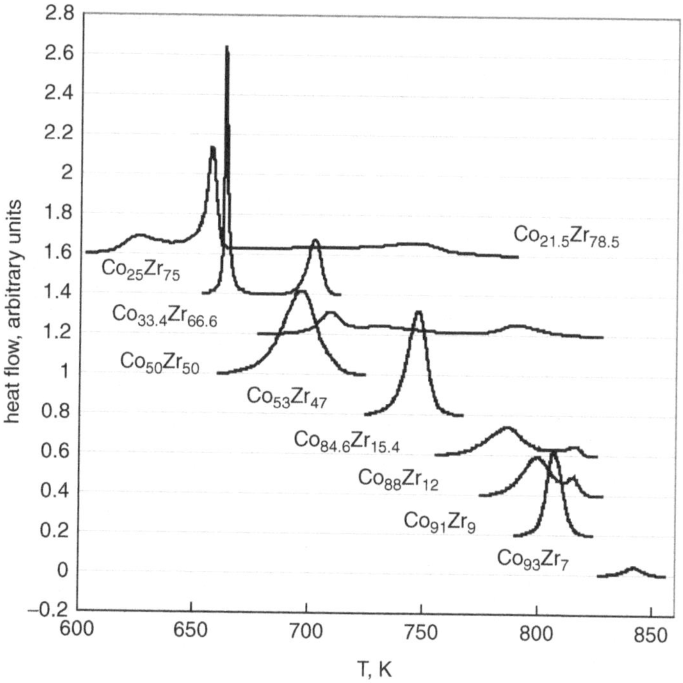

Fig. 6.1 Crystallization DSC peaks of all rapidly solidified $Co_{100\text{-}x}Zr_x$ metallic alloys at a heating rate 20 K/min. Perkin-Elmer DSC 2C has been used

Figure 6.1 shows the crystallization peaks of all $Co_{100\text{-}x}Zr_x$ glassy alloys, studied at heating rate of 20 K/min.

The onset temperature of crystallization T_{on} as a function of the Zr content at three different heating rates decreases smoothly along with the decreasing of the Zr content. Deviations from this behavior demonstrate the alloys $Co_{93}Zr_7$ and $Co_{50}Zr_{50}$ – Fig. 6.2. The same is valid for the temperature T_{max} of the maximum of the first crystallization peak. In order to find an explanation of these deviations, a structural phase analysis of all rapidly solidified $Co_{100\text{-}x}$ Zr_x alloys was carried out with the aid of X-ray diffraction, Fig. 6.3. The conclusion was made that the deviations of calorimetric data for the $Co_{93}Zr_7$ and $Co_{50}Zr_{50}$ samples are most probably due to their mixed amorphous, nanocrystalline, and microcrystalline structures. It was confirmed for the $Co_{93}Zr_7$ alloy also by the transmission electron microscopy (TEM) study [24].

Tables 6.1 and 6.2 present in a systematic way all calorimetric (DSC) data for all rapidly solidified $Co_{100\text{-}x}Zr_x$ metallic alloys. The method of Kissinger [27] was used for the performance of kinetic analysis of the crystallization process. The parameter ln (q/T^2_{max}) vs. $1000/T_{max}$ was plotted. From the slopes of these linear dependences, the activation energies of different crystallization stages were calculated. They are presented in Table 6.2.

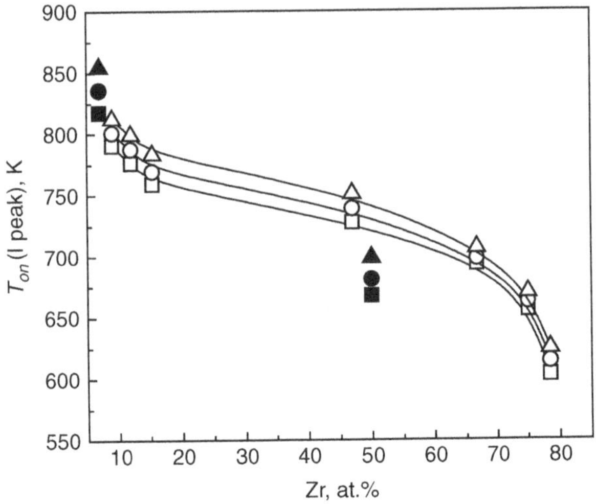

Fig. 6.2 Dependence of T_{on} on the Zr content of alloys studied, at heating rates of 10(□), 20 (○), and 40(Δ) K/min, respectively. The solid symbols (■,●, ▲) denote the values of these characteristic temperatures for the case of partially crystallized $Co_{93}Zr_7$ and $Co_{50}Zr_{50}$ alloy compositions, respectively

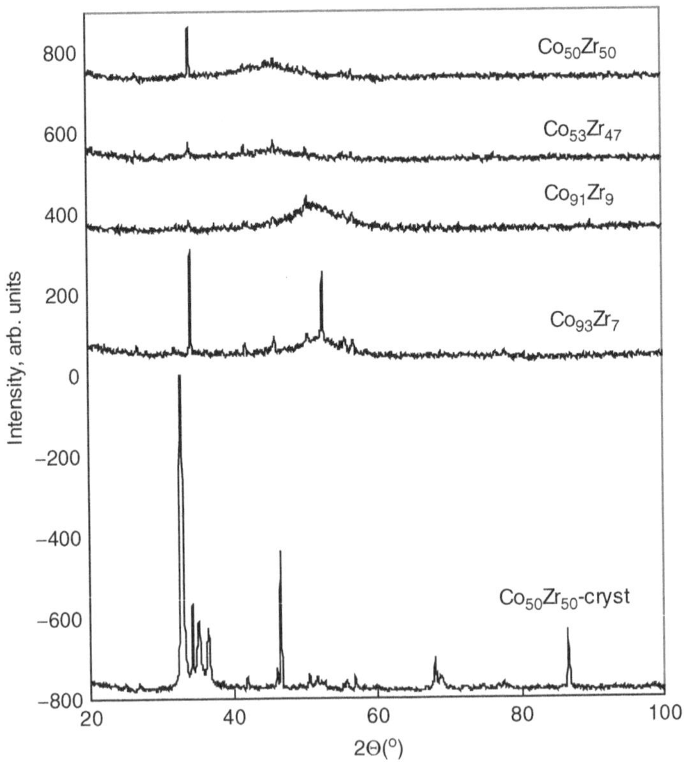

Fig. 6.3 Diffraction patterns of as-quenched $Co_{93}Zr_7$, $Co_{91}Zr_9$, $Co_{50}Zr_{50}$, and $Co_{53}Zr_{47}$ compositions, respectively, together with the diffraction pattern of a crystallized $Co_{50}Zr_{50}$ sample

Table 6.1 The number of crystallization peaks, onset temperature of crystallization T_{on}, overall crystallization enthalpy ΔH_{tot}, glass transition temperature T_g, and the viscosity value at T_g, $\eta(T_g)$, for all rapidly solidified $Co_{100-x}Zr_x$ metallic alloys at heating rates of 10, 20, and 40 K/min

Alloy composition and the number of crystallization peaks	Heating rate q, K/min	T_{on}, K	ΔH_{tot}, cal/g	T_g, K	$\eta(T_g)$, Pa s
$Co_{93}Zr_7$	10	817.7	−0.9		
1 peak	20	835.7	−1.0	806	5.47×10^{12}
	40	854.0	−1.4		
$Co_{91}Zr_9$	10	790.3	−10.0		
1 peak	20	800.8	−12.0	790	7.18×10^{11}
	40	811.9	−12.3		
$Co_{88}Zr_{12}$	10	776.2	−12.9		
2 peaks	20	787.3	−13.2	775	3.19×10^{11}
	40	798.9	−13.8		
$Co_{84.6}Zr_{15.4}$	10	759.0	−15.1		
2 peaks	20	769.2	−13.3	710	2.14×10^{12}
	40	783.0	−13.0		
$Co_{53}Zr_{47}$	10	727.5	−19.8		
1 peak	20	738.2	−19.0	722	3.29×10^{11}
	40	750.7	−16.7		
$Co_{50}Zr_{50}$	10	667.6	−27.8		
1 peak	20	681.1	−26.2	668	7.55×10^{11}
	40	698.4	−26.0		
$Co_{33.4}Zr_{66.6}$	10	693.7	−13.4		
3 peaks	20	700.2	−12.8	688	5.90×10^{10}
	40	706.5	−12.7		
$Co_{25}Zr_{75}$	10	655.7	−13.4		
2 peaks	20	662.6	−15.6	633	4.55×10^{10}
	40	669.7	−16.0		
$Co_{21.5}Zr_{78.5}$	10	603.0	−25.7		
3 peaks	20	614.0	−29.1	588	1.81×10^{11}
	40	624.0	−22.1		

The viscosity temperature dependence of all $Co_{100-x}Zr_x$ amorphous alloys studied at a heating rate of 20 K/min is shown in Fig. 6.4 and published in Refs. [28–34].

The experimentally determined viscosity values are plotted with graphical symbols in Fig. 6.5 for four of the amorphous alloys studied. The fitting curves of the nonequilibrium and the (quasi)equilibrium viscosity, respectively, are calculated according to the (▶ Eqs. 5.8) and (▶ 5.17). The steeper lines a, b, c, and d represent the (quasi)equilibrium viscosity (▶ Eq. 5.9) of the different alloy compositions. The cross point of the nonequilibrium and (quasi)equilibrium viscosity lines corresponds to the glass transition temperature T_g of the different alloy compositions at a heating rate of 20 K/min. After reaching the onset temperature of crystallization, the viscosities begin to grow up. This is clearly demonstrated in Fig. 6.6a–c for the case glassy alloys with one, two, or three stages of

Table 6.2 Temperature of the peak maximum T_{max} and activation energy for crystallization for the different crystallization stages Q_{cr}

Alloy	q, K/min	T_{max} of I, II, and five crystallization peaks, respectively, K	Q_{cr} of I, II, and five crystallization stages, respectively, kJ/mol
	10	823.8	
$Co_{93}Zr_7$	20	842.3	183
	40	865.4	
	10	796.4	
$Co_{91}Zr_9$	20	806.8	345
	40	817.5	
	10	788.5; 803.1	
$Co_{88}Zr_{12}$	20	800.6; 815.4	289.5; 312.9
	40	812.9; 826.5	
	10	775.2; 804.9	
$Co_{84.6}Zr_{15.4}$	20	787.4; 817.6	269.6; 281.8
	40	800.5; 831.0	
	10	737.9	
$Co_{53}Zr_{47}$	20	749.1	293.7
	40	758.4	
	10	683.8	
$Co_{50}Zr_{50}$	20	697.0	195.2
	40	710.9	
	10	702.2; 721.3; 780.1	
$Co_{33.4}Zr_{66.6}$	20	709.6; 733.4; 791.1	396.2; 328.3; 310.4
	40	716.4; 740.5; 802.4	
	10	657.1; 691.7	
$Co_{25}Zr_{75}$	20	663.9; 703.0	339.4; 240.3
	40	671.6; 714.3	
	10	616.6; 647.0; 734.5	
$Co_{21.5}Zr_{78.5}$	20	626.1; 657.3; 745.5	241.3; 217.4; 252.9
	40	635.8; 669.0; 758.3	

crystallization, respectively. The glassy alloys $Co_{21.5}Zr_{78.5}$, $Co_{25}Zr_{75}$, and $Co_{50}Zr_{50}$ are used as examples. The two very important viscous flow characteristics – the glass transition temperature T_g and the viscosity value $\eta(T_g)$ at T_g are given in Table 6.1.

The FVM parameters of viscous flow, determined via multiparameter nonlinear regression analysis of the viscosity experimental data, are given in Table 6.3.

The determination of the viscosity temperature dependence of metallic glasses under nonisothermal experimental conditions gives the opportunity to make important for their practical application conclusions. It gives an information about the glass forming ability of metallic melts by using the Angell's fragility number m_A (Eq. 6.2) as well as the indexes of Moynihan m_M (Eq. 6.3) and m'_M (Eq. 6.4), respectively. The three types of melt fragility indexes are presented in Table 6.4.

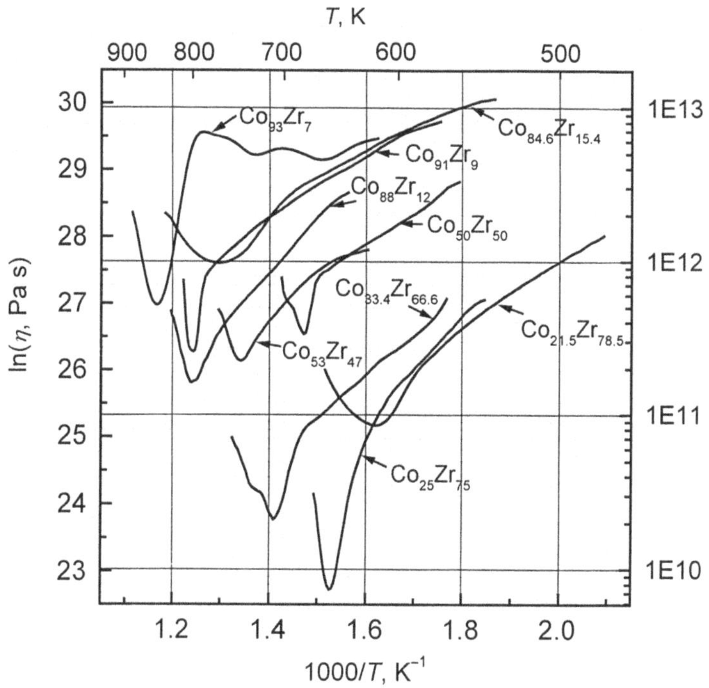

Fig. 6.4 Viscosity temperature dependence of all $Co_{100-x}Zr_x$ alloys studied

Fig. 6.5 Viscosity temperature dependence for four of the $Co_{100\text{-}x}Zr_x$ glassy alloys: ○, □, △, and ◇ – experimentally determined viscosity values; a, b, c, d – *lines* represent the (quasi) equilibrium viscosities

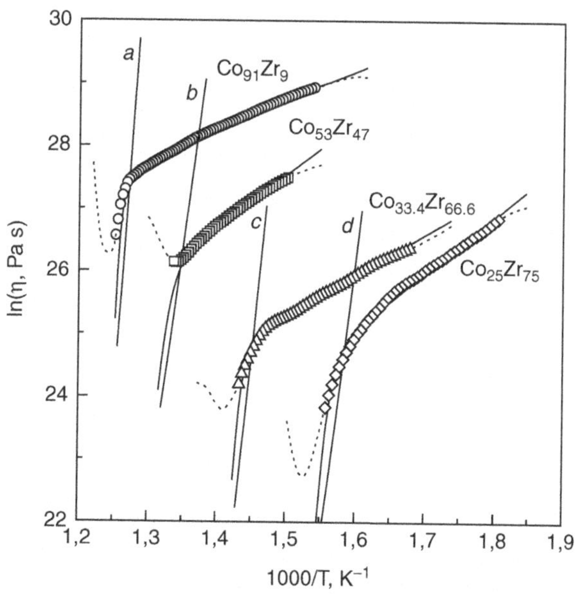

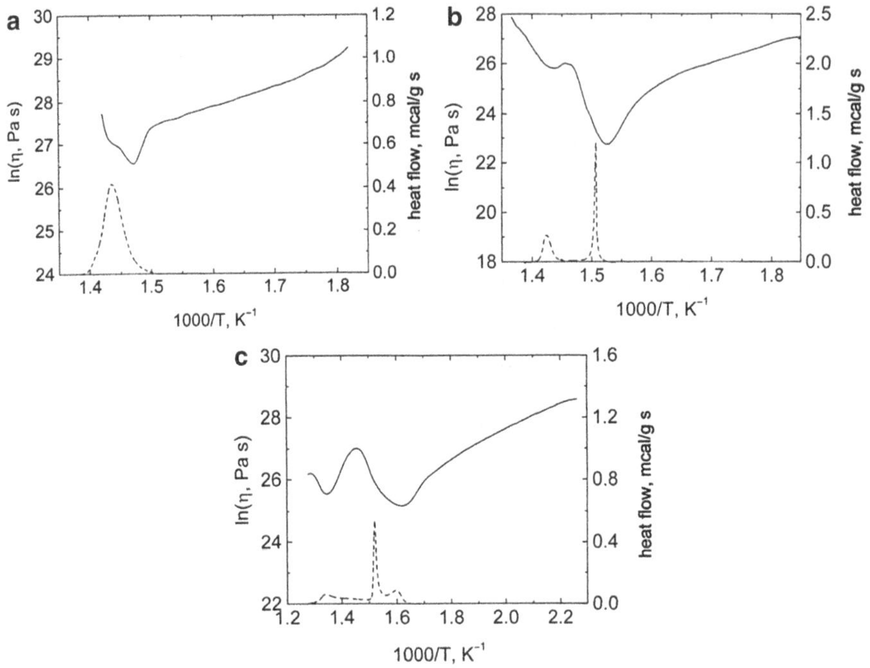

Fig. 6.6 Correlation between the viscous flow (—) and crystallization processes (– –) for: (**a**) $Co_{50}Zr_{50}$ alloy with one-stage crystallization, (**b**) $Co_{25}Zr_{75}$ alloy with two-stage crystallization, and (**c**) $Co_{21.5}Zr_{78.5}$ alloy with three-stage crystallization. The heating rate used was 20 K/min

Figure 6.7 shows the *Zr* concentration dependence of these indexes for the case of the rapidly solidified $Co_{100\text{-}x}Zr_x$ amorphous metallic alloys. As is seen, the melt fragility indexes m_A and m'_M derived on the basis of FVM concepts coincide practically. On their basis it can be concluded that practically all alloy compositions, except that of lowest *Zr* content – $Co_{93}Zr_7$ and $Co_{91}Zr_9$, show melt fragility numbers and GFA, which is between 90 and 40. It could be concluded that most of the alloys in the Co-Zr system belong to the relatively good metallic glass formers. The classical melt fragility number m_M of Moynihan deviates drastically as compared to the value and tendency with the other two fragility indexes derived on the basis of FVM concepts. This is why the index m_M should be used cautiously and after stipulations.

Crystallization, Viscosity, and Glass Forming Ability of Rapidly Solidified Amorphous $Cu_{100\text{-}x}Zr_x$ Alloys

Six $Cu_{100\text{-}x}Zr_x$ amorphous alloys (x in the range 35.7–60 at.%) were prepared via chill block melt spinning (CBMS) method under low pressure helium atmosphere. Their crystallization and viscous flow behavior was studied with the aid of

Table 6.3 FVM parameters η_o, $c_{f,o}$, Q_η, v_r, Q_r, B, and T_o for the $Co_{(100\text{-}x)}Zr_x$ glassy alloys studied

Parameters	Dimension	Zr, at. %								
		7.0	9.0	12.0	15.4	47.0	50.0	66.6	75.0	78.5
ν	1/s	–	1.4 E30	1.8 E24	6.6 E27	9.9 E21	8.5 E23	5.7 E24	9.4 E19	4.5 E23
Q_r	kJ/mol	–	209	193	227	184	151	150	101	130
$c_{f,o}$		–	2.5 E-4	3.7 E-4	3.8 E-4	6.9 E-6	6.1 E-6	1.7 E-7	6.1 E-6	7.6 E-7
T_o	K	–	598	550	505	475	452	500	455	430
Q_η	KJ/mol	–	270	281	280	270	204	220	191	190
B	K	–	7850	7200	6300	6700	6730	6640	5500	4970
η_o	Pa s/K	–	1.1 E-27	3.9 E-24	1.3 E-24	4.7 E-22	1.4 E-21	1.3 E-24	5.4 E-22	5.2 E-22

Table 6.4 The calculated according to Eqs. (6.2), (6.3), and (6.4), melt fragility numbers m_A, m_M, and m'_M, respectively

Alloy	m_M	m'_M	m_A
$Co_{91}Zr_9$	146	87	90
$Co_{88}Zr_{12}$	126	65	66
$Co_{84.6}Zr_{15.4}$	24	60	66
$Co_{53}Zr_{47}$	89	52	54
$Co_{50}Zr_{50}$	103	40	42
$Co_{33.4}Zr_{66.6}$	115	70	72
$Co_{25}Zr_{75}$	43	58	63
$Co_{21.5}Zr_{78.5}$	45	62	67

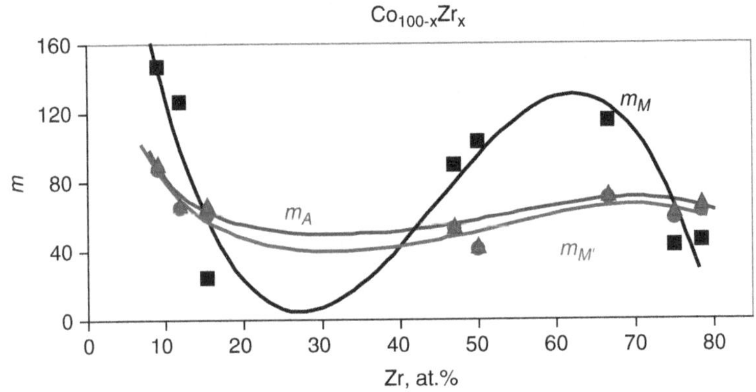

Fig. 6.7 *Zr* concentration dependence of melt fragility indexes m_M, m'_M, and m_A

Perkin-Elmer DSC 2C and Perkin Elmer TMS 2 devices, respectively [20, 35, 36]. The viscous flow temperature dependencies at a heating rate of 20 K/min were interpreted on the basis of the free-volume model. The DSC and TMS data were used to determine the melt fragility number m of Angell in three different ways as a function of alloy composition. It has been shown that the fragility number goes over a maximum and has a minimum at alloy composition very near to $Cu_{64}Zr_{36}$, in good agreement with the results of Donghua Xu et al. [37] and Wang D et al. [38].

The $Cu_{100\text{-}x}Zr_x$ glassy alloys studied were X-ray amorphous. The research experimental methods were already extensively discussed in the previous chapter (▶ Chap. 3) of the book. The results obtained have been published in Refs. [20, 35, 36]. The crystallization behavior was studied at three different constant heating rates, 10, 20, and 40 K/min, respectively. The viscosity temperature of the glassy alloys studied, depending on the alloy composition, was determined at a heating rate of 20 K/min. A good correlation between the onset temperatures of crystallization and the viscosity temperature dependences has been established.

Figure 6.8 shows the DSC crystallization peaks of all $Cu_{100\text{-}x}Zr_x$ glassy alloys studied at a heating rate of 20 K/min. The thermal stability of the amorphous alloys grows up along with the increasing *Zr* content. Table 6.5 presents in a systematic way all DSC data for the $Cu_{100\text{-}x}Zr_x$ glassy alloys studied by heating rates of 10, 20,

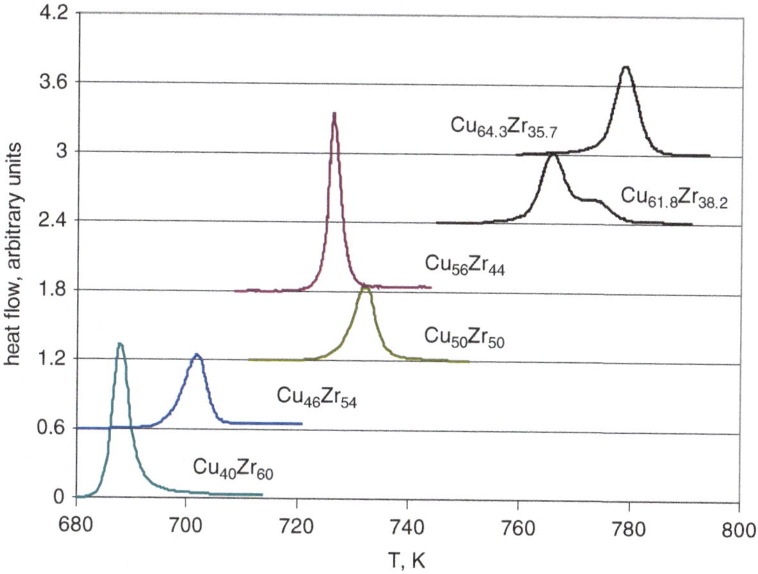

Fig. 6.8 DSC crystallization peaks of all $Cu_{100\text{-}x}Zr_x$ glassy alloys studied at a heating rate of 20 K/min

Table 6.5 The number of crystallization DSC peaks, the onset temperature of crystallization T_{on}, the temperature of peak maxima T_{max}, crystallization enthalpy ΔH_{tot}, and the activation energy of the different crystallization stages Q_{cr}. of all $Cu_{100\text{-}x}Zr_x$ glassy alloys studied

Alloy	q, K/min	$T_{on}{}^{I}$, K	$T_{max}{}^{I}$, K	$T_{on}{}^{II}$, K	$T_{max}{}^{II}$, K	ΔH^{I}, cal/g	ΔH^{II}, cal/g	Q_{cr}, kJ/mol
$Cu_{40}Zr_{60}$	10	677.9	680.4	–	–	−15.17	–	
1 peak	20	685.3	688.7	–	–	−16.77	–	297
	40	695.2	698.1	–	–	−16.89	–	
$Cu_{46}Zr_{54}$	10	690.7	695.2	728.9	738.4	−13.1	−1.26	
2 peaks	20	696.5	701.6	743.7	752.5	−13.29	−0.97	362,240
	40	705.1	710.3	756.2	764.0	−14.93	−0.68	
$Cu_{50}Zr_{50}$	10	716.7	721.7	–	–	−11,98	–	
1 peak	20	725.4	729.9	–	–	−12.57	–	343
	40	734.0	739.0	–	–	−13,42	–	
$Cu_{56}Zr_{44}$	10	715.1	717.8	–	–	−13.11	–	
1 peak	20	724.1	726.2	–	–	−13.76	–	353
	40	731.4	734.4	–	–	−14.55	–	
$Cu_{61.8}Zr_{38.2}$	10	751.1	756.4	–	767.2?	−13.77	–	
Twin peak	20	761.6	765.6	–	773.2	−14.94	–	354,455
	40	771.6	774.8	–	781.9	−14.97	–	
$Cu_{64.3}Zr_{35.7}$	10	767.0	770.6	–	–	−11.98	–	
1 peak	20	775.0	778.8	–	–	−12.56	–	436
	40	781.1	786.1	–	–	−13.00	–	

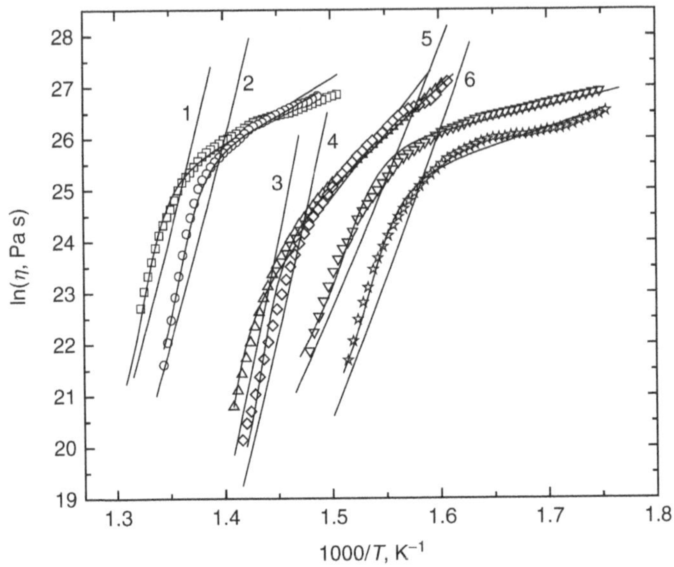

Fig. 6.9 Measured (symbols) and fitted (solid lines) viscosity temperature dependences of $Cu_{00\text{-}x}Zr_x$ glassy alloys at a heating rate of 20 K/min: □- $Cu_{64.3}Zr_{35.7}$, ○- $Cu_{61.8}Zr_{38.2}$, △ – $Cu_{56}Zr_{44}$, ◇- $Cu_{50}Zr_{50}$, ▽- $Cu_{46}Zr_{54}$, and ★- $Cu_{40}Zr_{60}$. The *curves* 1–6, respectively, present the (quasi)equilibrium viscosities η_{eq}

and 40 K/min. The method of Kissinger [27] was used for the kinetic analysis of the glassy alloy crystallization. The corresponding crystallization activation energies are given in Table 6.5. Figure 6.9 shows the experimentally determined and the calculated according to (▶ Eqs. 5.8) and (▶ 5.17) viscosity temperature dependences of the $Cu_{100\text{-}x}Zr_x$ glassy alloys. On the same figure are presented also the temperature dependences of the (quasi)equilibrium viscosities calculated according to (▶ Eq. 5.9) by using the FVM parameters obtained by the regression analysis of the viscosity experimental data. All FVM parameters, together with T_g and $\eta(T_g)$, are given in Table 6.6.

Table 6.7 shows the glass forming ability of the $Cu_{100\text{-}x}Zr_x$ glassy alloys presented by the FVM modified melt fragility numbers of Angell m_A, according to Eq. (6.2), the classical melt fragility number of Moynihan m_M, according to Eq. (6.3), and the FVM modified index of Moynihan m'_M, according to Eq. (6.4).

Figure 6.10 presents the melt fragility numbers, given in Table 6.7, as a function of the Zr content. It can be again concluded that the melt fragility indexes m_A and m'_M, calculated according to the concepts of FVM, coincide practically. The values of the classical melt fragility index of Moynihan are systematically lower by about 33 % as compared to the melt fragility indexes m_A and m'_M and should not be taken into account.

From the above considered follows that the Cu-Zr glassy alloys with Zr concentration around and close to 40 at. % Zr possess very good glass forming ability, as

Table 6.6 FVM parameters of viscous flow determined by nonlinear multiparameter regression analysis of the experimental viscosity temperature dependences of all $Cu_{100-x}Zr_x$ glassy alloys studied at heating rate of 20 K/min

Parameters	Dimensions	Zr, at. %					
		35.7	38.2	44	50	54	60
ν	1/s	2.8E+15	1.1E+17	4E+16	8E+16	2E+14	4.9E+15
Q_r	KJ/mol	116	135	95	102	96	98
$c_{f,o}$		2.5E-07	2.9E-07	8E-08	7.8E-07	7E-05	1.7E-05
T_o	K	529	510	481	460	409	400
Q_η	KJ/mol	238	248	334	275	166	171
B	K	4290	4297	5620	5484	4477	5036
η_o	Pa s/K	1.1E-18	1.5E-19	4E-30	4.2E-25	3E-14	3.4E-16
T_g	K	734	716	692	676	642	632
$\eta(T_g)$	Pa s	8.4e10	1.4e11	1.4e10	3.6e10	1.1e11	7.4e10

Table 6.7 The calculated according to Eqs. (6.2), (6.3), and (6.4), melt fragility numbers m_A, m_M, and m'_M of the glassy $Cu_{100\text{-}x}Zr_x$ alloys studied

Alloy	m_M	m'_M	m_A
$Cu_{64.3}Zr_{35.7}$	36	45	49
$Cu_{61.8}Zr_{38.2}$	31	45	49
$Cu_{56}Zr_{44}$	43	59	63
$Cu_{50}Zr_{50}$	27	51	55
$Cu_{46}Zr_{54}$	24	33	36
$Cu_{40}Zr_{60}$	24	36	39

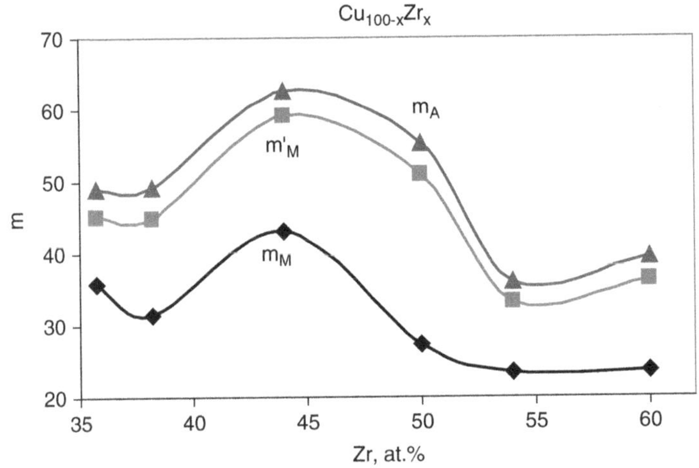

Fig. 6.10 Zr concentration dependence of the melt fragility numbers m_A, m_M, and m'_M, for the $Cu_{100\text{-}x}Zr_x$ glassy alloys

well as the glassy alloys with Zr concentration higher than 54 at. %. These alloy compositions are expected to allow the production of bulk Cu-Zr glassy samples.

Donghua Xu et al. [37] and D. Wang et al. [38] have also experimentally established that $Cu_{100\text{-}x}$ Zr_x is a metallic system, which relatively easy can be subjected to vitrification. In the concentration range $x = 34–40$ at. %, bulk amorphous samples can be produced. This is in good agreement with the results of our study.

It follows from the above stated that:

- For a first time it is theoretically shown and experimentally proven that the free-volume model quantitatively describes the viscosity temperature dependence of all kinds of metallic glasses under nonisothermal, experimental conditions by a constant heating rate and adequately takes into account the structural relaxation phenomena around the glass transition.
- It is again convincingly shown that the proposed method for precise viscosity determination under nonisothermal conditions works perfectly.

- The exact viscosity measurements of glassy metals allow the determination of FVM parameters describing its temperature dependence. In this way it is possible to calculate the melt fragility number of Angell and to estimate the glass forming ability (GFA) of metallic alloys.

References

1. Battezzati L (2004) Mater Sci Eng A375–377:60
2. Battezzati L, Greer AL (1987) Int J Rapid Solid 3:130
3. Peker A, Johnson WL (1993) Appl Phys Lett 63:2342
4. Ma H, Xu J, Ma E (2003) Appl Phys Lett 83:2793
5. Angell CA (1985) Strong and fragile liquids. In: Ngai K, Wright GB (eds) Relaxation in complex systems. US Department of Commerce, Springfield
6. Ito K, Moynihan CT, Angell CA (1999) Nature 398(4):492
7. Battezzati L (2005) Mater Trans 46:2915
8. Spaepen F (1980) In: Balian R, Kléman M, Poirer JP (eds) Physics of defects, Les Houches 1980, session XXXV. NorthHolland, Amsterdam, p 135
9. Gladstone S, Laidler KJ, Eyring H (1941) The theory of rate processes. McGraw-Hill, New York, p 480
10. Van den Beukel A, Huizer E, Mulder AL, van der Zwaag S (1986) Acta Metall 34:483
11. Taub AI, Spaepen F (1980) Acta Metall 28:1781
12. Koebrugge GW, Sietsma J, van den Beukel A (1992) Acta Metall Mater 40:753
13. Tsao SS, Spaepen F (1985) Acta Metall 33:881
14. Duine PA, Sietsma J, van den Beukel A (1992) Acta Metall Mater 40:743
15. Van den Beukel A, Sietsma J (1990) Acta Metall Mater 38:383
16. Russew K, Zappel BJ, Sommer F (1995) Nonisothermal viscous flow behaviour of $Pd_{40}Ni_{40}P_{20}$ glassy alloy considered as a free volume related phenomenon. Scr Met Mater 32:271
17. Moynihan CT (1993) J Am Ceram Soc 76:1081
18. Russew K, Stojanova L (1993) Viscous flow behaviour and thermal stability of $Ni_{100-x}P_x$ metallic glasses from $Ni_{84}P_{16}$ to $Ni_{79}P_{21}$. Mater Lett 17:199
19. Stojanova L, Russew K (1995) Thermal stability and viscous flow of $Ni_{100-x}P_x$ and $Fe_{100-x}P_x$ metallic glasses. In: Proceedings of 8th international congress on metallurgy and materials, Istanbul, June 1995, p 1421
20. Russew K, Stojanova L, Yankova S et al (2009) Thermal behavior and melt fragility number of Cu100-xZrx glassy alloys in terms of crystallization and viscous flow. In: 13th international conference on rapidly quenched and metastable materials'2008, Drezden. Journal of Physics: conference series 2009, vol 144, p 012094
21. Savalia RT, Tewari R, Dey GK, Banerjee S (1996) Acta Metall Mater 44:57
22. Buschow KHJ (1983) Acta Metall Mater 31:155
23. Jergel M, Vlasak G, Duhaj P (1989) Phys Status Solidi (a) 111:597
24. Dyakova V, Kamenova TZ, Varga LK et al (2005) Thermal and mechanical properties of rapidly solidified amorphous and nanocrystalline Co-Zr alloys of high Co content. In: XVI national conference on non-destructive testing, NDT-2005, Sozopol. Scientific communications of STU machine building vol XII no 1(79), p 299
25. Kim YH, Inoue A, Masumoto T (1990) Mater Trans JIM 31:747
26. Bakonyi I, Mehner F, Rapp M et al (1995) Zs Metallkde 86:619
27. Kissinger HE (1957) Anal Chem 29:1702
28. Russew K, Stojanova L, Yankova S (2006) Free volume model interpretation of rheological properties of $Co_{(100-x)}$ Zr_x metallic glasses from $Co_{93}Zr_7$ to $Co_{21.5}Zr_{78.5}$. J Mater Sci Technol 14(3):144

29. Yankova S, Stojanova L, Russew K et al (2006) Crystallization behavior of $Co_{100-x}Zr_x$ amorphous metallic alloys. In: XXI national conference on non-destructive testing, NDT-2006, Sozopol. Scientific communications of STU machine building vol XIII no 3(86), p 119
30. Yankova S, Stojanova L, Russew K (2006) Rheological behavior of $Co_{100-x}Zr_x$ amorphous alloys. In: XXI national conference on non-destructive testing, NDT-2006, Sozopol. Scientific communications of STU machine building vol XIII no 3(86), p 113
31. Russew K, Stojanova L (2006) Crystallization, thermomagnetic gravimetry and viscous flow features of $Co_{100-x}Zr_x$ rapidly solidified alloys. In: Proceedings of V intern congress mechanical engineering technologies'06, Varna, Sept 2006, Sec II, p 24
32. Russew K, Stojanova L, Yankova S (2007) Viscous flow behaviour and thermal stability of Co100-xZrx metallic glasses from $Co_{91}Zr_9$ to $Co_{21.5}Zr_{78.5}$. In: Balabanova E, Dragieva I (eds) Proceedings of conference on nanoscience and nanotechnology'7. Heron Press, Sofia, p 133
33. Russew K, Stojanova L, Varga LK et al (2009) Glass forming ability and thermal behavior of binary Co-Zr amorphous alloys. In: Proceedings of national conference on metal science and novel materials'2008, Sofia, p 91
34. Russew K, Stojanova L, Varga LK, Fazakas E, Yankova S (2009) Glass forming ability and thermal behavior of binary Co-Zr amorphous alloys. J Mater Sci Technol 17(1):29
35. Stojanova L (2009) Relation between viscous flow and mechanical properties of Co-Zr and Cu-Zr amorphous alloys In: Proceedings of national conference on metal science and novel materials'2008, Sofia, p 96
36. Yankova S, Stojanova L, Varga LK et al (2008) Viscous flow behavior and mechanical properties of $Cu_{100-x}Zr_x$ amorphous alloys. In: XXV national conference on non-destructive testing NDT-2008, Sozopol. Scientific communications of STU machine building vol XV no 2(105), p 324
37. Donghua X, Boonrat L, Gang D, Johnson WL, Garland C (2004) Acta Mater 52:2621
38. Wang D, Li Y, Sun BB, Sui ML, Lu K, Ma E (2004) Appl Phys Lett 84(20):4029

7 Bend Stress Relaxation of Amorphous Metallic Alloys Under Isothermal and Nonisothermal Heating Conditions

Abstract

A model analysis about the relation between the viscosity and the relaxation of bend stresses of ribbonlike amorphous metallic alloys, both under isothermal and nonisothermal conditions, is presented. Additionally, an experimental proof of the theoretical considerations based on the free volume model (FVM) is given. The experimental proof upon a study of the nonisothermal viscous flow behavior of $Fe_{40}Ni_{40}Si_6B_{14}$ amorphous metallic glass, in the high temperature range around its glass transition temperature, is performed. Additionally a study of the (nonisothermal and isothermal) relaxation of bend stresses in the same ribbonlike alloy at considerably lower temperatures is carried out. It is shown that the FVM interpretation of the two kinds of experimental data is possible with practically one and the same common set of FVM parameters. A new method for quantitative experimental separation of viscous flow caused plastic and anelastic deformations, respectively, is developed. It is shown that the relaxation of the anelastic deformation can be very well described by the empirical equation of Kohlrausch–Williams–Watts.

Keywords

Glassy metals • Ribbonlike amorphous alloys • Bend stress • Relaxation • Isothermal • Nonisothermal • Irreversible/reversible relaxation of bend stresses • Anelastic deformation • Viscosity • Equation of Kohlrausch–Williams–Watts

This book section aims to present a model analysis about the relation between the viscosity and the relaxation of bend stresses of ribbonlike amorphous metallic alloys both under isothermal and nonisothermal conditions. Additionally, an experimental proof of the theoretical considerations based on the free volume model (FVM) is given. The experimental proof is focused on a study of the nonisothermal viscous flow behavior of $Fe_{40}Ni_{40}Si_6B_{14}$ amorphous metallic glass, in the high temperature range around its glass transition temperature, and a study of the

K. Russew, L. Stojanova, *Glassy Metals*,
DOI 10.1007/978-3-662-47882-0_7

(nonisothermal and isothermal) relaxation of bend stresses in the same ribbonlike alloy at considerably lower temperatures.

The experimental study of bend stress relaxation in ribbonlike amorphous metallic alloys requires the application of high bending stresses of 500–600 MPa order of magnitude and can be successfully carried out only at temperatures substantially lower than the glass transition temperature. At such temperatures the tensile test viscosity determinations are practically not possible. This is due to the circumstance that the applied low tensile stresses of 10–30 MPa in the case of tensile creep viscosity tests cause an extremely low elongation rate of the ribbonlike amorphous samples at lower temperatures, which is in the range of the experimental accuracy of measurements or even lower. On the other hand the relaxation of bend stresses at temperatures close to the glass transition temperature proceeds so fast that it is impossible to be monitored experimentally. This is why the combining of these two methods for studying of viscous flow behavior of glassy metals provides the possibility to extend substantially the temperature range of investigation. The possibility to overcome the shortcomings of both methods and to make use of their advantages is demonstrated. The limitations of applicability of FVM for study of structural relaxation phenomena in glassy metals, concerning the range of applied experimental stresses and temperatures, respectively, have been established.

Relaxation of Bend Stresses in Ribbonlike Amorphous Metallic Alloys: Model Considerations

By coiling of amorphous metallic ribbon into a restraining ring of radius r_o, it develops a stress profile. For a thin ribbon the maximal value of bending stress σ_0 at the surface of the ribbon [1–5] is given by

$$\sigma_o = E\frac{d}{2r_o}, \tag{7.1}$$

where E denotes the Young modulus and d is the ribbon thickness. The initial deformation is fully elastic and is defined as

$$\varepsilon_0 = \frac{\sigma_0}{E} = \frac{d}{2r_o}. \tag{7.2}$$

By application of a restraining ring, the bend radius remains constant, i.e., the overall initial deformation ε_o keeps constant also (see Fig. 7.1). Under isothermal or nonisothermal annealing, overall plastic deformation $\varepsilon(t)$ or $\varepsilon(T, q)$ for isothermal or nonisothermal conditions of the experiment appears, consisting of a component of plastic deformation $\varepsilon_f(t)$ or $\varepsilon_f(T_F, q)$ due to the viscous flow and anelastic deformation $\varepsilon_a(t)$ or $\varepsilon_a(T_F, q)$. The anelastic component of overall deformation is due to processes with nature not well understood so far. As a result the initial bending stress σ_o decreases to σ_l, which value depends on the annealing time t and

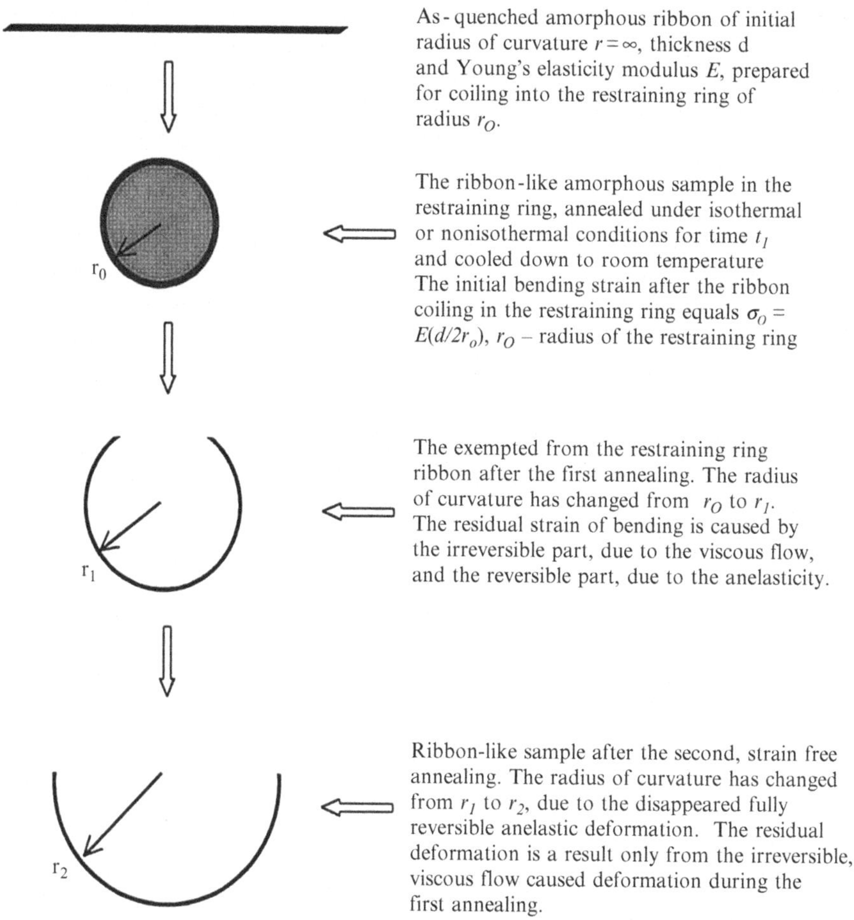

Fig. 7.1 Scheme of the ribbonlike amorphous sample shape changes during the experimental procedure for the determination of the fully reversible and fully irreversible anelastic and viscous flow-caused deformations

temperature or on the combination of final temperature of heating T_F and heating rate q, under isothermal and nonisothermal conditions, respectively. Hence, it follows that

$$\varepsilon_0 = \frac{\sigma(t)}{E} + \varepsilon_f(t) + \varepsilon_a(t), \tag{7.3a}$$

for isothermal experimental conditions, and

$$\varepsilon_0 = \frac{\sigma(T_F, q)}{E} + \varepsilon_f(T_F, q) + \varepsilon_a(T_F, q), \tag{7.3b}$$

for nonisothermal constant heating rate conditions. Here $\sigma(t)/E$ and $\sigma(T_F, q)/E$ denote the elastic deformation as a function of the annealing time t at the different constant temperatures, and $\sigma(T_F, q)/E$ is the elastic deformation as a function of the heating rate q and the final temperature of annealing T_F, under nonisothermal conditions. Provided that after annealing time t_I at a constant temperature T_I the ribbon is taken out from the restraining ring and allowed to spring back, it changes its radius of curvature from r_o to a new value r_I (Fig. 7.1), which depends on the time of annealing t_I, due to the residual overall strain $\varepsilon_f(t_I) + \varepsilon_a(t_I)$. Under nonisothermal conditions the residual overall strain will be dependent on the combination of final temperature T_{FI} and heating rate $q - \varepsilon_f(T_{FI}, q) + \varepsilon_a(T_{FI}, q)$. The dependence of the residual deformations ε_f and ε_a is related to the initial radius r_o and the resulting r_I radius of curvature, respectively, shown as follows [1, 6]:

$$\frac{\varepsilon_f(t_1) + \varepsilon_a(t_1)}{\varepsilon_o} = \frac{r_o}{r_1} \tag{7.4a}$$

and

$$\frac{\varepsilon_f(T_{F1}, q_1) + \varepsilon_a(T_{F1}, q_1)}{\varepsilon_o} = \frac{r_o}{r_1}. \tag{7.4b}$$

By using Eq. (7.2), Eqs. (7.3a, 7.3b) and Eqs. (7.4a, 7.4b), the dependence of bend stress relaxation σ/σ_o on the experimental conditions according to Luborsky et al. can be derived [3–5]:

$$\frac{\sigma(t_1)}{\sigma_o} = 1 - \frac{r_o}{r_1}, \quad \text{and} \tag{7.5a}$$

$$\frac{\sigma(T_{F1}, q_1)}{\sigma_o} = 1 - \frac{r_o}{r_1}. \tag{7.5b}$$

Following the considerations of Luborsky et al. when considering the bend stress relaxation under isothermal experimental conditions, from the equation of Newton describing the relation between the applied tensile stress and the strain rate and Eq. (7.5a), Luborsky et al. obtained for the irreversible bend strain caused by the viscous flow:

$$\frac{\varepsilon_f(t)}{\varepsilon_o} = \frac{E}{3\eta_o \exp\left(\frac{Q_\eta}{RT}\right)} \int_o^t \left(1 - \frac{r_o}{r_1(t')}\right) \frac{dt'}{t'}. \tag{7.6a}$$

Equation (7.6a) does not take into account the presence of an anelastic strain.

For nonisothermal constant heating rate conditions, the relation of Newton for the stress dependence of the viscous flow rate can be integrated to

$$\varepsilon_f(T_F, q) = \frac{1}{q}\int_{T_B}^{T_F} \frac{\sigma(T', q)}{3\eta(T', q)} dT', \quad \text{and}$$
$$\frac{\varepsilon_f(T_F, q)}{\varepsilon_o} = \frac{E}{q}\int_{T_B}^{T_F} \left(1 - \frac{r_o}{r_1}\right) \frac{1}{3\eta(T, q)} dT, \tag{7.6b}$$

where T_F denotes the maximal temperature of annealing reached during the nonisothermal constant heating rate q of annealing and T_B – the starting temperature of the nonisothermal heat treatment. The term $\eta(T, q)$ represents the temperature dependence of the sample viscosity by a constant heating rate q in a temperature range considerably below the glass transition temperature T_g and is defined by (▶ Eq. 5.20).

In order to eliminate the contribution of the anelastic strain ε_a from the overall residual strain and to make possible to deal only with the caused by the viscous flow part ε_f of the residual deformation, one has to perform additional second annealing for a time t_2 and temperature T_2, by a full absence of bending stresses. As far as the anelastic deformation is fully reversible by definition, it should disappear, provided that the second annealing time t_2 is sufficiently long for the chosen temperature T_2 of second annealing. As a result, the radius of curvature should increase to a value r_2 (Fig. 7.1) [1, 6]. The new radius is determined only by the experimental conditions of the first annealing in the restraining ring and can be used for a quantitative evaluation of the irreversible plastic deformation $\varepsilon_f(t_1)$ or $\varepsilon_f(T_{F1}, q)$. By isothermal or nonisothermal annealing, the deformations due to the viscous flow only are

$$\frac{\varepsilon_f(t_1)}{\varepsilon_o} = \frac{r_o}{r_2} \quad \text{and} \tag{7.7a}$$

$$\frac{\varepsilon_f(T_{F1}, q_1)}{\varepsilon_o} = \frac{r_o}{r_2}. \tag{7.7b}$$

After rearranging Eqs. (7.4a, 7.4b) and (7.5a, 7.5b), it follows for the anelastic part of bend deformation that

$$\frac{\varepsilon_a(t_1)}{\varepsilon_o} = r_o\left(\frac{1}{r_1} - \frac{1}{r_2}\right) \quad \text{and} \tag{7.8a}$$

$$\frac{\varepsilon_a(T_{F1}, q_1)}{\varepsilon_o} = r_o\left(\frac{1}{r_1} - \frac{1}{r_2}\right). \tag{7.8b}$$

Equations (7.5a, 7.5b) and (7.8a, 7.8b) provide the possibility for quantitative evaluation of both time and heating rate-dependent parts of the overall strain and their further FVM interpretation. Maddin et al. [2] and Taub and Luborsky [4, 5] have tried to describe the bend stress relaxation of ribbonlike amorphous metallic alloys on the assumption that the strains caused by the elastic deformation decrease only as a result of the viscous flow, taken place during the annealing. Should it be true, however, the anelastic strain components $\varepsilon_a(t)$ and/or $\varepsilon_a(T_F, q)$ should be simply considered as negligible, and the time derivative of Eqs. (7.3a) and (7.3b) should reduce to

$$\frac{d\sigma(t)}{dt} = -E\dot{\varepsilon}_f = -E\frac{\sigma(t)}{3\eta(t)} \quad \text{or} \tag{7.9a}$$

$$\frac{d\sigma(T_F,q)}{dt} = -E\dot{\varepsilon}_f = -E\frac{\sigma(T_F,q)}{3\eta(T_F,q)}. \tag{7.9b}$$

The integration of Eq. (7.9a) for isothermal experimental conditions results in

$$\ln\left(\frac{\sigma(t)}{\sigma_o}\right) = -\frac{E}{3k_r\eta_o T\exp\left(\frac{Q_\eta}{RT}\right)}\ln\left(1 + k_r c_{f,o} t\right), \tag{7.10a}$$

where $k_r = \nu_r\, exp(-Q_r\,/RT)$ denotes the rate constant of relaxation.

For nonisothermal conditions of the bend stress relaxation experiment under constant heating rate q, the solution of Eq. (7.9b) is

$$\ln\left(\frac{\sigma(T_F,q)}{\sigma_o}\right) = -\frac{E}{3q\eta_o}\int_{T_B}^{T_F}\left[c_{f,o}^{-1} + \frac{\nu_r}{q}\int_{T_B}^{T}\exp\left(-\frac{Q_r}{RT'}\right)dT'\right]^{-1}\left[T\exp\left(\frac{Q_\eta}{RT}\right)\right]^{-1}dT \tag{7.10b}$$

where T_B and T_F are the starting temperature of heating and the maximal temperature reached before performing the rapid cooling, interrupting the experiment.

The usage of Eqs. (7.10a) and (7.10b) as an approximated theoretical description of bend stress relaxation under isothermal and nonisothermal experimental conditions is substantially improved by Russew et al. [1, 6, 7], via replacing the term $(1\text{-}r_o/r_1)$ in Eqs. (7.5a) and (7.5b) with experimentally determined decrease of bend stresses σ_2/σ_o, due to the viscous flow only. This term is determined from the ribbon radius of curvature r_2 based on the expressions:

$$\frac{\sigma_2(t)}{\sigma_o} = 1 - \frac{r_o}{r_2} \quad \text{and} \tag{7.11a}$$

$$\frac{\sigma_2(T_F,q)}{\sigma_o} = 1 - \frac{r_o}{r_2}, \tag{7.11b}$$

where r_2 denotes the sample radius of curvature after the second, stress-free annealing of the amorphous sample. As discussed above, this stress-free annealing fully eliminates the contribution of the anelastic deformation ε_a to the overall strain ε_o.

The degree of anelastic deformation (see Eq. (7.8a)) under isothermal experimental conditions can be described by the empirical equation of Kohlrausch–Williams–Watts (KWW) [8–10]:

$$\frac{\varepsilon_a(t)}{\varepsilon_o} = \exp\left[-\left(\frac{t}{\tau_r}\right)^b\right], \tag{7.12}$$

where the temperature dependence of the time of relaxation τ_r is given by

$$\tau_r = \tau_o \exp\left(\frac{Q_r}{RT}\right). \tag{7.13}$$

Q_r in Eq. (7.13) denotes the activation energy of relaxation. The constants τ_o and b are empirical. The constant τ_o is expected to be approximately equal to the reciprocal of the Debye frequency $\nu_D \sim 10^{13}\ s^{-1}$. The parameter b is the so-called KWW-stretched or fraction exponent. Its value varies within 0.1–1.

Under nonisothermal experimental conditions, taking into account that $(T - T_B) = qt$, the KWW Eq. (7.12) transforms to

$$\frac{\varepsilon_a(T,q)}{\varepsilon_o} = \exp\left[-\left(\frac{(T - T_B)/q}{\tau_o \exp(Q_r/RT)}\right)^b\right], \tag{7.14}$$

where T_B is the starting temperature of annealing and q is the heating rate.

Equation (7.14) can be expressed as a function of the time t and the heating rate q used:

$$\frac{\varepsilon_a(t,q)}{\varepsilon_o} = \exp\left[-\left(\frac{t}{\tau_o \exp\left[\frac{Q_r}{R(qt+T_B)}\right]}\right)^b\right]. \tag{7.15}$$

With fraction exponent $0.1 < b < 1$, the KWW equation gives a good description of existing experimental data on relaxation of inorganic glasses and polymers [11]. Most of the attempts to explain the Kohlrausch equation so far [8] have tried to give its direct derivation from complicated models for stress-induced flow into the relaxation kinetics [12]. It turns out that in fact the physical meaning of the KWW equation itself should be attributed to the models to which it is a mathematical approximation.

Structural Relaxation in $Fe_{40}Ni_{40}Si_6B_{14}$ Glassy Alloy

The next section presents the analysis of the relation between the viscosity and the relaxation of bend stresses in $Fe_{40}Ni_{40}Si_6B_{14}$ glassy alloy, both under isothermal and nonisothermal experimental conditions. The analysis is based on the concepts of the free volume model.

The experimental study is limited to:

- Study upon the high-temperature viscous flow behavior of the $Fe_{40}Ni_{40}Si_6B_{14}$ glassy alloy via direct creep viscosity determinations under constant heating rate annealing
- Lower-temperature isothermal determinations of the bend stress relaxation of the same glassy alloy at different annealing temperatures
- Lower-temperature nonisothermal determinations of the bend stress relaxation of the glassy alloy at different heating rates and different end heating temperatures

Direct Measurement of the High-Temperature Viscous Creep Under Nonisothermal Conditions

The specimens tested were amorphous $Fe_{40}Ni_{40}Si_6B_{14}$ ribbons with cross-sectional dimensions 0.68 mm × 0.024 mm produced by single roller quenching from the melt. The nonisothermal viscosity measurements were performed by using a Perkin-Elmer TMS-2 thermomechanical analyzer with a specially designed home-made silica glass assembly for high-temperature creep measurements described in detail in ▶ Chap. 3 (▶ Fig. 3.3). Before starting the experiments, the temperature accuracy of ± 2 K of the TMS-2 at the heating rate used was calibrated using the abrupt increase in the length change of 4 mm × 2 mm × 0.2 mm plates of pure Sn, Pb, and Zn at their melting points. All measurements were carried out in a protective atmosphere using purified Ar as a purging gas. The elongation of 4 mm long pieces of the ribbon studied at a constant heating rate of 20 K/min was monitored up to temperatures higher than the onset temperature of crystallization (740 K) with an accuracy better than 1 μm. Two measurements were performed at different applied loads σ_1 and σ_2 of 31 and 92 MPa, respectively. The sample viscosity was calculated based on the equation of Newton – ▶ Chap. 3, Eq. (3.5).

Figure 7.2 shows the experimentally obtained temperature dependence of the viscosity of the glassy alloy studied, together with the best fit curve obtained on the basis of (▶ Eq. 5.8) and (▶ Eq. 5.17) at a heating rate of 20 K/min. The values of the fitting parameters η_0, $c_{f,0}$, Q_η, ν_r, Q_r, B, and T_0 are given in Table 7.1 (A). Figure 7.2 also shows the temperature dependence of the metastable quasi-equilibrium viscosity η_{eq}, calculated based on (▶ Eq. 5.9) using the obtained fitting parameters. The intersection of both viscosity temperature dependencies corresponds to the dynamic glass transition temperature T_g (~729 K) of the alloy studied at a heating rate of 20 K/min.

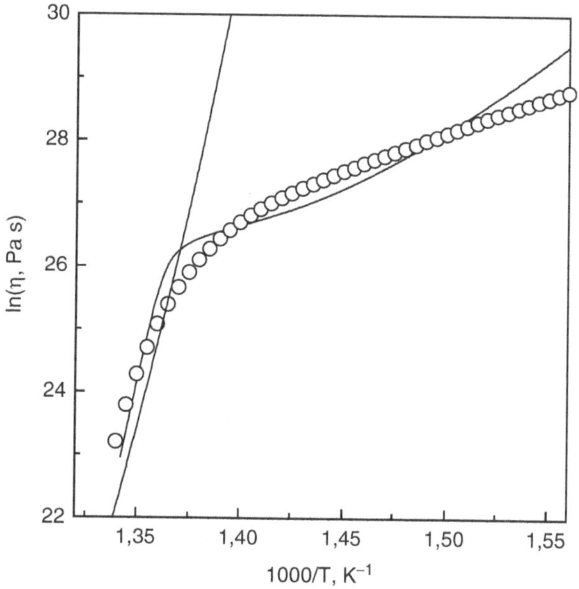

Fig. 7.2 Experimentally determined viscosity temperature dependence of $Fe_{40}Ni_{40}Si_6B_{14}$ amorphous ribbon, measured at a heating rate of 20 K/min – ○. The solid curves represents the calculated non-equilibrium viscosity temperature dependence $\eta(T)$. The steeper solid line represents the calculated quasi-equilibrium viscosity temperature dependence $\eta_{eq}(T)$

Study upon the Relaxation of Bend Stresses via Isothermal and Nonisothermal Measurements

Isothermal Bend Stress Relaxation Measurements

The isothermal bend stress relaxation experiments upon $Fe_{40}Ni_{40}Si_6B_{14}$ amorphous ribbons were carried out at four constant temperatures – 550 K, 575 K, 600 K, and 625 K. The starting temperature of heating T_B was 330 K. The isothermal sample annealing was carried out in Ar atmosphere in a Perkin-Elmer DSC-2 micro-furnace. The calorimeter sample pan was used as a restraining ring of radius 3.25 mm. The value of the Young modulus of the amorphous samples was accepted to be 150 GPa [13]. The sample ribbon was subjected to initial elastic stress via its dense coiling in contact to the internal calorimeter sample pan wall. Based on Eq. (7.1), it follows that the elastic stress arising into the ribbon after its positioning into the restraining ring equals to $\sigma_o = 560$MPa. At every temperature of isothermal annealing T_F 5, different times of annealing were chosen for the monitoring of the resulted degree of relaxation process. The coiled ribbon samples were heated up to the chosen temperature of annealing T_F with a heating rate of 320 K/min, i.e., fast enough in order to restrict to a great extent the transition stage of changing temperature until the constant experimental temperature was reached. As mentioned above, for each temperature of isothermal annealing, five different annealing times were chosen, before the sample was subjected to rapid cooling down to room temperature. After taking out the samples from the restraining ring, they were let to spring back and the resulting new sample radius of curvature r_I was evaluated, as it is described in [2–5]. This procedure was repeated five times with five different samples, and the mean radius value of all samples was evaluated [14]. After the first annealing a second fully stress-free sample

Table 7.1 Summarized FVM and KWW parameters for the three different types of experiments upon ribbonlike amorphous $Fe_{40}Ni_{40}Si_6B_{14}$ alloy, obtained from FVM-based regression analysis and the KWW equation

	Type of experiment					
	A	B	C	D	E	F
Model parameter, dimension	Viscosity	Isothermal bend stress relaxation	Nonisothermal bend stress relaxation	Isothermal nonreversible bend stress deformation	Isothermal reversible anelastic bend stress deformation	Nonisothermal reversible anelastic bend stress deformation
E, Pa	–	1.5×10^{11}	1.5×10^{11}	1.5×0^{11}	–	–
η_o, Pa s/K	1.7×10^{-20}	3.28×10^{-21}	3.28×10^{-21}	3.28×10^{-21}	–	–
$\nu_{r,}$, 1/s	1.2×10^{23}	3.09×10^{24}	3.07×10^{24}	3.09×10^{24}	–	–
Q_r, kJ/mol	177	177	177	177	284,0	278.0
Q_η , kJ/mol	224	224	224	224		–
$c_{f.o}$	3.6×10^{-12}	1.23×10^{-11}	1.23×10^{-11}	1.23×10^{-11}		–
$1/\tau_o$, 1/s	–	–	–	–	5.6×10^{23}	1.3×10^{22}
B	3612	–	–	–	–	–
T_0, K	603	–	–	–	–	–
b	–	–	–	–	0.19	0.15

(A) Model parameters η_0, Q_η, ν_r, $c_{f,0}$, B, and T_0 obtained on the basis of direct viscosity determinations
(B) Model parameters η_0, ν_r, and $c_{f,0}$ determined on the basis of isothermal bend stress relaxation determinations $\ln(\sigma_2\,(t)/\sigma_0)$. Q_η and Q_r are fixed to be 177 and 224 kJ/mol, respectively; the Young modulus is fixed to be equal to 150 GPa
(C) Model parameters η_0, ν_r, and $c_{f,0}$ determined on the basis of nonisothermal bend stress relaxation $\ln(\sigma_2\,(T,q)/\sigma_0)$. Q_η and Q_r are fixed to be 177 and 224 kJ/mol, respectively; the Young modulus is fixed to be equal to 150 GPa
(D) Model parameters η_0, ν_r, and $c_{f,0}$ determined on the basis of isothermal nonreversible bend stress deformation $\varepsilon_f(t)/\varepsilon_0$. Q_η and Q_r are fixed to be 177 and 224 kJ/mol, respectively; the Young modulus is fixed to be equal to 150 GPa
(E) KWW – model parameters Q_r, $1/\tau_0$, and b determined on the basis of isothermal fully reversible anelastic bend stress relaxation
(F) KWW – model parameters Q_r, $1/\tau_0$, and b determined on the basis of nonisothermal fully reversible anelastic bend stress relaxation

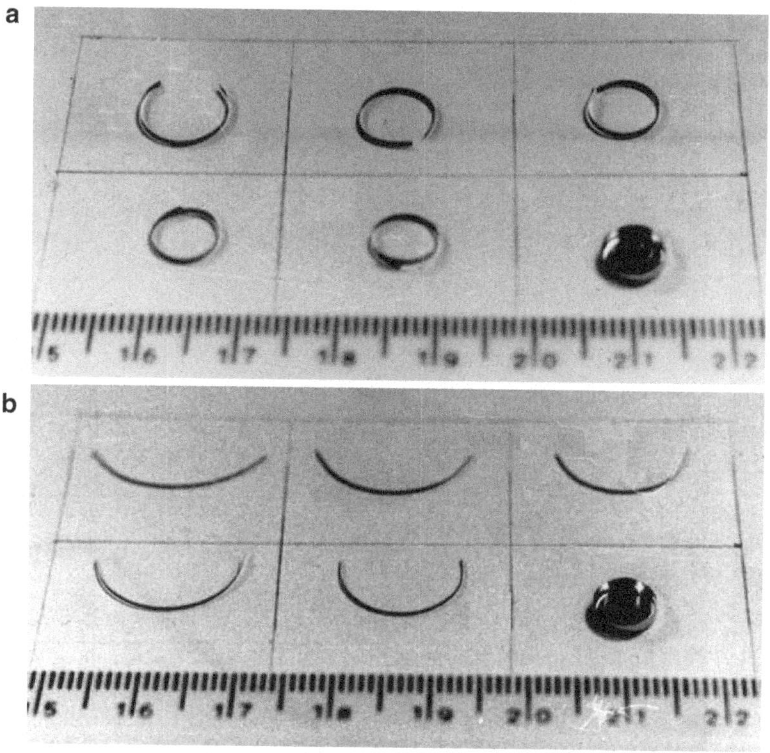

Fig. 7.3 Photographic images of bend stress-relaxed $Fe_{40}Ni_{40}Si_6B_{14}$ amorphous ribbons. (**a**) After the first isothermal anneal at temperature 575 K at five different annealing times: 1–5 min, 2–20 min, 3–60 min, 4–180 min, and 5–480 min; 6-standard DSC sample pan, used as a restraining ring. The sequence is from left to right. (**b**) The same samples after the second stress-free anneal after heating by a heating rate of 10 K/min up to 643 K and cooling down to ambient temperature

annealing was carried out in order to eliminate the contribution of the anelastic strain to the overall strain. For this purpose additional annealing of the partially relaxed specimens in a separate furnace under zero stress conditions was carried out, with a heating rate of 10 K/min up to 643 K and cooling down to ambient temperature. After cooling down, the new radius r_2 of curvature was measured. The contribution of the deformation ε_f caused by the viscous flow only was calculated based on Eq. (7.7a).

Figure 7.3a shows the photographic images of $Fe_{40}Ni_{40}Si_6B_{14}$ amorphous ribbonlike samples after subjecting them to bend stress relaxation at 575 K for the indicated times of isothermal annealing. The samples after the second, fully stress-free annealing are shown in Fig. 7.3b. The resulted increase in the sample radius of curvature is obvious, thus an indication of a real presence of "frozen-in" anelastic sample deformation ε_a after the first annealing.

Figure 7.4 shows the degree of bend stress relaxation $\ln(\sigma_l(t)/\sigma_o)$ of amorphous $Fe_{40}Ni_{40}Si_6B_{14}$ samples after the first anneal, as a function of annealing time t under isothermal experimental conditions at 550 K, 575 K, 600 K, and 625 K.

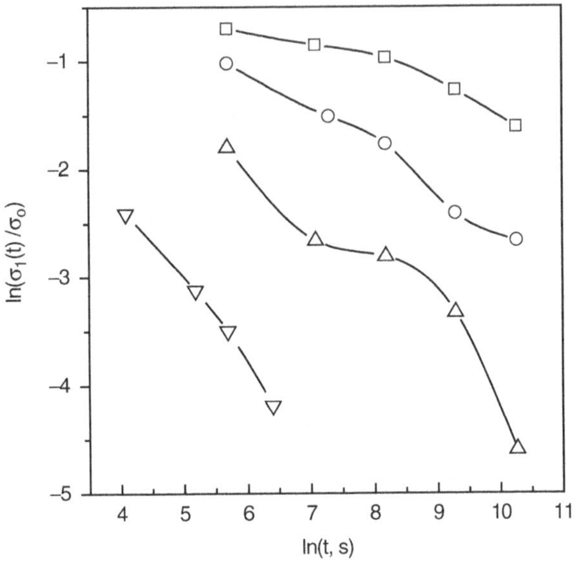

Fig. 7.4 Degree of bend stress relaxation $\ln(\sigma_l(t)/\sigma_o)$ of amorphous $Fe_{40}Ni_{40}Si_6B_{14}$ ribbonlike samples after isothermal annealing at temperatures: □, 550 K; o, 575 K; Δ, 600 K; and ∇, 625 K. *Solid lines* are guides for the eye

Figure 7.5 represents the degree of bend stress relaxation $\ln(\sigma_2(t)/\sigma_o)$ of amorphous $Fe_{40}Ni_{40}Si_6B_{14}$ samples due to the viscous flow after the second stress-free annealing, as function of the first annealing time at constant temperatures 550 K, 575 K, 600 K, and 625 K. The symbols present the bend stress values calculated by using the radii of curvature r_2 based on Eq. (7.11a).

The final temperature of the second stress-free annealing was 643 K. As is seen, the degree of bend stress relaxation decreases substantially. The regression analysis of the experimental data was carried out on the basis of Eq. (7.10a) by fixed values of the activation energies Q_η and Q_r. The values used were taken from the regression analysis of the viscosity experimental data. The obtained values of the three FVM parameters η_o, ν_r, and $c_{f,o}$, together with fixed values of the activation energies, are given in Table 7.1 (B). They coincide practically with the regression FVM parameters obtained from the experimental data of the direct viscosity creep determinations under constant heating rate conditions and with the results from the bend stress relaxation experiments under isothermal conditions [14]. All these FVM parameters are shown in Table 7.1. Figure 7.6 shows the ratio $\varepsilon_f(t)/\varepsilon_o$ corresponding to the relaxation of bend stresses calculated with the aid of Eq. (7.7a). The solid lines show the best fit curves based on Eq. (7.6a). The obtained values of the three FVM parameters η_o, ν_r, and $c_{f,o}$, together with fixed values of the activation energies, are given in Table 7.1 (D). As is seen, the contribution of the deformation $\varepsilon_f(t)/\varepsilon_O$ to the overall deformation increases along with increasing annealing time and also with the increasing of the constant annealing temperature. The saturation value of $\varepsilon_f(t)/\varepsilon_o = 1$ for fully relaxed samples is reached for samples annealed at 625 K.

The bend stress relaxation data obtained at four different constant temperatures under isothermal conditions were subjected to regression analysis simultaneously.

Fig. 7.5 Degree of bend stress relaxation $\ln(\sigma_2(t)/\sigma_o)$ of amorphous $Fe_{40}Ni_{40}Si_6B_{14}$ samples due to the viscous flow, after the second stress-free annealing, at constant temperatures: □, 550 K; o, 575 K; Δ, 600 K; and ∇, 625 K. The *symbols* present the bend stress values calculated by using the radii of curvature r_2 based on Eq. (7.11a). The *solid lines* represent $\ln(\sigma_2/\sigma_0)$ based on Eq. (7.10a)

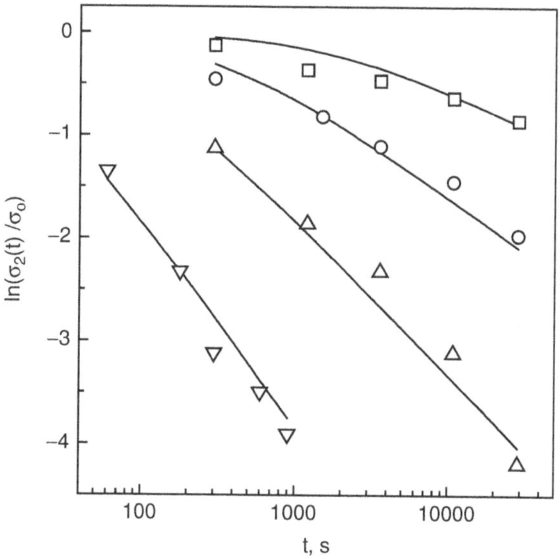

Fig. 7.6 Degree of the irreversible bend stress deformation $\varepsilon_f(t)/\varepsilon_o$ of amorphous $Fe_{40}Ni_{40}Si_6B_{14}$ ribbons, due to the viscous flow only, as a function of annealing time at four constant temperatures: □, 550 K; o, 575 K; Δ, 600 K; and ∇, 625 K. *Solid lines* – best fit, simultaneously calculated curves, based on Eq. (7.6a)

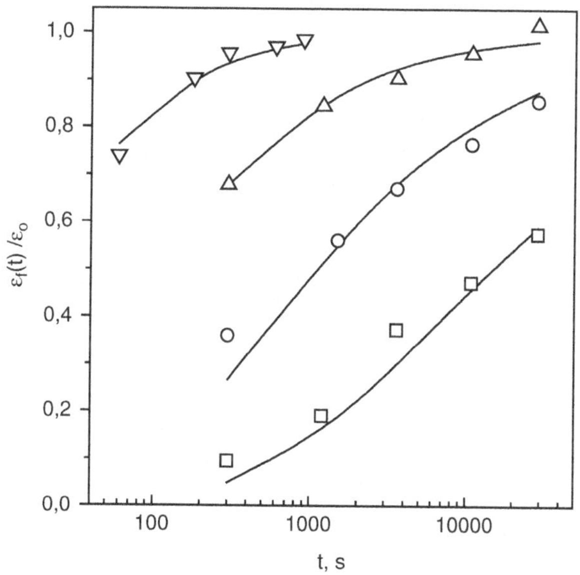

By carrying out the analysis, both activation energies Q_r and Q_η were fixed to be equal to the respective activation energies from the direct viscous flow determinations. In this way the number of the unknown regression coefficients was reduced to three. The regression analysis had as a result a combination of three model parameters η_0, ν_r, and $c_{f,0}$ given in Table 7.1. They coincide practically with the values of the same model parameters obtained from the experiments upon viscous flow and nonisothermal bend relaxation. Although the free volume model is mostly

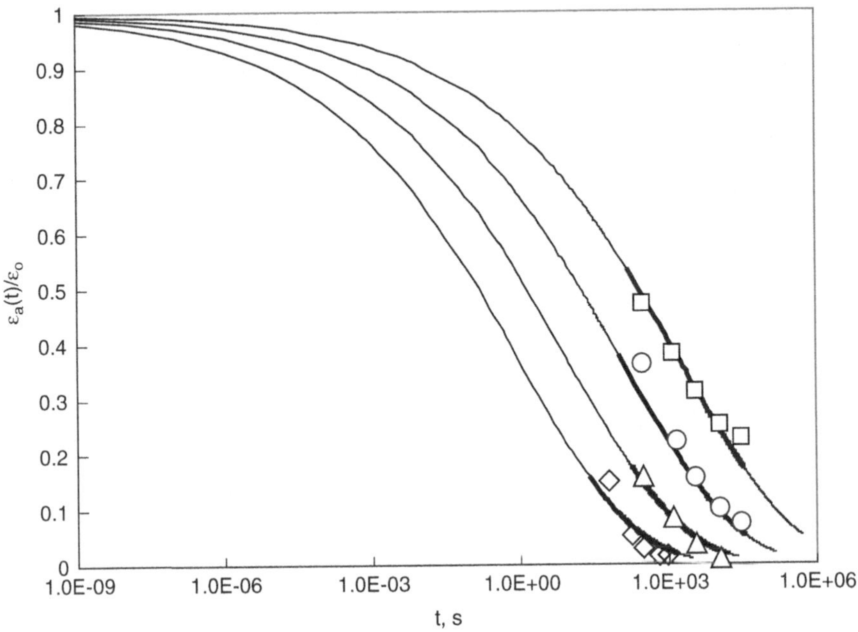

Fig. 7.7 Time dependence of the contribution of the fully reversible anelastic bend stress deformation $\varepsilon_a(t)/\varepsilon_o$ to the overall strain of amorphous $Fe_{40}Ni_{40}Si_6B_{14}$ ribbons at constant annealing temperatures: □, 550 K; o, 575 K; Δ, 600 K; and ◇, 625 K. The *solid lines* are the best fit curves based on the KWW equation (7.12)

criticized for the "too" great number of fitting parameters, here the three independent experiments presented with adequately interpreted, convincing, and uncontradictory common results give a proof for its physical consistency.

The contribution of the fully reversible anelastic deformation for the reduction of the bend stress level is substantial. This is more clearly seen in Fig. 7.7, showing the ratio $\varepsilon_a(t)/\varepsilon_0$ based on Eq. (7.8a). Figure 7.7 shows the relaxation of the reversible anelastic deformation under isothermal experimental conditions $\varepsilon_a(t)/\varepsilon_0$ as a function of time at different constant annealing temperatures. The solid lines present the best fit curves based on the equation of Kohlrausch–Williams–Watts – Eq. (7.12). The corresponding KWW model parameters are given in Table 7.1 (E). As is seen, the anelastic deformation decreases along with the increasing time of annealing and by sufficiently long annealing times almost vanishes.

Nonisothermal Bend Stress Relaxation Measurements

Nonisothermal bend stress relaxation tests were carried out at seven different constant heating rates from 0.31 to 20 K/min, using the starting temperature $T_B = 330$ K and heating up to five final temperatures T_F, viz., 550 K, 575 K, 600 K, 625 K, and 650 K. The nonisothermal annealing of the specimens was carried out in a protective purified Ar atmosphere with the aid of a Perkin-Elmer DSC-2 by using the sample pan of the calorimeter as a restraining ring with a radius of curvature $r_0 = 3.25$ mm. Using $E = 150$ GPa [13], it follows from Eq. (7.1) that $\sigma_0 = 560$ MPa. The ribbons were subjected to bend stress by coiling them into the sample pans, annealed with different heating rates q from T_B to each final temperature T_F, and subsequently rapidly cooled down to T_B and allowed to spring back. Spring-back measurements of coiled $Fe_{40}Ni_{40}Si_6B_{14}$ amorphous ribbons were performed as described in [2–5]. The radius of curvature r_1 was measured for each T_F and q, followed by additional annealing of the partially relaxed specimens in a separate furnace under zero stress conditions with a heating rate of 10 K/min up to 650 K and cooling down to ambient temperature. The resulting new radius of curvature r_2 was measured, from which the amount of deformation caused by viscous flow was calculated using Eq. (7.7b)].

In Fig. 7.8a–c photographic images of the $Fe_{40}Ni_{40}Si_6B_{14}$ ribbons used for the bend stress relaxation measurements are shown after the first anneal at a heating rate q of 20 K/min up to the final temperatures T_F indicated in the figure. The same specimens after the second anneal at zero stress are shown in Fig. 7.8a′–c′. The increase in the radius of curvature of the ribbons after the second anneal is apparent, indicating the presence of the frozen-in anelastic contribution after the first annealing treatment.

Figure 7.9 shows the ratio $\varepsilon_f(T_F,q)/\varepsilon_0$ reflecting the relaxation of bend stresses calculated based on Eq. (7.7b) [1]. The solid lines give the best fit curves based on Eq. (7.6b). It can be seen that at all final temperatures of heating, the viscous flow contribution $\varepsilon_f(T_F,q)/\varepsilon_0$ increases with increasing T_F and with decreasing rate of heating, almost reaching, at $T_F = 650$ K, a saturation value of unity for a fully relaxed specimen. The data from all bend stress relaxation experiments for all final temperatures of heating T_F were fitted simultaneously, fixing the values of the activation energies Q_r and Q_η, to be equal to the values of the corresponding activation energies from the direct viscosity measurements, thus reducing the number of fitting parameters to three. The fitting procedure yields a set of fitting parameters η_0, $c_{f,0}$, and ν_r which are given in Table 7.1(b). It can be seen that under these fitting conditions, the values of the three free fitting parameters coincide practically with the corresponding values from the viscosity measurements. Most of the criticism of the free volume model is based on its many adjustable parameters [15]. Our present results show that the theoretical description of the irreversible part of relaxation of bend stresses under nonisothermal conditions on the basis of the free volume model is physically meaningful and self-consistent.

The degree of bend stress relaxation, $\ln(\sigma_1(T_F, q)/\sigma_0)$, after the first annealing treatment of $Fe_{40}Ni_{40}Si_6B_{14}$ is given in Fig. 7.10, and the degree of bend stress relaxation, $\ln(\sigma_2(T_F, q)/\sigma_0)$, due to viscous flow only was obtained after the second anneal under zero stress conditions shown in Fig. 7.11. For the calculation of ln $(\sigma_2(T_F, q)/\sigma_0)$ based on Eq. (7.11b), only the experimentally obtained r_2 radii of curvature were used. The experimental data were fitted with Eq. (7.10b) with the activation energies Q_η and Q_r fixed to the values obtained from the viscosity measurement. The values of the three free fitting parameters η_0, $c_{f,0}$, and ν_r obtained, together with the fixed values of Q_η and Q_r, are shown in Table 7.1(C). They practically coincide with the fitting parameter values from the viscosity and

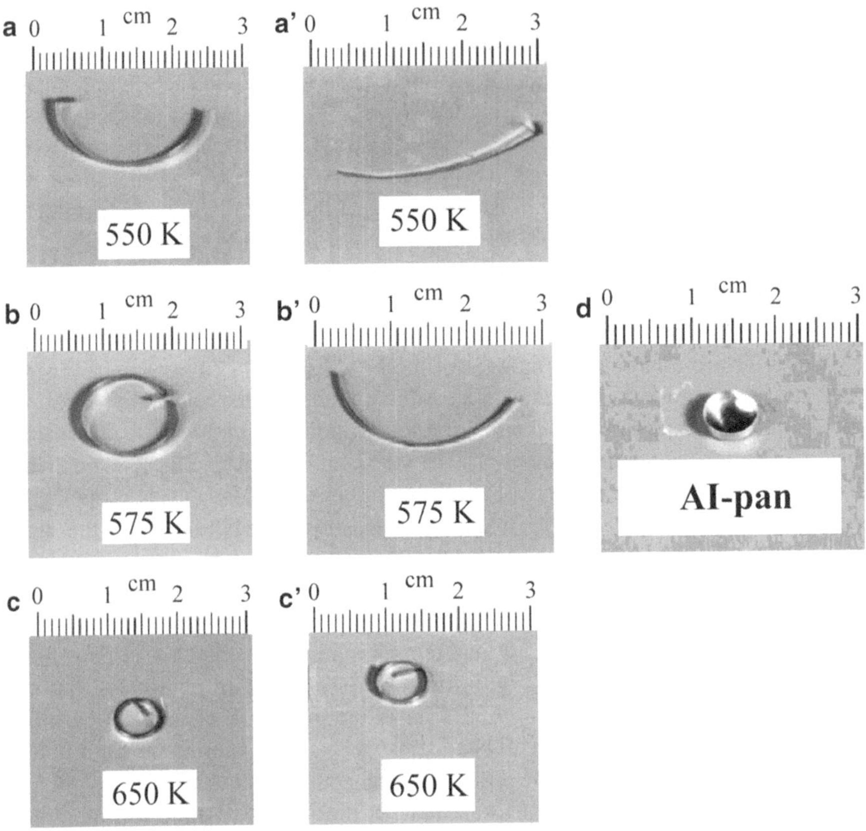

Fig. 7.8 (**a**)–(**c**): Photographic images of bend stress-relaxed $Fe_{40}Ni_{40}Si_6B_{14}$ amorphous ribbons after the first anneal at a heating rate of 20 K/min up to the final temperatures T_F indicated, (**a′**)–(**c′**). The same ribbons after the second stress-free annealing at a heating rate of 10 K/min up to 650 K and cooling down to room temperature. (**d**) Aluminium sample pan used as a restraining ring during the bend stress relaxation measurements

viscous flow measurements given in Table 7.1 (A) and (B). It turns out that Eq. (7.10b) can be used as a good approximation describing the viscous flow contribution by the relaxation of bend stresses in the alloy studied. It is clear from Fig. 7.11 and Table 7.1 that the use of Eq. (7.10b) as an approximation to Eq. (7.6b) is valid when considering the stress relaxation due to viscous flow only.

The contribution of the anelastic strain to the stress reduction during the first anneal is considerable. This is seen even more clearly in Fig. 7.12, which shows the ratio $\varepsilon_a(T_F,q)/\varepsilon_0$ calculated based on Eq. (7.8b), with r_2 values determined from the smoothed curves in Fig. 7.11. At lower final temperatures of heating, the anelastic strain contribution increases rapidly with decreasing heating rate, passes over a maximum, and decreases with increasing T_F. This behavior can be explained as follows. Since the total amount of anelastic strain that can take place is proportional to the actual stress level, the actual amount of anelastic strain first increases with

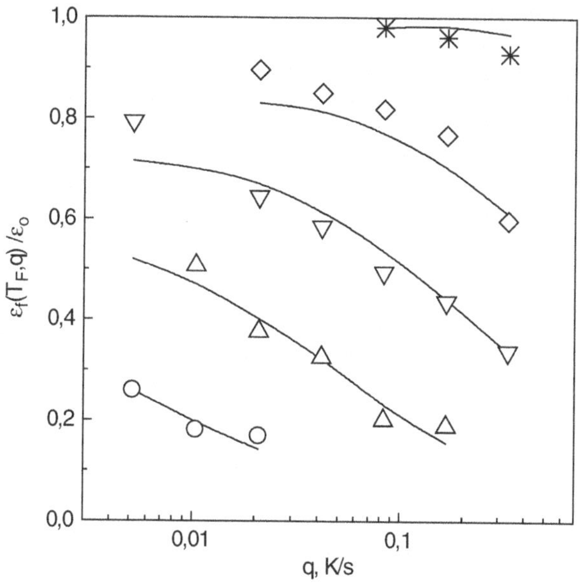

Fig. 7.9 Normalized viscous flow contribution $\varepsilon_f(T_F,q)/\varepsilon_o$ to the overall strain during bend stress relaxation of $Fe_{40}Ni_{40}Si_6B_{14}$ as a function of heating rate q and final temperature of heating T_F by first anneal: *, 650 K; ◇, 620 K; ∇, 600 K; Δ, 575 K; and o, 550 K. The data are fitted simultaneously based on Eq. (7.6b)

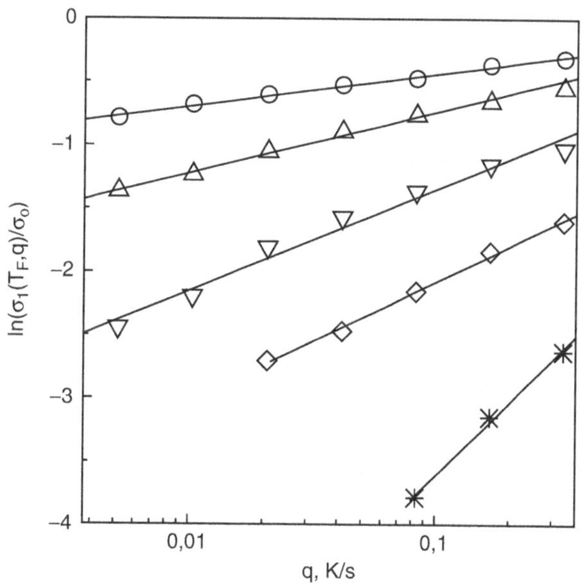

Fig. 7.10 The degree of bend stress relaxation ln $(\sigma_1(T_F,q)/\sigma_0)$ in $Fe_{40}Ni_{40}Si_6B_{14}$ after first anneal as a function of heating rate q and final temperatures T_F: *, 650 K; ◇, 620 K;∇, 600 K; Δ, 575 K; and o, 550 K. Solid lines are guides to the eye

decreasing heating rate at lower temperatures. With increasing final temperatures of heating and further decreasing heating rates, however, the amount of viscous flow increases, lowering the stress level, and the actual amount of anelastic strain then decreases. At a final temperature of 650 K, the anelastic strain contribution equals to zero, i.e., it has been annealed out already during the first anneal of stressed ribbons. This circumstance shows unambiguously the fully reversible nature of anelastic strain.

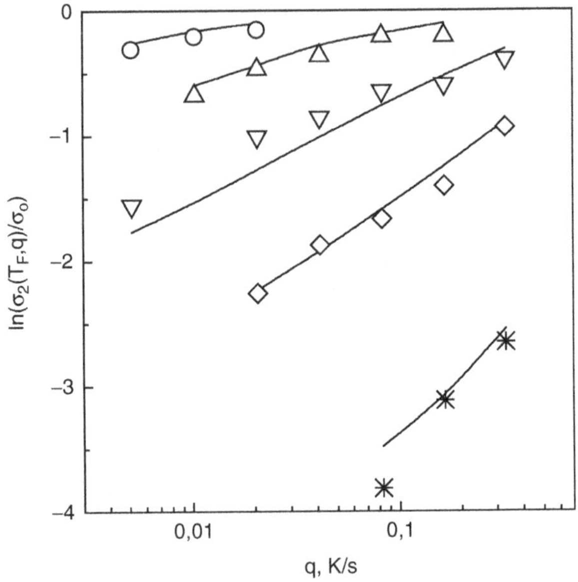

Fig. 7.11 The degree of bend stress relaxation ln $(\sigma_2(T_F,q)/\sigma_0)$ after the second anneal at zero stress as a function of heating rate q and final temperatures T_F indicated above. *Solid lines* are calculated values of ln (σ_2/σ_0) based on Eq. (7.10b)

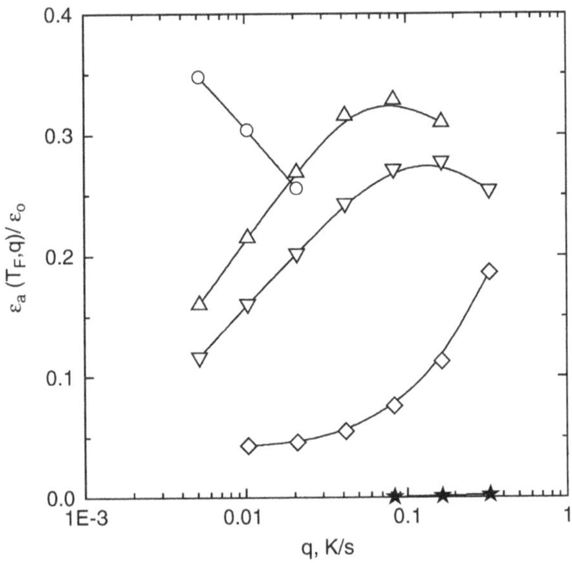

Fig. 7.12 Heating rate dependence of the normalized anelastic strain contribution $\varepsilon_a(T_F,q)/\varepsilon_0$ to overall strain during bend stress relaxation of $Fe_{40}Ni_{40}Si_6B_{14}$ calculated based on Eq. (7.8b). Final temperatures of first anneal T_F: *, 650 K; ◇, 620 K; ∇, 600 K; Δ, 575 K; and o, 550 K. *Solid lines* are guides to the eye

Considering the experimental results presented, it is clear that the free volume model describes in an adequate way both the kinetics of structural changes during the directly measured high-temperature viscous flow around the glass transition temperature T_g and the low-temperature irreversible part of relaxation of bend stresses in $Fe_{40}Ni_{40}Si_6B_{14}$ glassy ribbons with practically one and the same set of intrinsic structural relaxation parameters η_0, $c_{f,0}$, Q_η, ν_r, Q_r, B, and T_0 governing the

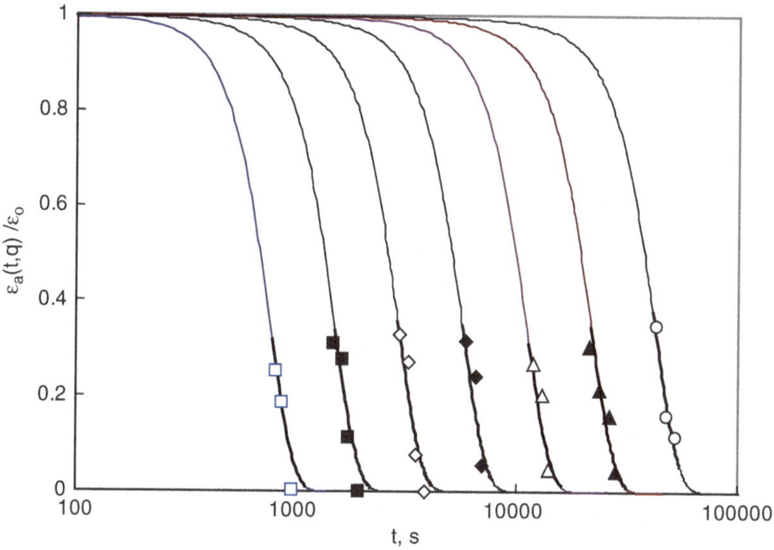

Fig. 7.13 Time dependence of the nonisothermal degree of anelastic bend stress deformation $\varepsilon_a(t,q)/\varepsilon_o$ in $Fe_{40}Ni_{40}Si_6B_{14}$ glassy alloy at different heating rates: □, 20 K/min; ■, 10 K/min; ◇, 5 K/min; ◆, 2.5 K/min; Δ, 1.25 K/min; ▲, 0.63 K/min; and ○, 0.31 K/min. The *solid lines* are the best fit curves based on the KWW Eq. (7.15)

atomic mobility in amorphous metals. The parameter values are physically acceptable and meaningful except for the value of the attempt frequency ν_r being several orders of magnitude higher than the Debye frequency. In this respect the discussion given in the work of Duine et al. [16] seems to be quite useful and fully applicable to the results presented in this article. Later on a special discussion will be dedicated to the physical significance of FVM parameters.

An important result of the present study is the possibility of distinguishing, both theoretically and experimentally, between the viscous flow and anelastic strain contributions taking place during the low-temperature relaxation of bend stresses. Although several authors have examined the effect of structural changes on the kinetics of anelastic relaxation of metallic glasses [17–22] during creep experiments, no systematic investigation of the anelastic relaxation under isothermal and nonisothermal conditions has been reported. It is generally accepted that the anelastic strain is governed by a spectrum of activation energies, making the theoretical description both under nonisothermal and isothermal conditions quite complicated.

The time dependence of the nonisothermal degree of anelastic bend stress deformation $\varepsilon_a(t,q)/\varepsilon_o$ in $Fe_{40}Ni_{40}Si_6B_{14}$ glassy alloy at different heating rates is shown in Fig. 7.13. The solid lines present the best fit curves based on the equation of Kohlrausch–Williams–Watts – Eq. (7.15). The corresponding KWW model parameters are given in Table 7.1 (F). It is obvious from Fig. 7.13 that Eq. (7.15) adequately describes the relaxation of fully reversible anelastic deformation in $Fe_{40}Ni_{40}Si_6B_{14}$ glassy alloy under nonisothermal conditions.

A generalization discussion about the physical significance of the FVM parameters will be presented on the example of the amorphous alloy Al_{25} La_{50} Ni_{25} in ▶ Chap. 8.

References

1. Russew K, de Hey P, Sietsma J, van den Beukel A (1997) Viscous flow of amorphous $Fe_{40}Ni_{40}Si_6B_{14}$ glassy alloy studied by direct creep measurements and relaxation of bend stresses under nonisothermal conditions. Acta Mater 45(5):2129
2. Maddin R, Masumoto T (1972) Mater Sci Eng 9:153
3. Luborsky F, Walter J (1978) Mater Sci Eng 35:255
4. Taub A, Luborsky F (1981) Acta Metall 29:1939
5. Taub A, Luborsky F (1982) Mater Sci Eng 56:157
6. Russew K, Sommer F (1998) Determination of the viscosity, thermal expansion, heat capacity and bend stress relaxation temperature dependence of glassy alloys – a tool for understanding of relaxation phenomena. Int J Non-Equilb Process 11:3
7. Russew K, Zappel BJ, Sommer F (1995) Nonisothermal viscous flow behaviour of $Pd_{40}Ni_{40}P_{20}$ glassy alloy considered as a free volume related phenomenon. Scr Mater 24:271
8. Dobreva A, Gutzow I, Schmelzer J (1997) J Non-Cryst Solids 209:257
9. Weiss M, Moske M, Samwer K (1996) Appl Phys Lett 69(21):3200
10. Ulfert W, Kronmüller H (1996) J Phys IV France 06(C8):C8–617
11. Mazurin OV, Leko VK (1986) Vitrification. Nauka, Leningrad (in Russian)
12. Jonscher AK (1983) Dielectric relaxation in solids. Chelsea Dielectrics Press, London, p 5
13. ASM International Handbook Committee (1990) ASM handbook, vol 02. Metal Park
14. Koleva E, Stojanova L, Russew K (2001) Thermomechanical study of relaxation phenomena in $Fe_{40}Ni_{40}Si_6B_{14}$ metallic glass using creep, bend stress and anelastic strain relaxation measurements under isothermal conditions. J Mater Sci Technol 9(2):108
15. Limoge Y (1990) Acta Metall Mater 38:1733
16. Duine PA, Sietsma J, van den Beukel A (1992) Acta Metall Mater 40:743
17. Taub AI (1981) Mater Sci 16:3887
18. Chen HS (1980) Rep Prog Phys 43:353
19. Berry BS (1978) In: Gilman JJ, Leamy HJ (eds) Metallic glasses. ASM, Metals Park, p 161
20. Leusink GJ, van den Beukel A (1988) Acta Metall 36:3019
21. De Hey P, Sietsma J, van den Beukel A (1996) J Non-Cryst Solids 205–207:696
22. Van den Beukel A (1992) Phys Stat Solidi (a) 129:49

Structural Relaxation in Various Amorphous Metallic Alloys Studied via Direct Viscosity and Bend Stress Relaxation Measurements

8

Abstract

This book section is a continuation of those presented in Chap. 7, theoretical bend stress relaxation considerations with respect to their relation to the viscosity. The chapter provides additional experimental proof with the aid of viscosity and bend stress relaxation measurements on $Pd_{77.5}Cu_6Si_{16.5}$, $Al_{25}La_{50}Ni_{25}$, and $Al_{7.5}Cu_{17.5}Ni_{10}Zr_{65}$ glassy alloys.

The FVM interpretation of the two kinds of experimental data is possible with practically one and the same set of FVM parameters. A new method for quantitative experimental separation of viscous flow caused plastic and anelastic deformations, respectively, is developed. It is shown that the relaxation of the anelastic deformation, both under isothermal and nonisothermal conditions, can be very well described by the empirical equation of Kohlrausch-Williams-Watts.

Keywords

Glassy metals • Amorphous metallic alloys • $Pd_{77.5}Cu_6Si_{16.5}$ • $Al_{25}La_{50}Ni_{25}$ • $Al_{7.5}Cu_{17.5}Ni_{10}Zr_{65}$ glassy alloys • Viscous flow • Bend stress relaxation • Irreversible/reversible relaxation • Anelastic deformation • Equation of Kohlrausch-Williams-Watts

This book section is a continuation of those presented in ▶ Chap. 7, theoretical bend stress relaxation considerations with respect to their relation to the viscosity. It provides additional experimental proof with the aid of viscosity and bend stress relaxation measurements on $Pd_{77.5}Cu_6Si_{16.5}$, $Al_{25}La_{50}Ni_{25}$, and $Al_{7.5}Cu_{17.5}Ni_{10}Zr_{65}$ glassy alloys.

K. Russew, L. Stojanova, *Glassy Metals*,
DOI 10.1007/978-3-662-47882-0_8

The Amorphous Metallic Alloy $Pd_{77.5}Cu_6Si_{16.5}$

Direct Viscosity Measurements Under Nonisothermal Conditions

The amorphous metallic $Pd_{77.5}Cu_6Si_{16.5}$ ribbons were produced according to the CBMS method. Their cross section was $0.38 \times 0.042\ mm^2$.

The nonisothermal constant heating rate viscosity determinations were carried out with the aid of Perkin-Elmer TMS-2 silica glass dilatometer at a heating rate of 20 K/min by using the assembly for creep measurements, as described in ▶ Chap. 2 (▶ Fig. 2.3). The initial sample length was fixed to be 4 mm. The sample elongation was monitored in the temperature range 550–660 K with accuracy better than ±1 μm. The temperature was measured with an accuracy of ±2 K. The sample viscosity was calculated according to the equation of Newton – ▶ Chap. 3, Eq. (3.3).

Figure 8.1 shows the experimentally obtained viscosity temperature dependence of the $Pd_{77.5}Cu_6Si_{16.5}$ amorphous alloy studied, together with the best-fit curve, calculated by using the combination of (▶ Eqs. 5.8) and (▶ 5.17) (▶ Chap. 5).

The values of the FVM regression coefficients η_o, Q_η, ν_r, $c_{f,o}$, Q_r, B, and T_o are presented in Table 8.1(A). Figure 8.1 shows also the temperature dependence of the quasi-equilibrium viscosity η_{eq}, calculated according to the (▶ Eq. 5.9), ▶ Chap. 5 by using the FVM model parameters from Table 8.1(A). The point of intersection of both curves represents the glass transition temperature T_g (645 K) of the amorphous alloy studied at a heating rate of 20 K/min. According to Sestak [1], the glass transition temperature of the amorphous metallic alloy $Pd_{83}Si_{17}$ is 620 K and of $Pd_{70.5}$ $Ni_{12}.5Si_{17}$ - T_g =666 K, respectively, i.e., quite similar to the amorphous alloy studied.

Bend Stress Relaxation Study Under Isothermal Conditions

The isothermal study of bend relaxation was carried at different constant temperatures – 420, 460, 500, 530, and 560 K, respectively, starting from 330 K. The annealing of the specimens was carried out in a protective purified Ar atmosphere with the aid of a Perkin-Elmer DSC 2 by using the sample pan of the calorimeter as a restraining ring with a radius of curvature r_0 = 3.25 mm. Using E = 108 GPa [2], it follows from (▶ Eq. 7.1), ▶ Chap. 7 that σ_0 = 646MPa. The ribbons were subjected to bend stress by coiling them into the sample pans, annealed with heating rates 320 K/min from 330 K to each final constant temperature, mentioned above. After the isothermal annealing the samples were subsequently rapidly cooled down to ambient temperature and allowed to spring back. Spring-back measurements of the coiled amorphous ribbons were performed as described in [3–6]. This procedure was repeated 5 times with 5 different samples and the mean radius value of all samples was evaluated [7]. After the first annealing, a second, fully stress-free sample annealing was carried out in order to eliminate the contribution of the anelastic strain to the overall strain. For this purpose additional annealing of the partially relaxed specimens in a separate furnace under zero stress conditions was

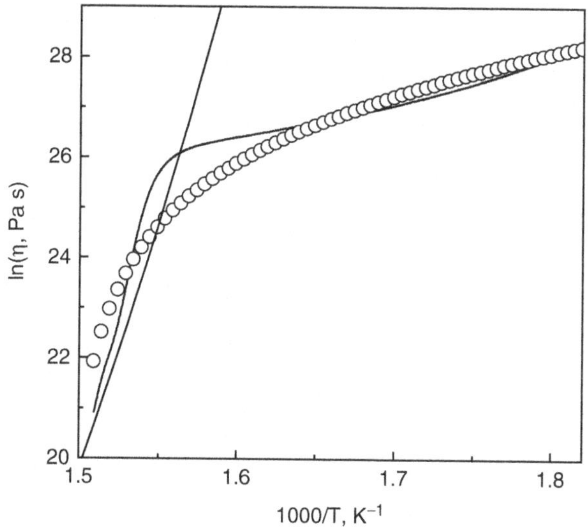

Fig. 8.1 Viscosity temperature dependence of ribbonlike amorphous $Pd_{77.5}Cu_6Si_{16.5}$ metallic alloy at a heating rate of 20 K/min. The solid curve represents the FVM best-fit curve. The steeper line presents the temperature dependence of the quasi-equilibrium viscosity – (▶ Eq. 5.9), ▶ Chap. 5

Table 8.1 FVM parameters determined from the different kinds of experiments – direct nonisothermal creep viscosity determinations and isothermal bend stress relaxation for ribbonlike $Pd_{77.5}Cu_6Si_{14}$ amorphous samples

	Kind of experiment		
FVM parameters, dimension	(A) Viscosity	(B) Isothermal bend stress relaxation	(C) Isothermal bend strain relaxation
E, Pa	–	1×10^{11}	1×10^{11}
η_o, Pa s/K	2.6×10^{-20}	6.84×10^{-20}	6.84×10^{-20}
ν_r, 1/s	5.3×10^{21}	2.97×10^{22}	2.97×10^{22}
Q_r, kJ/mol	100	100	100
Q_η, kJ/mol	155	155	155
$c_{f.o}$	7.7×10^{-15}	2.5×10^{-11}	2.5×10^{-11}
B	5712	–	–
T_o, K	480	–	–
T_g, K	645	–	–
$\eta(T_g)$, Pa s/K	2×10^{11}	–	–

(A) FVM parameters η_o, Q_η, ν_r, $c_{f.o}$, Q_r, B, and T_o obtained from direct viscosity determination according to (▶ Eqs. 5.8) and (▶ 5.17) (▶ Chap. 5)

(B) FVM parameters η_o, ν_r, $c_{f.o}$, obtained from the bend stress experimental data $\ln(\sigma_2(t)/\sigma_0)$ according to (▶ Eq. 7.10a), ▶ Chap. 7. Q_η and Q_r are fixed to the values from the viscosity determinations. The Young's modulus E is fixed to 100 GPa

(C) FVM parameters η_o, ν_r, $c_{f.o}$, obtained from the bend strain experimental data $\varepsilon_f(t)/\varepsilon_0$ according to (▶ Eq. 7.6a), ▶ Chap. 7

carried out, with a heating rate of 10 K/min up to 563 K and sample cooling down to ambient temperature. After cooling down, the new radius r_2 of curvature was measured. The contribution of the deformation $\varepsilon_f/\varepsilon_o$ caused by the viscous flow only was calculated according to (▶ Eq. 7.7a), ▶ Chap. 7.

Fig. 8.2 Isothermal bend stress relaxation $\ln(\sigma_1(t)/\sigma_0)$ of ribbonlike amorphous $Pd_{77.5}Cu_6Si_{16.5}$ samples after the first annealing as a function of time at: □ – 420 K, o – 460 K, Δ – 500 K, ∇ – 30 K, ◇ – 560 K

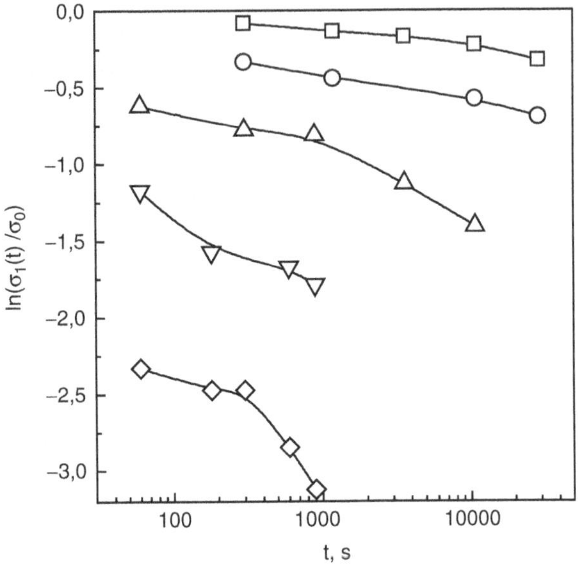

The free volume model considerations presented in the forgoing book sections were again subjected to a rigorous proof. For this purpose simultaneous nonlinear multiparameter regression analysis of all bend stress relaxation experimental results gained at all constant temperatures was carried out with a result – common combination of FVM model-fitting parameter values.

Figure 8.2 shows the bend stress relaxation results $\ln(\sigma_1(T)/\sigma_0)/t$ for $Pd_{77.5}Cu_6Si_{16.5}$ glassy alloy after the first isothermal annealing at 420 K, 460 K, 500 K, 530 K, and 560 K, respectively. The solid lines are guides for the eye. As is seen, the degree of bend stress relaxation increases along with increasing the time of annealing and the bend stresses decrease along with increasing the annealing temperature.

Figure 8.3 shows the degree of bend stress relaxation $\ln(\sigma_2/\sigma_0)$ after the second fully stress-free anneal of the samples as a function of time of the first annealing. The residual deformation is due only to the viscous flow taken place during the first anneal. The graphic symbols denote the bend stress relaxation values calculated according to (▶ Eq. 7.11a), ▶ Chap. 7 by using the radii of curvature r_2. The solid lines are calculated values of $\ln(\sigma_2/\sigma_0)$ according to (▶ Eq. 7.10a), ▶ Chap. 7. All curves are fitted simultaneously. As is seen, along with increasing the annealing time, the degree of bend stress relaxation increases and bend stresses decrease. The same effect is observed along with increasing the annealing temperature. The decrease of bend stresses after the second stress-free anneal is quite visible. The regression analysis of experimental data was carried out on the basis (▶ Eq. 7.10a) by fixed values for the activation energies Q_η and Q_r, taken from the regression analysis of the data from the direct nonisothermal viscosity determination.

Figure 8.4 shows the degree of bend stress deformation $\varepsilon_f(t)/\varepsilon_0$ as a function of annealing time of ribbonlike samples of amorphous $Pd_{77.5}Cu_6Si_{16.5}$ samples due to the viscous flow during the first annealing. The solid lines represent the best-fit

Fig. 8.3 Bend stress relaxation $\ln(\sigma_2(t)/\sigma_0)$ as a function of annealing time of ribbonlike samples of amorphous $Pd_{77.5}Cu_6Si_{16.5}$ samples due to the viscous flow during the first annealing at temperatures: □ – 420 K, o – 460 K, Δ – 500 K, ∇ – 530 K, ◇ – 560 K

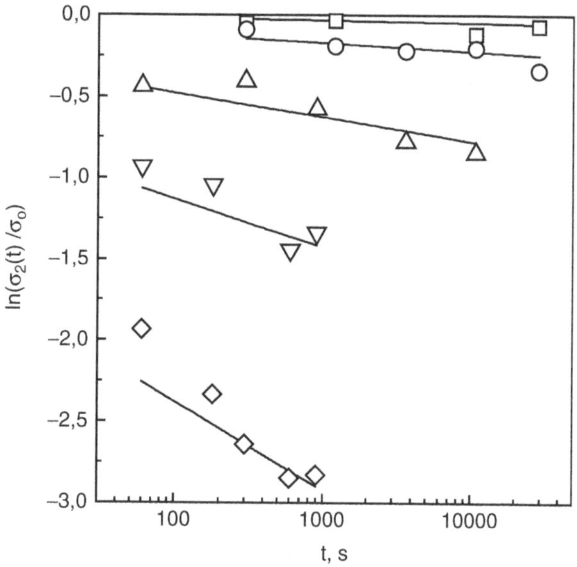

Fig. 8.4 Degree of bend stress deformation $\varepsilon_f(t)/\varepsilon_0$ as a function of annealing time of ribbonlike samples of amorphous $Pd_{77.5}Cu_6Si_{16.5}$ samples due to the viscous flow during the first annealing at temperatures: □ – 420 K, o – 460 K, Δ – 500 K, ∇ – 530 K, ◇ – 560 K. The solid lines represent the best-fit curves according (▶ Eq. 7.6a), ▶ Chap. 7. All curves are fitted simultaneously – Table 8.1

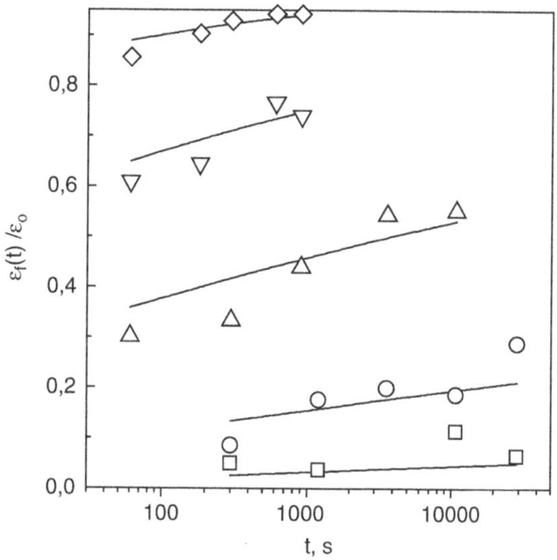

curves according (▶ Eq. 7.6a). All curves are fitted simultaneously. The values of the FVM parameters η_o, ν_r, $c_{f,o}$ together with the fixed values of Q_η and Q_r are given in Table 8.1(B). They coincide practically with the FVM parameter values, determined from the direct creep viscosity determinations.

It is clear again that, similar to the case of the amorphous $Fe_{40}Ni_{40}Si_6B_{14}$ system, the usage of (▶ Eq. 7.6a), ▶ Chap. 7 is rightful by considering the relaxation of bend stresses due only to the viscous flow, i.e., after the second stress-free anneal.

The Amorphous Metallic Alloy $Al_{7.5}Cu_{17.5}Ni_{10}Zr_{65}$

The amorphous $Al_{7.5}Cu_{17.5}Ni_{10}Zr_{65}$ ribbons were produced using the melt-spinning apparatus at the Max-Planck-Institut fur Metallforschung. The ribbon production was carried out in a low-pressure (1.3×10^4 Pa He atmosphere) chamber; the gap between the chilling copper wheel and the silica glass crucible with a planar slot was 3 mm. The surface velocity of the copper wheel was about 20 m/s. The resulting ribbon studied was 2.1 mm wide by 0.036 mm thick. Viscosity measurements of the glassy alloys studied were performed with the aid of a Perkin-Elmer TMS-2 silica glass dilatometer using the experimental approach described in detail in ▶ Chap. 2 (Fig. 2.3).

Direct Viscosity Determination Under Nonisothermal Conditions

Figure 8.5 shows the viscosity temperature dependence of $Al_{7.5}Cu_{17.5}Ni_{10}Zr_{65}$ glassy alloy obtained at heating rates of 5, 10, and 20 K/min, respectively [8, 9]. Experimental data were fitted using the combination of (▶ Eq. 5.8) and (▶ Eq. 5.17), ▶ Chap. 5 and the FVM parameters are given in Table 8.2. The temperature dependence of η_{eq} calculated using the fitting parameter values obtained is also shown. The temperature of intersection of the η_{eq} curve with the dynamic viscosity temperature dependence corresponds to the reciprocals of the dynamic glass transition temperature T_g at the heating rates used.

Study of Bend Stress Relaxation Under Isothermal and Nonisothermal Conditions

Isothermal Bend Stress Relaxation Measurements

Isothermal and nonisothermal bend stress and anelastic strain relaxation of melt-spun ribbons of the $Al_{7.5}Cu_{17.5}Ni_{10}Zr_{65}$ bulk metallic glass was studied at 4 constant temperatures and 6 constant heating rates using a Perkin-Elmer DSC 2 as a heat treatment equipment. The equipment and the theoretical approach are described in [8–10], respectively.

The isothermal study of bend relaxation was carried at 4 different constant temperatures–450 K, 475 K, 500 K, and 520 K, starting from 330 K. The annealing of the specimens was carried out in a protective purified Ar atmosphere with the aid of a Perkin-Elmer DSC 2 by using the sample pan of the calorimeter as a restraining ring with a radius of curvature $r_0 = 3.25$ mm. Using $E = 6.2\times10^9$ Pan [2], it follows from (▶ Eq. 7.1), ▶ Chap. 7 that $\sigma_0 \sim 485$ MPa. The ribbons were stressed by coiling them into the sample pans, annealed with heating rates 320 K/min from 330 K to each final constant temperature, mentioned above. After the isothermal annealing the samples were subsequently rapidly cooled down to ambient temperature and allowed to spring back. Spring-back measurements of the coiled amorphous ribbons were performed as described in [3–6]. This procedure was repeated 5 times with

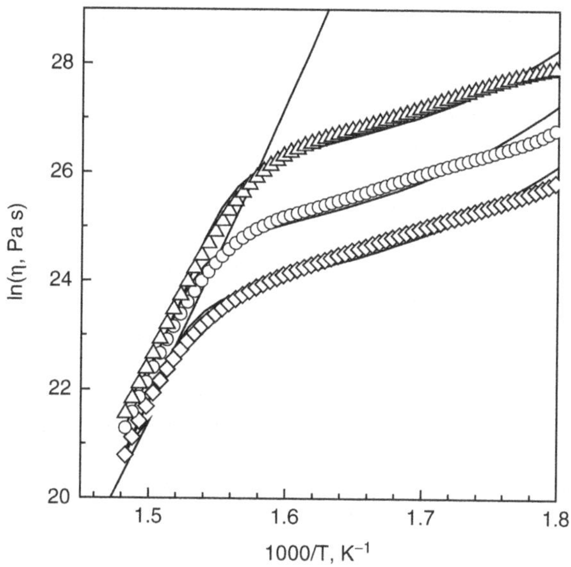

Fig. 8.5 Viscosity temperature dependence of $Al_{7.5}Cu_{17.5}Ni_{10}Zr_{65}$ glassy alloy at constant heating rates of: ◇ – 20 K/min, o – 10 K/min, Δ – 5 K/min

Table 8.2 FVM parameters of the viscosity temperature dependence of $Al_{7.5}Cu_{17.5}Ni_{10}Zr_{65}$ glassy alloy obtained at heating rates of 5, 10, and 20 K/min, respectively

	Viscosity		
FVM parameters	5 K/min	10 K/min	20 K/min
η_o, Pa s/K	2.6×10^{-18}	2.6×10^{-18}	2.6×10^{-18}
ν_r, 1/s	1.5×10^{19}	8.5×10^{18}	3.1×10^{18}
Q_r, kJ/mol	140.4	141.1	137.2
Q_η kJ/mol	195.5	196.0	195.9
$c_{f,o}$	3×10^{-9}	9×10^{-9}	3.1×10^{-8}
T_o, K	350	350	350
B, K	6497	6420	6252
T_g	635	647	665
$\eta(T_g)$, Pa s	1.3×10^{11}	2.6×10^{10}	6.5×10^{9}

5 different samples and the mean radius value of all samples was evaluated. After the first annealing, a second, fully stress-free sample annealing was carried out, in order to eliminate the contribution of the anelastic strain to the overall strain. For this purpose additional annealing of the partially relaxed specimens in a separate furnace under zero stress conditions was carried out, with a heating rate of 10 K/min up to 563 K and cooling down to ambient temperature. After cooling down, the new radius r_2 of curvature was measured.

The free volume model consideration presented in the forgoing book sections were again subjected to a rigorous proof. For this purpose simultaneous nonlinear multiparameter regression analysis of all bend stress relaxation experimental results gained at all constant temperatures was carried out with a result – common combination of FMM model-fitting parameter values.

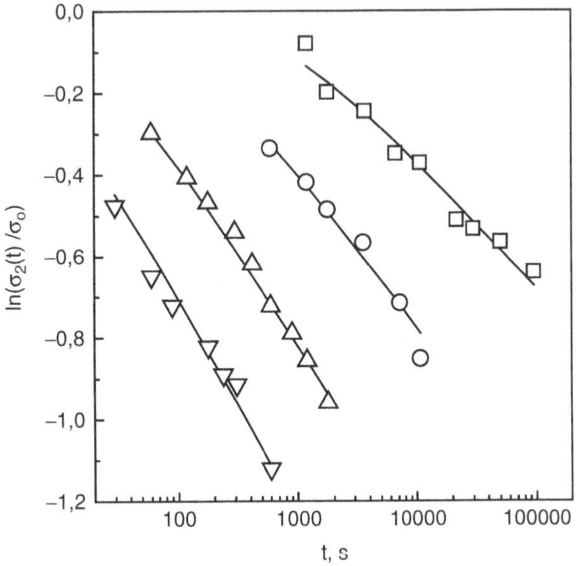

Fig. 8.6 $Al_{7.5}Cu_{17.5}Ni_{10}Zr_{65}$ glassy alloy. Time dependence of the degree of bend stress relaxation $\ln\sigma_2(t)/\sigma_0$, determined after the second stress-free anneal. The bend stress relaxation is caused by the viscous flow only, during the first anneal at temperatures : □ – 450 K, o – 475 K, Δ – 500 K, ∇ – 520 K

Figure 8.6 shows the degree of bend stress relaxation ln $(\sigma_2(t)/\sigma_0)$ after the second fully stress-free annealing of the samples as a function of the time and temperature of the first isothermal annealing for $Al_{7.5}Cu_{17.5}Ni_{10}Zr_{65}$ glassy alloy. The isothermal annealing was performed at 450 K, 475 K, 500 K, and 520 K, respectively. As is seen, the degree of bend stress relaxation increases along with increasing the time of annealing and the bend stresses decrease along with increasing the annealing temperature.

Figure 8.7 shows the experimentally determined time dependence of the normalized fully irreversible deformation contribution $\varepsilon_f/\varepsilon_o$ of viscous flow according to (▶ Eq. 7.7a), ▶ Chap. 7. The best-fit curves are calculated via multiparameter nonlinear regression analysis according (▶ Eq. 7.6a). The FVM parameter values are given in Table 8.3. Figure 8.8 shows the time dependence of the experimentally determined normalized contribution of the fully reversible anelastic deformation $\varepsilon_{an}/\varepsilon_o$ of $Al_{7.5}Cu_{17.5}Ni_{10}Zr_{65}$ glassy alloy at the different temperatures of the first anneal. The experimental data are fitted with the aid of KWW (▶ Eq. 7.12).

Nonisothermal Bend Stress Relaxation Measurements

Bend stress relaxation results of the $Al_{7.5}Cu_{17.5}Ni_{10}Zr_{65}$ glassy alloy under nonisothermal conditions were obtained using a Perkin-Elmer TMS-2 thermomechanical analyzer as described in detail in [10] at different heating rates ranging from 0.31 to 10 K/min and different final temperatures of heating T_F ranging from 425 to 520 K. The starting temperature of heating was $T_B = 315$ K.

Figure 8.9 shows the heating rate dependence of the degree of bend stress relaxation $\ln\sigma_2(T_F,q)/\sigma_o$ caused by the viscous flow only. The solid lines are

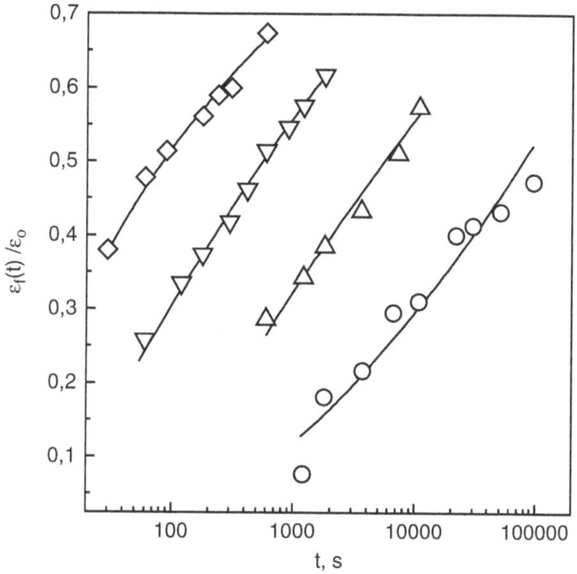

Fig. 8.7 Time dependence of the degree of irreversible bend stress deformation $\varepsilon_f(t)/\varepsilon_o$ of $Al_{7.5}Cu_{17.5}Ni_{10}Zr_{65}$ glassy alloy, caused by the viscous flow only, determined after the second fully stress-free anneal. Constant temperatures of first anneals: □ – 450 K, o – 475 K, Δ – 500 K, ∇ – 520 K

calculated on the basis of the regression analysis of experimental data according to (▶ Eq. 7.10b). Due to the structural relaxation taken place, the stress degree reduces considerably.

Figure 8.10 shows the heating rate dependence of the degree of fully irreversible bend stress deformation $\varepsilon_f(T_f, q)/\varepsilon_0$ up to the indicated final temperatures T_F of heating. The solid limes represent the best-fit curves calculates by the regression analysis on the basis of (▶ Eq. 7.6b).

Figure 8.11 shows the heating rate dependence of fully reversible anelastic deformation $\varepsilon_a(T_f, q)/\varepsilon_0$ as a function of the final temperature T_F of heating.

Figure 8.12 shows time dependence of the nonisothermal degree of anelastic bend stress deformation in $Al_{7.5}Cu_{17.5}Ni_{10}Zr_{65}$ glassy alloy at different heating rates according to (▶ Eq. 7.15).

Table 8.3 shows the FVM and KWW parameter values obtained by the regression analysis of the experimental data considered above.

As is seen, the values of the corresponding model parameters obtained from the different kinds of experiments coincide practically. This circumstance is a proof for the high degree of relevance of the theoretical models used [11–14].

The relaxation of the anelastic bend stress deformation is described quite well with the equation of Kohlrausch-Williams-Watts both under isothermal and nonisothermal experimental conditions.

As is seen, the study of bend stress relaxation is a suitable and successful method for determination of the reversible and irreversible relaxation phenomena in glassy metals. The free volume model describing the atomic mobility and relaxation phenomena in glassy metals can be successfully applied for consideration to both the viscosity and bend stress relaxation in amorphous metallic alloys.

Table 8.3 FVM- and KWW parameters for $Al_{7.5}Cu_{17.5}Ni_{10}Zr_{65}$ glassy alloy obtained by the regression analysis of the experimental data from the different kinds of experiments

	Type of experiment				
Parameters, dimension	(A) Isothermal bend stress relaxation	(B) Nonisothermal bend stress relaxation	(C) Nonisothermal irreversible bend strain contribution	(D) Isothermal anelastic strain relaxation	(E) Nonisothermal anelastic Strain relaxation
E, Pa	6.2×10^{9}	6.2×10^{9}	6.2×10^{9}	–	–
η_0, Pa s/K	9×10^{-17}	8.7×10^{-17}	8.5×10^{-17}	–	–
ν_r, 1/s	2.5×10^{21}	8.37×10^{20}	8.37×10^{20}	–	–
$\boldsymbol{Q_r}$, kJ/mol	137.5	135.2	135.2	161.1	170.0
$\boldsymbol{Q_\eta}$, kJ/mol	156.1	157.4	161.1	–	–
$c_{f,0}$	5×10^{-9}	4.7×10^{-10}	4.7×10^{-10}	–	–
$1/\tau_o$, 1/s	–	–	–	1.28×10^{17}	1.05×10^{18}
b	–	–	–	0.179	0.1

(A) FVM parameters η_o, ν_r, $c_{f,o}$, Q_η, and Q_r obtained from the regression analysis of the $\ln(\sigma_2(t)/\sigma_0)$ experimental data
(B) FVM parameters η_o, ν_r, $c_{f,o}$, Q_η, and Q_r obtained from the regression analysis of the $\ln(\sigma_2(T_F,q)/\sigma_o)$ experimental data
(C) FVM parameters η_o, ν_r, $c_{f,o}$, Q_η, and Q_r obtained from the bend strain experimental data $\varepsilon_f(t)/\varepsilon_o$
(D) KWW parameters Q_r, $1/\tau_0$, and b, obtained from the regression analysis of the $\varepsilon_a(t)/\varepsilon_0$ experimental data
(E) KWW parameters Q_r, $1/\tau_0$, and b, obtained from the regression analysis of the $\varepsilon_a(T, q)/\varepsilon_0$ experimental data

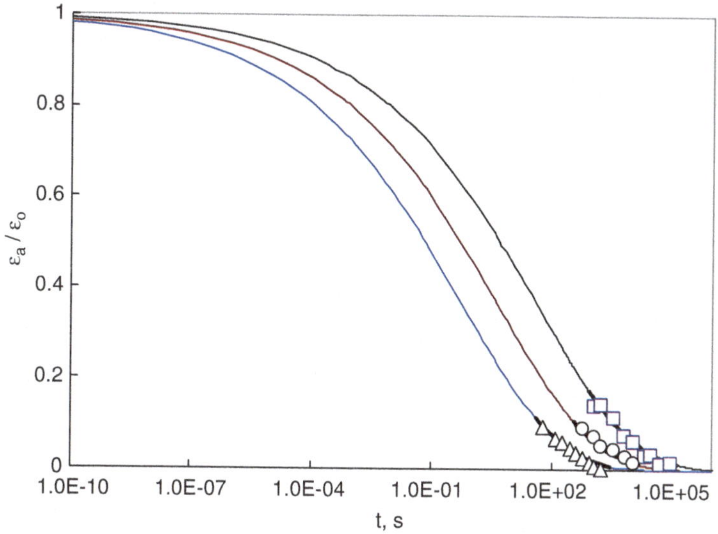

Fig. 8.8 Time dependence of the degree of isothermal reversible anelastic deformation $\varepsilon_\alpha(t)/\varepsilon_O$ of $Al_{7.5}Cu_{17.5}Ni_{10}Zr_{65}$ glassy alloy at temperatures of first anneal: □ – 450 K, o – 475 K, Δ – 500 K

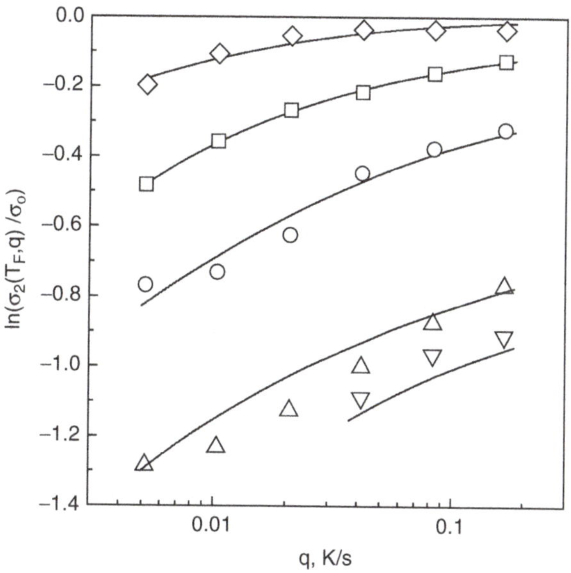

Fig. 8.9 Bend stress relaxation $\ln(\sigma_2(T_F,q)/\sigma_o)$ due to the viscous flow only, of glassy $Al_{7.5}Cu_{17.5}Ni_{10}Zr_{65}$ alloy under nonisothermal experimental conditions. The final temperatures T_F of heating are ◇ – 425 K, □ – 450 K, o – 475 K, Δ – 500 K, ∇ – 520 K

The Amorphous Metallic Alloy $Al_{25}La_{50}Ni_{25}$

The amorphous $Al_{25}La_{50}Ni_{25}$ ribbons were produced using the melt-spinning apparatus at the Max-Planck-Institut fur Metallforschung. The ribbon production was carried out in a low-pressure (1.3×10^4 Pa) He atmosphere chamber; the gap

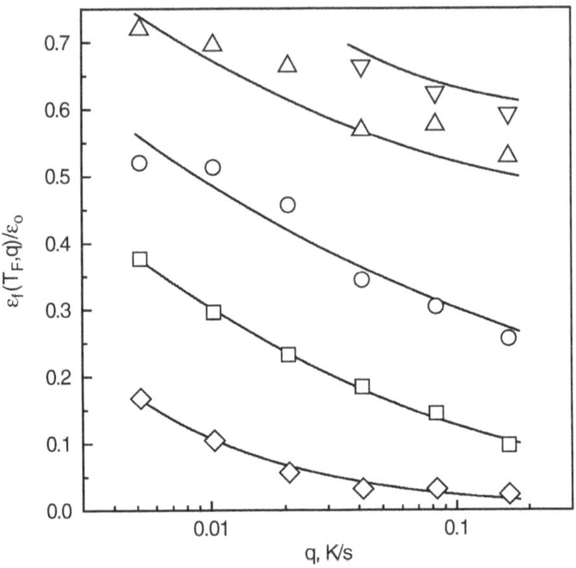

Fig. 8.10 Degree of bend deformation $\varepsilon_f(T_f, q)/\varepsilon_0$ of glassy $Al_{7.5}Cu_{17.5}Ni_{10}Zr_{65}$ alloy under nonisothermal experimental conditions up to final temperatures T_F: ◇ – 425 K, □ – 450 K, o – 475 K, Δ – 500 K, ∇ – 520 K

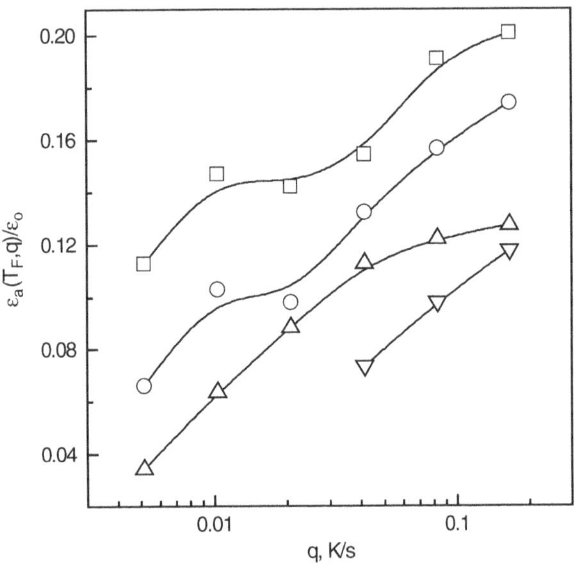

Fig. 8.11 Heating rate dependence of the degree of fully reversible anelastic deformation $\varepsilon_a(T_F, q)/\varepsilon_o$ of glassy $Al_{7.5}Cu_{17.5}Ni_{10}Zr_{65}$ alloy as a function of heating rate, up to final temperatures T_F: □ – 450 K, o – 475 K, Δ – 500 K, ∇ – 520 K. Solid lines are guides for the eye

between the chilling copper wheel and the silica glass crucible with a planar slot was 3 mm. The surface velocity of the copper wheel was about 20 m/s. The resulting ribbon studied was 2.5 mm wide by 0.038 mm thick. Viscosity measurements of the glassy alloys studied were performed with the aid of a Perkin-Elmer TMS-2 silica glass dilatometer using the experimental approach described in detail in ▶ Chap. 3 (▶ Fig. 3.3).

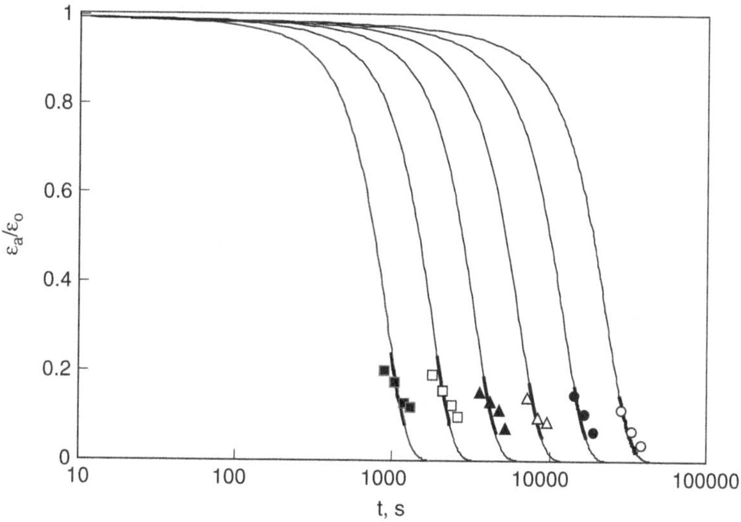

Fig. 8.12 Time dependence of the nonisothermal degree of anelastic bend stress deformation in $Al_{7.5}Cu_{17.5}Ni_{10}Zr_{65}$ glassy alloy at different heating rates: ■ – 10 K/min; □ – 5 K/min; ▲ – 2.5 K/min; Δ – 1.25 K/min; • – 0.63 K/min; ○ – 0.31 K/min; $T_b = 300$ K

Direct Viscosity Measurements Under Nonisothermal Conditions

Figure 8.13 shows the viscosity temperature dependence of $Al_{25}La_{50}Ni_{25}$ glassy alloy obtained at heating rates of 2.5, 10, and 20 K/min, respectively [15, 16]. Experimental data were fitted simultaneously using the combination of (▶ Eqs. 5.8) and (▶ 5.17), ▶ Chap. 5 and the fitting parameters are given in Table 8.4(A). The temperature dependence of η_{eq} calculated using the fitting parameter values obtained is also shown. The temperature of intersection of the η_{eq} curve with the dynamic viscosity temperature dependence corresponds to the reciprocals of the dynamic glass transition temperature T_g (485, 492, and 495 K) at the different heating rate used.

Bend Stress Relaxation Studies Under Isothermal and Nonisothermal Conditions

Bend stress relaxation results of the $Al_{25}La_{50}Ni_{25}$ glassy alloy both under nonisothermal and isothermal conditions were obtained using a Perkin-Elmer DSC 2 as described in detail in [10] at different heating rates ranging from 0.31 to 20 K/min using $T_B = 305$ K as the starting temperature of heating and different final temperatures of heating T_F ranging from 373 to 443 K or performing isothermal annealing for different periods of time at different fixed temperatures ranging from 393 to 443 K, respectively [15–17].

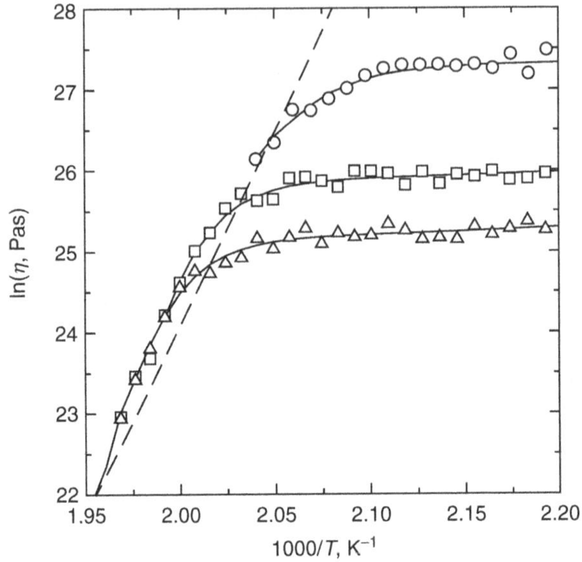

Fig. 8.13 Viscosity temperature dependence of glassy $Al_{25}La_{50}Ni_{25}$ alloy at heating rates of: o – 2.5 K/min; □ – 10 K/min; Δ – 20 K/min; broken line – temperature dependence of quasi-equilibrium viscosity η_{eq}

Isothermal Bend Stress Relaxation Measurements

Figures 8.14 and 8.15 show the dependences of bend stress relaxation $\sigma_2(t)/\sigma_o$ and the bend stress deformation $\varepsilon_f(t)/\varepsilon_o$, respectively, on the time of isothermal annealing at different constant temperatures [12, 13]. The experimental points are determined by using (▶ Eq. 7.11a) and (▶ Eq. 7.7a), ▶ Chap. 7, respectively. The best-fit curves are obtained by (▶ Eq. 7.10a), and (▶ Eq. 7.6a), ▶ Chap. 7.

The FVM parameters determined by the regression analyses are given in Table 8.4.

Figure 8.16 shows the experimentally determined time dependence of the fully reversible anelastic contribution to the overall bend deformation $\varepsilon_{an}(t)/\varepsilon_o$ as a result of the isothermal annealing at different constant temperatures.

The values of $\varepsilon_a/\varepsilon_o$, shown in the Fig. 8.16 were obtained using r_2 values calculated from the fitting curves in Fig. 8.15. The KWW fitting curves are obtained by (▶ Eq. 7.12), and the fitting parameters are given in Table 8.4(D).

Nonisothermal Bend Stress Relaxation Measurements

Figure 8.17 shows the nonisothermal bend stress relaxation of $Al_{25}La_{50}Ni_{25}$ glassy alloy as a heating rate dependence of the degree of bend stress reduction $\sigma_2(T_F,q)/\sigma_0$ due to the viscous flow contribution only [15, 16].

Figure 8.18 presents the nonisothermal bend stress relaxation of $Al_{25}La_{J0}Ni_{25}$ glassy alloy as a heating rate dependence of the normalized reversible, anelastic strain contribution $\varepsilon_{an/}\varepsilon_0$ for final temperatures T_f of 373 K, 393 K, 413 K,423 K, and 443 K. Figure 8.19 shows the nonisothermal bend stress relaxation of $Al_{25}La_{J0}Ni_{25}$ glassy alloy as a heating rate dependence of the normalized reversible, anelastic strain contribution $\varepsilon_{an/}\varepsilon_0$ for different final temperatures. The free volume

Table 8.4 FVM and KWW model parameters obtained from direct viscosity determinations and isothermal and nonisothermal bend stress relaxation measurements of $Al_{25}La_{50}Ni_{25}$ glassy alloy

	Type of experiment				
	(A)	(B)	(C)	(D)	(E)
Model parameter, dimension	Viscosity	Isothermal bend stress relaxation	Nonisothermal bend stress relaxation	Isothermal reversible anelastic bend stress deformation	Nonisothermal reversible anelastic bend stress deformation
q, K/min	2.5, 10, 20	–	0.31 ÷ 20	–	0.31 ÷ 20
η_o, Pa s/K	1.1×10^{-17}	1.1×10^{-17}	1.1×10^{-17}	–	–
ν_r, 1/s	8.6×10^{21}	8.6×10^{21}	8.6×10^{21}	–	–
Q_r, kJ/mol	134.4	134.4	134.4	166.3	176.9
Q_η, kJ/mol	149.6	149.6	149.6	–	–
$c_{f,o}$	1.8×10^{-7}	1.8×10^{-7}	1.8×10^{-7}	–	–
B, K	3770	–	–	–	–
T_0, K	320	–	–	–	–
E, Pa	–	85×10^9	85×10^9	–	–
b	–	–	–	0.33	0.19
$1/\tau_o$, 1/s	–	–	–	$1.3\text{x}10^{20}$	$9.1\text{x}10^{20}$
T_g, K	485, 492, 495	–	–	–	–
$\eta(T_g)$, Pa s	$3 \times 10^{11}, 1\times10^{11}, 6\times10^{10}$	–	–	–	–

(A) FVM parameters η_o, Q_η, ν_r, $c_{f,o}$, Q_r, B, and T_o obtained from direct viscosity determination according (▶ Eqs. 5.8) and (▶ 5.17), ▶ Chap. 5
(B) FVM parameters η_o, ν_r, $c_{f,o}$, Q_η, and Q_r obtained from the regression analysis of the $\ln(\sigma_2(t)/\sigma_0)$ experimental data
(C) FVM parameters η_o, ν_r, $c_{f,o}$, Q_η, and Q_r obtained from the regression analysis of the $\ln(\sigma_2(T_F,q)/\sigma_o)$ experimental data
(D) KWW parameters Q_r, $1/\tau_0$, and b, obtained from the regression analysis of the $\varepsilon_a(t)/\varepsilon_0$ experimental data
(E) KWW parameters Q_r, $1/\tau_0$, and b, obtained from the regression analysis of the $\varepsilon_a(T, q)/\varepsilon_0$ experimental data

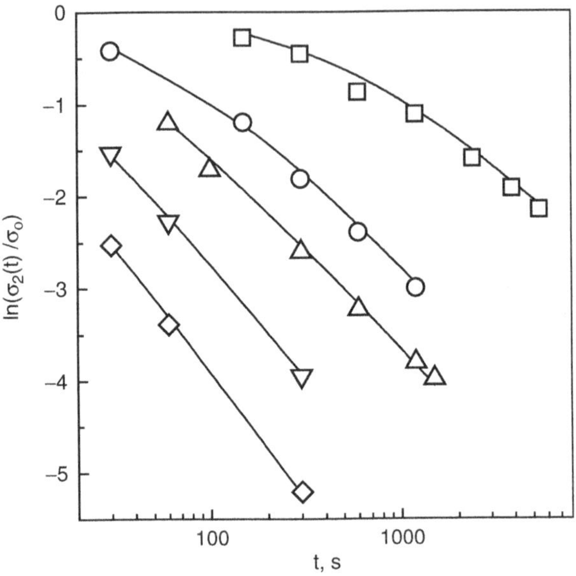

Fig. 8.14 Isothermal bend stress relaxation of glassy $Al_{25}La_{50}Ni_{25}$ alloy. Experimentally determined bend stress reduction $\sigma_2(t)/\sigma_o$, due only to the viscous flow at constant temperatures: □ – 393 K; o – 413 K; Δ – 423 K; ∇ – 433 K; ◇ – 443 K. Solid lines – best-fit curves according (▶ Eq. 7.10a), ▶ Chap. 7

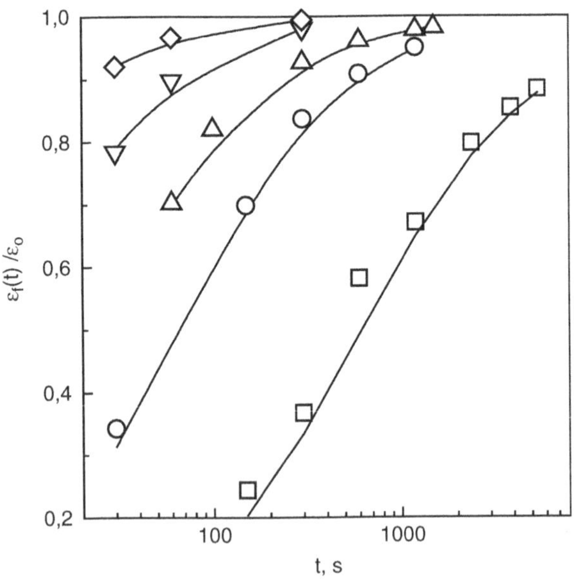

Fig. 8.15 Isothermal bend stress deformation $\varepsilon_f(t)/\varepsilon_o$ of $Al_{25}La_{50}Ni_{25}$ glassy alloy at temperatures: □ – 393 K; o – 413 K; Δ – 423 K; ∇ – 433 K; ◇ – 443 K. Solid lines – best-fit curves according to (▶ Eq. 7.6a), ▶ Chap. 7

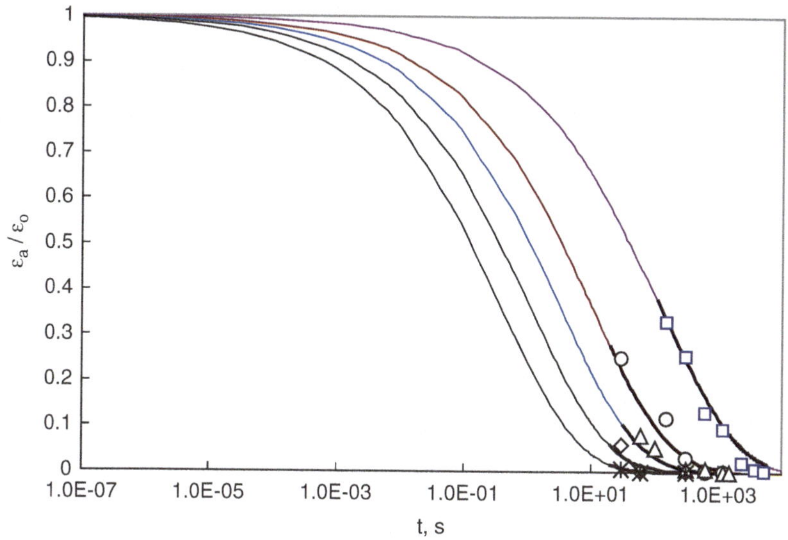

Fig. 8.16 Isothermal bend stress relaxation of glassy $Al_{25}La_{50}Ni_{25}$ alloy. Experimentally determined fully reversible anelastic deformation contribution to the overall strain $\varepsilon_{an}(t)/\varepsilon_o$ as a result of annealing at: □ – 393 K; o – 413 K; Δ – 423 K; ◇ – 433 K; * – 443 K

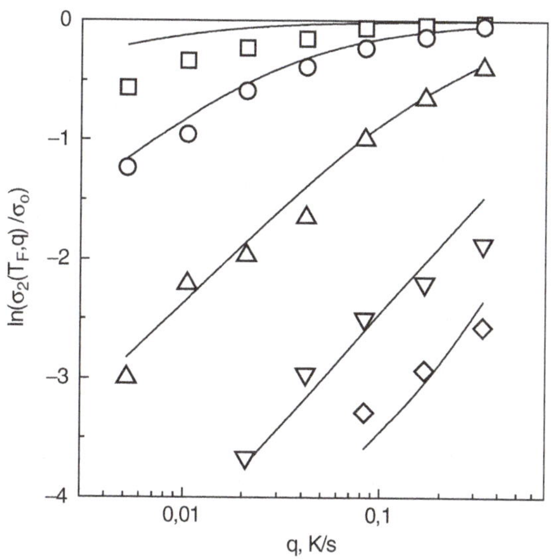

Fig. 8.17 Nonisothermal bend stress relaxation of glassy $Al_{25}La_{50}Ni_{25}$ alloy. Heating rate dependence of the normalized stress reduction $\sigma_2(T_F,q)/\sigma_o$ for final temperature of heating T_F : □ – 373 K; o – 393 K; Δ – 413 K; ∇ – 423 K; ◇ – 443 K. The best-fit curves are calculated according to (▶ Eq. 7.10b), ▶ Chap. 7

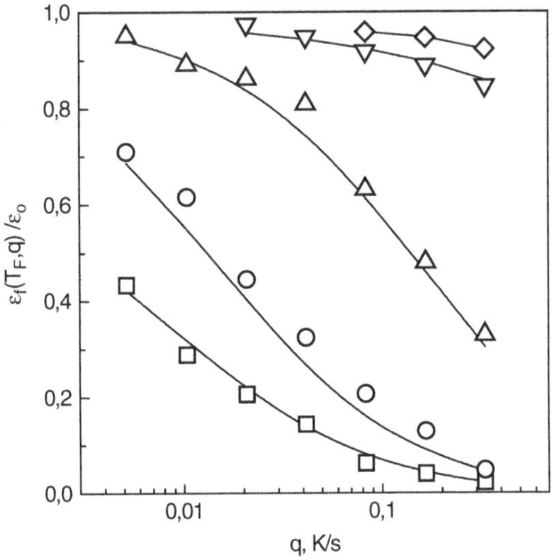

Fig. 8.18 Heating rate dependence of the normalized deformation contribution $\varepsilon_f/\varepsilon_o$, due to the viscous flow only, for final temperatures of heating T_F: □ – 373 K; o – 393 K; Δ – 413 K; ∇ – 423 K; ◇ – 443 K. The best-fit curves are calculated according to (▶ Eq. 7.6b), ▶ Chap. 7

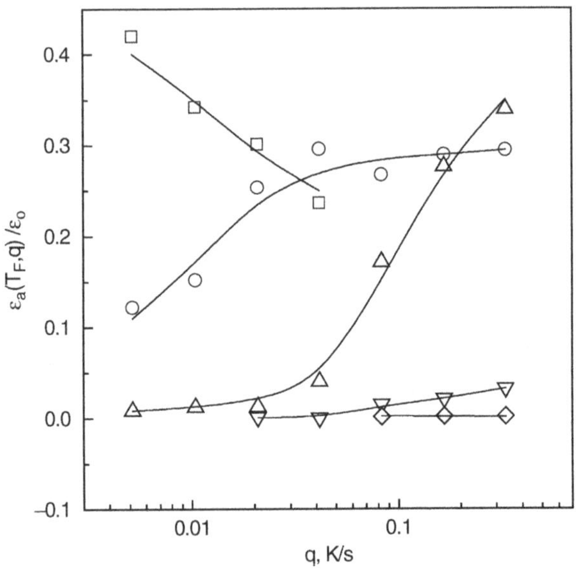

Fig. 8.19 Nonisothermal bend stress relaxation of $Al_{25}La_{J0}Ni_{25}$ glassy alloy as a heating rate dependence of the normalized reversible, anelastic strain contribution $\varepsilon_{an}/\varepsilon_0$ for final temperatures T_F of: O – 373 K; □ – 393 K; Δ – 413 K; ∇ – 423 K; ◇ – 443 K. The solid curves are guides for the eye

model parameters for the isothermal and nonisothermal bend stress relaxation are given in Table 8.4(B) and (C).

Figure 8.20 shows the time dependence of the nonisothermal degree of anelastic bend stress deformation in $Al_{25}La_{50}Ni_{25}$ glassy alloy at different heating rates according to (▶ Eq. 7.15), ▶ Chap. 7.

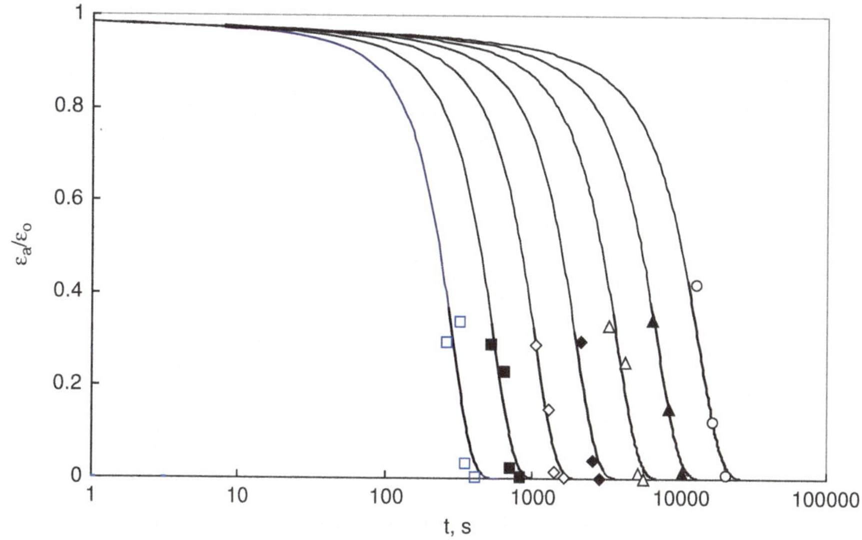

Fig. 8.20 Time dependence of the nonisothermal degree of anelastic bend stress deformation in $Al_{25}La_{50}Ni_{25}$ glassy alloy at different heating rates: □ – 20 K/min; ■ – 10 K/min; ◇ – 5 K/min; ♦ – 2.5 K/min; Δ – 1.25 K/min; ▲ – 0.63 K/min; ○ – 0.31 K/min. T_b = 305 K

Physical Significance of the FVM Model Parameters

The physical significance of the FVM parameters will be generally considered and applicable for all glassy alloys studied and presented in this book on the basis of the experimentally determined values for the $Al_{25}La_{50}Ni_{25}$ glassy alloy. Depending on the chemical composition of the different amorphous metallic alloys presented in this book, they differ somewhat from alloy to alloy, remaining however with comparable order of magnitude. This is why the discussion presented here is applicable for all cases of studied glassy alloys.

The Viscosity Pre-exponential Factor η_0 The value of η_0 obtained from viscosity and bend stress relaxation experiments for $Al_{25}La_{50}Ni_{25}$ glassy alloy is 1.2×10^{-17} Pa s/K. The physical significance of this model parameter (as well as of all model parameters involved in the combination of (▶ Eqs. 5.8) and (▶ 5.3), ▶ Chap. 5 representing the temperature dependence of the (quasi)equilibrium viscosity η_{eq}) can be demonstrated by calculating the viscosity value of the alloy studied at the liquidus temperature T_L (=950 K) with the aid of (▶ Eqs. 5.8) and (▶ 5.3), ▶ Chap. 5. This results in a value for $\eta(T_L) = 1.7 \times 10^{-3}$ Pa s. This value is considerably lower than the results obtained for the amorphous PdSi alloy ($\eta(T_m)$ = 0.5 to 1 Pa s), which is well known as a very easy glass-forming system [18], and is closer to the viscosity of glass-forming alloys based on Fe, Co, and Ni, whose viscosity at T_L is known to be [19, 20], around 10^{-2} to 10^{-3} Pa s. So at least semiquantitatively, the value of η_0, although very low, can be considered as acceptable.

The Initial Concentration of Frozen-in Structural (Flow) Defects $c_{f,o}$ The nature of structural defects in the amorphous state is far from complete understanding as yet. Nevertheless, there exist some modeling results of atomic transport properties which are strongly related to the amount and the nature of structural defects. Model calculations have been carried out using reverse Monte Carlo simulation of the structure of the metallic glasses [21, 22]. Duine et al. [22], have shown by their model calculations that for the $Pd_{52}Ni_{32}P_{16}$ amorphous alloy, a very low initial defect concentration of 6×10^{-9} for the as-quenched state and 4×10^{-16} for the relaxed state is to be expected. The value of 1.8×10^{-7} for $c_{f,0}$ in the $Al_{25}La_{50}Ni_{25}$ glassy alloy is in the range of these model calculations. The glass transition temperature during the melt-spinning process can be estimated according to (▶ Eq. 5.3) using $c_{f,0}$ as equal to $c_{f,e}$ at T_g. The obtained temperature of 565 K is a realistic T_g value in comparison to the dynamic glass transition temperature (496 K) at a heating rate of 10 K/min for the glassy alloy studied. The obtained initial concentration $c_{f,0}$ of structural (flow) defects seems therefore to be quite realistic.

The Activation Energy for Viscous Flow Q_η The value of obtained from the viscosity and bend stress relaxation measurements of $Al_{25}La_{50}Ni_{25}$ glassy alloy is 149.6 kJ/mol. It should in a first approximation coincide with the apparent activation energy of crystallization [23], of the amorphous alloy studied which was determined by us to be 225 kJ/mol after the method of Kissinger [24]. The Kissinger equation, however, works on the assumption that crystallization occurs via Arrhenius-type thermally activated processes. Taking into account the temperature dependence of the (quasi)equilibrium viscosity η_{eq}, it becomes however clear that the apparent activation energy for viscous flow should be higher due to the Vogel-Fulcher-Tammann term $\exp(B/(T- T_0))$.

The Attempt Frequency ν_r The value of 8.6×10^{21} 1/s obtained for $Al_{25}La_{50}Ni_{25}$ glassy alloy is in agreement with the results of other authors, e.g., [25–27]. This value is very high as compared to the expected Debye frequency (10^{13} 1/s). There exist some more or less convincing explanations [27, 28], but this model parameter value still awaits an explanation. B.S. Berry [27], for example, proposes the following considerations:

According to the FVM, the temperature dependence of the viscosity of amorphous bodies is presented by (see ▶ Eq. 5.8) ▶ Chap. 5

(a) $\eta = T\frac{k\Omega_0}{a^2\nu_r}\exp\left(\frac{Q_\eta}{RT}\right)\frac{1}{c_f}$, where Ω_0 is the atomic volume for dense random packing (DRP), a is the mean atomic diameter, and ν_r is the so-called apparent relaxation frequency. The experimentally determined value of ν_r is usually about 10^{20} 1/s. Let us suppose that the activation energy of viscous flow is weakly temperature dependent and decreases along with increasing the temperature, according to the relation

(b) $Q_\eta = Q_\eta^0(1 - \Theta T)$, where Q_η^0 is the activation energy for viscous flow at $T = 0$ K, and Θ, 1/K, is a characteristic coefficient for a given kind of amorphous body (amorphous metallic alloy). Inserting equation (b) into equation (a), one obtains that by $\Theta = 6\times10^{-4}$ 1/K (very weak temperature dependence of Q_η) and by Debye frequency value $\nu_r^D = 10^{13}$ 1/s, the value of the apparent relaxation frequency ν_r should be about 10^{20} 1/s.

Keeping in mind that the order of magnitude of the reciprocal of the constant τ_o in the empirical KWW equation is expected to be around the order of magnitude of ν_r, the considerations concerning ν_r are valid for the Kohlrausch-Williams-Watts empirical constant τ_o also.

The Activation Energy for Relaxation Q_r The value of the activation energy for the rate constant of defect concentration changes was determined to be 134.4 kJ/mol. This value is reasonable when compared with the activation energy of vacancy concentration changes in crystals [29].

The Model Parameters B and T_0 As discussed in [23], in the framework of the FVM, the model parameter B is presented as $B \cong \gamma\, \alpha_v^{-1} \upsilon^*/\Omega_0$, where $\alpha_v = 3\alpha_l$ is the thermal coefficient of volume expansion, and Ω_0 is the atomic volume for dense random packing (DRP). The thermal coefficient of linear expansion α_l of the alloy studied was experimentally determined to be 2×10^{-5} 1/K. Inserting $B = 3770$ K for $Al_{25}La_{50}Ni_{25}$ glassy alloy which is a result of the fitting procedure and setting the free volume overlapping factor γ to be 0.75 (according to FVM $0.5 < \gamma < 1$), one obtains for the ratio υ^*/Ω_0 the value of 0.3 which is physically quite realistic; hence, the value of the model parameter B is physically realistic too. Another possibility to demonstrate the physical significance of B and T_0 parameter values is to insert them into (▶ Eq. 5.3), ▶ Chap. 5 and to calculate the concentration of defects at T_L (950 K) of the alloy studied. This results in $c_{f,e} = 2.5 \times 10^{-3}$. As pointed out by Seeger ([29] in Chap. 7), the concentration of vacancies in metals at their melting point is approximately 1×10^{-4}, in fairly good agreement with our estimations.

The Young's Modulus E The value 85.5 GPa obtained from the ratio E/η_0 in (▶ Eqs. 7.10a) and (▶ 7.10b), ▶ Chap. 7 corresponds very well to the Young's modulus of $Al_{25}La_{50}Ni_{25}$ glassy alloy, 77 GPa, roughly estimated from the Young's moduli of the pure components (E_{Al}, 25 GPa; E_{La}, 38.4 GPa; E_{Ni}, 207 GPa) [2].

The conclusion can be drawn that:

A new, simple method for the evaluation of bend stress relaxation in metallic glasses under nonisothermal constant heating rate conditions is developed which can be successfully applied for the study of both reversible and irreversible relaxation phenomena in amorphous metallic alloys.

The method provides the possibility of separating experimentally the time and heating rate-dependent viscous flow and anelastic strain contributions to the bend stress relaxation in order to consider the experimental data on the basis of separate theoretical considerations. The irreversible part of the bend stress relaxation, governed by viscous flow, is adequately described by the free volume model giving a quantitative link between the low- and high-temperature viscous flow phenomena in glassy alloy and can be described by a single set of parameters. The relaxation of the fully reversible anelastic deformation can be quantitatively described by the stretched function (empirical equation of Kohlrausch-Williams-Watts). It becomes clear that all experimental data obtained by studying the viscosity temperature dependence, and the isothermal and nonisothermal bend stress relaxation of glassy alloys, can be conveniently and convincingly presented in terms of the FVM.

References

1. Sestak J (1984) Thermo-physical properties of solids. Academia Prague
2. Metals handbook, 9th ed, vol 2, American Soc for Metals, Metal Park
3. Maddin R, Masumoto T (1972) Mater Sci Eng 9:153
4. Luborsky F, Walter J (1978) Mater Sci Eng 35:255
5. Taub A, Luborsky F (1981) Acta Metall 29: 1939
6. Taub A, Luborsky F (1982) Mater Sci Eng 56: 157
7. Koleva E, Russew K, Stojanova L (2001) Isothermal study of relaxation phenomena in $Pd_{77.5}Cu_6Si_{16.5}$ metallic glass. J Mater Sci Technol 9(3):171–179
8. Russew K, Stojanova L, Sommer F (1999) Viscosity and thermal expansion of the $Al_{7.5}Cu_{17.5}Ni_{10}Zr_{65}$ bulk metallic glass. In: J. Wood et al (eds) European Congress on advanced materials and processes EUROMAT'99, Munich, 27-30.09.1999, Proceedings, Volume 8, Materials development and processing–bulk amorphous materials, undercooling and powder metallurgy. WILEY-VCH, pp 65–70
9. Russew K , Sommer F (1999) Relaxation phenomena in the $Al_{7.5}Cu_{17.5}Ni_{10}Zr_{65}$ bulk metallic glass studied by bend stress and anelastic strain relaxation under isothermal and non-isothermal conditions. In: J. Wood et al (eds) European Congress on Advanced Materials and Processes EUROMAT'99, Munich, 27-30.09.1999, Proceedings, Volume 8, Materials development and processing–bulk amorphous materials, undercooling and powder metallurgy, WILEY-VCH, pp 55–64
10. Russew K et al (1997) Viscous flow of amorphous $Fe_{40}Ni_{40}Si_6B_{14}$ glassy alloy studied by direct creep measurements and relaxation of bend stresses under nonisothermal conditions. Acta Mater 45(No. 5):2129–2137
11. Chen HS (1980) Rep Prog Phys 43:353
12. Berry BS (1978) Metallic glasses. American Society of Metals, Cleveland, p 161
13. Leusink GJ, van den Beukel A (1988) Acta Metall 36: 3019
14. De Hey P, Sietsma J, van den Beukel A (1996) J Non-Cryst Solids 205–207: 696
15. Russew K (1997) Free volume related viscous flow, relaxation of bend stresses, thermal expansion and heat capacity of glassy metals: theory and experiment. Mater Sci Eng A226–228:779–783
16. Russew K, Stojanova L, Koleva E (1997) Structural defects and bend stress relaxation in $Al_{25}La_{50}Ni_{25}$ glassy alloy under isothermal and nonisothermal conditions. In: 12th national conference on non-destruct 7e testing, NDT-97, Sozopol, Scientific Communications of SNU Machine Building, Vol 7, 9–12 Jun 1997, pp 177–182

17. Russew Kr, Sommer F (1998) Determination of the viscosity, thermal expansion, heat capacity and bend stress relaxation temperature dependence of glassy alloys–a tool for understanding of relaxation phenomena. Int J Nonequilib Process 11: 3–23
18. Chen HS, Turnbull D (1968) J Chem Phys 48:531
19. Barfteld RN, Kitchener JA (1955) J Iron Steel Inst 180: 324
20. Cavalier G (1960) Physical chemistry of metallic solutions and intermetallic compounds, Vol II, Chem Publ, 54
21. McGreevy RL, Pusztal L (1988) Mol Simul 1:359
22. Duine PA, Sietsma J, van den Beukel A (1995) Defect and diffusion forum, 117/118, 91
23. Russew K, Stojanova L (1990) Viscous flow of $Fe_{82}B_{18}$ amorphous alloy under continuous heating conditions. Mater Sci Eng A123:59
24. Kissinger HE (1702) Anal Chem 29:1957
25. Tsao SS, Spaepen F (1985) Acta Metall 33:881
26. Duine PA, Sietsma J, van den Beukel A (1992) Acta Metall Mater, 40,743
27. Berry BS (1982) Scripta Metall 16:1407
28. van den Beukel A (1991) Phys Stat Sol (a) 128: 285
29. Seeger A, Schumacher D, Schilling W, Diehl J (eds) (1970) Vacancies and interstitials in metals. North Holland, Amsterdam

Thermal Expansion and Heat Capacity of Amorphous Metallic Alloys: Interpretation of Their Anomalous Behavior in the Framework of FVM

9

Abstract

A free volume model analysis upon the thermal expansion and specific heat relaxation peculiarities of glassy metals has been carried out. It has been shown, both theoretically and experimentally, that the processes of viscous flow and relaxation caused by anomalies in thermal expansion and specific heat of glassy metals are governed by one and the same mechanism of structural changes based upon annihilation and/or production of atomic excess free volume.

Keywords

Glassy metals • Amorphous metallic alloys • Structural relaxation • Viscosity • Thermal expansion • Specific heat • Free volume model • Glass transition temperature

Anomalies in Thermal Expansion of Amorphous Metallic Alloys: Model Consideration

It was shown in ▶ Chaps. 5, ▶ 6, and ▶ 7, both theoretically and experimentally, that an unambiguous relation between the irreversible structural relaxation (annihilation and/or production of free volume) and the temperature dependence of viscosity of glassy metals exists. This is true for viscosity determined both directly via high-temperature creep measurements and indirectly via bend stress relaxation measurements in the temperature range considerably lower than the glass transition temperature. It is apparent, however, that the annihilation or production of free volume should be accompanied by "anomalous" negative or positive deviations $\Delta L_f(T, q)$, from the linear temperature dependence of the thermal expansion of the metallic glasses [1]. The annihilation and/or production of free volume should be directly reflected by "anomalous" negative or positive deviations $\Delta L_f(T, q)$ from the simple linear dependence of the glassy metal sample length $L_o(T)$ upon the temperature T by heating the sample at constant heating rate q. This is shown schematically in Fig. 9.1.

K. Russew, L. Stojanova, *Glassy Metals*,
DOI 10.1007/978-3-662-47882-0_9

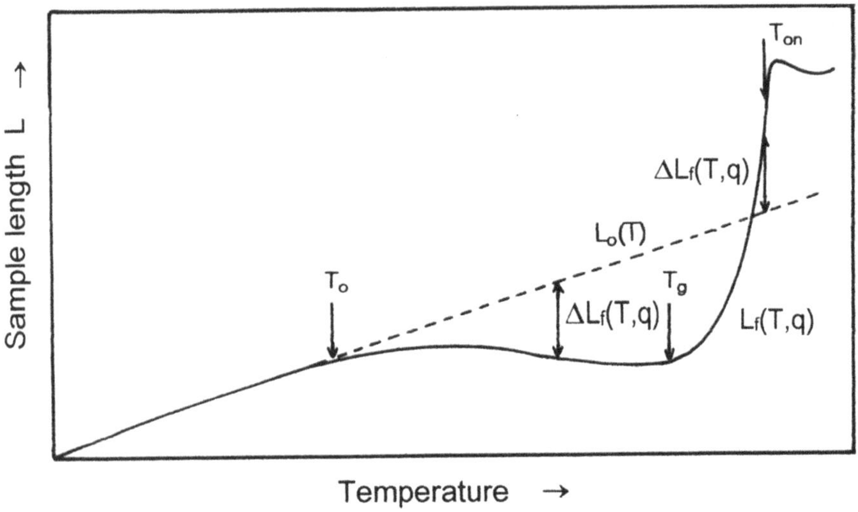

Fig. 9.1 Scheme of the glassy metal peculiarities in thermal expansion

The low-temperature linear temperature dependence of sample length $L_o(T)$ is described by the equation

$$L_0(T) = L_0(T_B)\left[1 + \alpha_l^0(T - T_B)\right], \tag{9.1}$$

where T_B is the starting temperature of heating, $L_o(T_B)$ is the initial sample length at T_B, α_l^o is the coefficient of linear thermal expansion in the low temperature range $T_B \div T_o$, and T_o is the ideal glass temperature. The deviation $\Delta L_f(T, q)$ from the real temperature dependence of the sample length $L_f(T, q)$, comparing it to the extrapolated to higher temperatures $L_o(T)$, is obviously equal to

$$\Delta L_f(T, q) = L_f(T, q) - L_0(T). \tag{9.2}$$

Taking into account that the reduced free volume x [2–5] correlates to the real mean free volume υ_f according to the expression $x = \upsilon_f / \gamma\upsilon^*$, its change along with increasing temperature can be presented through the observed length changes as follows [1]:

$$x(T, q) - x_0 = \frac{L_f^3(T, q) - L_0^3(T)}{\gamma\upsilon^* N}, \tag{9.3}$$

where N is the number of atoms in a cube of the glassy alloy studied with edge length $L_0(T_B)$ at temperature $T = T_B$. Obviously, N can be presented as

$$N = \frac{L_0^3(T_B)}{V_{mol}} N_A, \tag{9.4}$$

where N_A is the Avogadro number, and V_{mol} is the molar volume of the glassy alloy studied. Taking into account the definition of $\Delta L_f(T, q)$ given in Eq. (9.2) and neglecting all terms containing $\Delta L_f^2(T, q)$ and $\Delta L_f^3(T, q)$ when expanding Eq. (9.3) into a power series, one obtains

$$x(T,q) - x_0 \cong 3L_0^2(T)\Delta L_f(T,q)\frac{V_{mol}}{\gamma v^* N_A L_0^3(T_B)}. \tag{9.5}$$

Taking into account the relation between the reduced free volume and the concentration of the flow defects $c_f = \exp\left(-\frac{1}{x}\right)$ [4], one obtains

$$3L_0^2(T)\Delta L_f(T,q)\frac{V_{mol}}{\gamma v^* N_A L_0^3(T_B)} = \frac{1}{\ln c_{f,0}} - \frac{1}{\ln c_f(T,q)}, \tag{9.6}$$

where $c_f(T, q)$ is defined by ▶ Eq. (5.17) (Chap. 5). Keeping in mind that at temperatures $T \leq T_o$ $x(T) = x_o$, one obtains the following temperature dependence for the experimentally observed sample length by heating the sample with constant heating rate q:

(a) $T < T_o$:

$$L_f(T,q) = L_0(T) = L_0(T_B)\left[1 + \alpha_l^0(T - T_B)\right]; \tag{9.7}$$

(b) $T > T_o$:

$$L_f(T,q) = L_0(T_B)\left[1 + \alpha_l^0(T - T_B)\right] + \frac{\gamma v^* N_A L_0(T_B)}{3V_{mol}\left[1 + \alpha_l^0(T - T_B)\right]^2} \times$$
$$\times \left[\frac{1}{\ln c_{f,0}} + \frac{1}{\ln\left(\frac{1}{c_{f,0}} - \int_{T_0}^{T} Q(\theta,q) e^{-\int_{T_0}^{\theta} P(\theta',q)d\theta'} d\theta\right) + \int_{T_0}^{T} P(\theta',q)\,d\theta'}\right], \tag{9.8}$$

where the functions $P(T,q)$ and $Q(T, q)$ were defined in ▶ Chap. 5. In a first approximation, the coefficient of linear thermal expansion α_l^o can be accepted as a temperature-independent constant, which is experimentally obtained. For the "anomaly" of thermal expansion $\Delta L_f(T, q)$, one obtains

$$\Delta L_f(T,q) = \frac{\gamma v^* N_A L_0(T_B)}{3V_{mol}\left[1+\alpha_l^0(T-T_B)\right]^2} \mathrm{x}$$
$$\mathrm{x}\left[\frac{1}{\ln c_{f,0}} + \frac{1}{\ln\left(\frac{1}{c_{f,0}} - \int_{T_0}^{T} Q(\theta,q)e^{-\int_{T_0}^{\theta} P(\theta',q)d\theta'} d\theta\right) + \int_{T_0}^{T} P(\theta',q)d\theta'}\right]. \tag{9.9}$$

Apparently, (▶ Eq. 5.17) is a cardinal keystone in the extended free volume model by the authors. By using this equation, it is possible to check whether an interrelation between the nonisothermally determined viscosity and the rheological peculiarities of the thermal expansion of glassy metals exists.

Viscous Flow and Thermal Expansion of Amorphous Metallic Alloys $Pd_{82}Si_{18}$ and $Fe_{25}Zr_{75}$

The present section aims to demonstrate experimentally that the kinetics of free volume changes during the structural relaxation of glassy alloys determines both the features of nonisothermal viscous flow and the peculiarities (the rheology) of their thermal expansion. This is demonstrated for the first time by the authors by a parallel investigation of both phenomena in the glassy alloys $Pd_{82}Si_{18}$ [6, 7] and $Fe_{25}Zr_{75}$ [8, 9]. The viscosity determinations were carried out by using a silica glass dilatometers Perkin-Elmer TMS-2 ($Pd_{82}Si_{18}$) and Heraeus TMA 500 ($Fe_{25}Zr_{75}$), at heating rates of 20 K/min and 10 K/min, respectively. The temperature and length accuracy were ±2 K and ±0.5 μm.

The temperature dependence of the sample length $L_f(T, q)$ of the alloys studied was determined at the same heating rates as by the viscosity determinations.

Figures 9.2a and 9.3a show the experimentally determined viscosity temperature dependencies of $Pd_{82}Si_{18}$ and $Fe_{25}Zr_{75}$ glassy metals, together with the best fit curves obtained by the regression analysis based on (▶ Eqs. 5.8) and (▶ 5.17) for heating rates of 20 and 10 K/min.

The FVM parameter values obtained by the regression analysis of the experimental data are given in Tables 9.1 and 9.2. Figures 9.2a and 9.3a show also the calculated viscosity temperature dependencies at heating rates of 2, 20, 200, and 2,000 K/min ($Pd_{82}Si_{18}$) and 1, 10, 100, and 1,000 K/min ($Fe_{25}Zr_{75}$). The calculated temperature dependencies, based on (▶ Eq. 5.9), of the quasi-equilibrium viscosities are also presented. As is seen from the prognostic viscosity temperature dependencies, at low heating rates (e.g., 2 K/min), the sample viscosity firstly goes up along with increasing temperature, goes over a maximum, and then decreases (see Fig. 9.2a). This is due to the structural relaxation, which at relatively

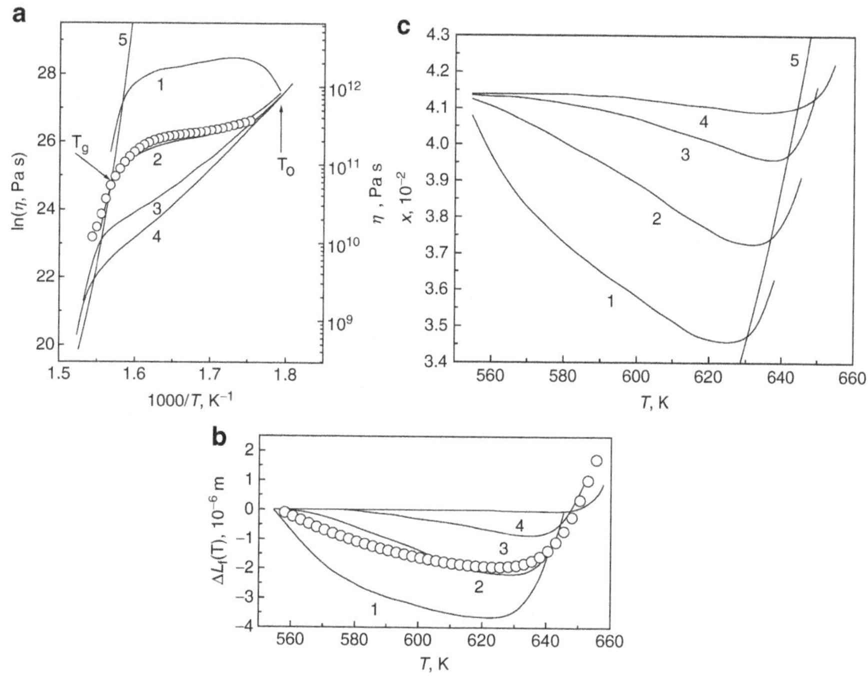

Fig. 9.2 The glassy $Pd_{82}Si_{18}$ alloy. (**a**) Experimentally determined (graphic symbols) and calculated viscosity temperature dependencies at heating rates 2, 20, 200, and 2,000 K/min (curves 1–4). (**b**) Experimental and calculated thermal expansion "anomalies" ΔL_f under the same as above heating conditions. (**c**) Calculated temperature dependencies of the reduced free volume x at the same heating rates as stated above

low temperatures and prolonged period of heating is first of all connected with free volume annihilation, as expected by the FVM. This structural relaxation effect becomes more and more difficult to be observed or even invisible along with increasing the heating rate value. Figures 9.2a and 9.3a are very illustrative and useful when discussing about the definition of glass transition temperature. It seems very judicious to accept the idea that the glass transition temperature T_g is actually $T_g(q)$, i.e., it depends on the heating rate q, and corresponds to the point of interception of the nonequilibrium and quasi-equilibrium viscosity temperature dependencies, at a given heating rate q.

The experimental results obtained by studying the thermal expansion of $Pd_{82}Si_{18}$ and $Fe_{25}Zr_{75}$ glassy alloys are, respectively, shown in Figures 9.2b and 9.3b. Obviously, they possess all specific peculiarities schematically shown in Fig. 9.1. The experimentally obtained data for $\Delta L_f(T,q)$ ($Pd_{82}Si_{18}$) and $L_f(T,q)$ ($Fe_{25}Zr_{75}$) are shown with graphic symbols. The best fit curves are presented with solid lines. The best fit curves were calculated based on Eqs. (9.8) and (9.9), by using the FVM parameter values given in Table 9.1.

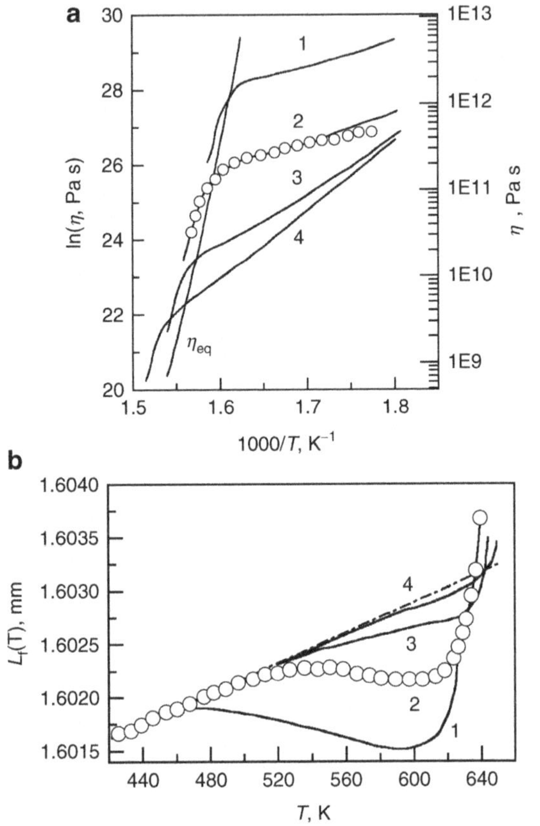

Fig. 9.3 The glassy $Fe_{25}Zr_{75}$ alloy. (**a**) Experimentally determined (graphic symbols) and calculated viscosity temperature dependencies at heating rates of 1, 10, 100, and 1,000 K/min (curves 1–4). (**b**) Experimentally determined (graphic symbols) and calculated sample length L_f at the same heating conditions

Table 9.1 FVM parameter values η_o, $c_{f,o}$, Q_η, v_r, Q_r, B, and T_o for the glassy alloys $Pd_{82}Si_{18}$ and $Fe_{25}Zr_{75}$ determined by the regression analysis of the nonisothermal viscosity experimental data at heating rate q

FVM parameter	Dimension	$Pd_{82}Si_{18}$ viscosity at q = 20/min	$Fe_{25}Zr_{75}$ viscosity at q = 10 K/min
η_o	Pa s/K	5×110^{-20}	5.9×110^{-20}
ν_r	1/s	6.8×110^{19}	7×110^{19}
Q_r	kJ/mol	120	89
Q_η	kJ/mol	191	154
$c_{f,o}$	–	2.54×10^{-11}	3.2×110^{-14}
B	K	2130	5749
T_0	K	557	460
T_g	K	638	628
$\eta(T_g)$	Pa s	6.5×110^{10}	1.3×110^{11}

Table 9.2 Experimentally determined values of thermal expansion coefficients α_l^o and estimated values of the parameters υ^*, V_{mol}, and γ for the glassy alloys $Pd_{82}Si_{18}$ and $Fe_{25}Zr_{75}$

Parameter	Dimension	$Pd_{82}Si_{18}$ thermal expansion at $q = 20$/min	$Fe_{25}Zr_{75}$ thermal expansion at $q = 10$ K/min
α_l^o	1/K	1.515×110^{-5}	5.42×110^{-6}
V_{mol}	m^3	9.6×1^{-6}	12.32×110^{-6}
υ^*	m^3	1.46×110^{-29}	2.24ь10^{-29}
γ	Dimensionless	0.78	0.78

The parameters shown in Table 9.2, together with the FVM parameters η_o, $c_{f,o}$, Q_η, ν_r, Q_r, B, and T_o determined by the nonisothermal viscosity determinations, were used to describe in a convincing way the thermal expansion experiments.

Figure 9.3c presents the temperature dependencies of the reduced free volume x at various heating rates, calculated by using (▶ Eq. 5.2), as well as the temperature dependence of the quasi-equilibrium free volume x_{eq} for the glassy alloy $Pd_{82}Si_{18}$.

The results presented in this book section demonstrate convincingly the excellent agreement between the calculated on the basis of FVM and experimentally obtained data for thermal expansion behavior of $Pd_{82}Si_{18}$ and $Fe_{25}Zr_{75}$ glassy alloys.

Conclusions

- The viscosity and thermal expansion measurements provide the possibility to interpret the experimental results on the basis of FVM.
- A new definition for the glass transition temperature of glassy metals is proposed by their heating from ambient temperature up to the temperatures of their devitrification. The glass transition temperature T_g is the temperature at which the relaxation tendency changes from annihilation to production of free volume.

Specific Heat Anomalies of Amorphous Metallic Alloys: Model Interpretation

A. van den Beukel and J. Sietsma [10] have supposed that the annihilation of free volume in glassy metals should be accompanied by energy (heat) release, while the production of free volume should be accompanied by energy (heat) absorption.

Supposing there is a direct proportionality between the system free energy change and free volume change of a given glassy alloy, an "anomalous" contribution ΔC_p to the specific heat C_p should appear (see Fig. 9.4), which can be presented as

$$\Delta C_p(T, q) = \beta \frac{dx}{dT}, \qquad (9.10)$$

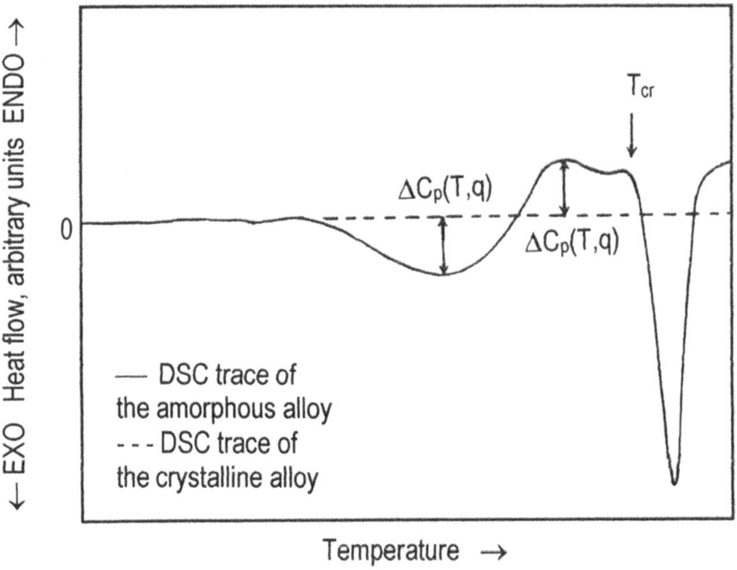

Fig. 9.4 Scheme of the peculiarities in the temperature dependence of glassy metal specific heat

where β is a material-specific coefficient of proportionality, which physical meaning should correspond to the energy consumption needed for the creation of a structural defect. Recollecting in mind (▶ Eq. 5.2), the relation between $\Delta C_p(T,q)$ and the concentration of structural defects in the temperature range considerably lower than the glass transition temperature T_g (see ▶ Eq. 5.19) is given by

$$\Delta C_{p,\,low}(T,q) = \beta \frac{c_{f,\,low}(T,q)Q(T,q)}{\left(\ln c_{f,\,low}(T,q)\right)^2}. \tag{9.11}$$

In the temperature range in the vicinity of the glass transition temperature (see ▶ Chap. 5), $\Delta C_p(T,q)$ is

$$\Delta C_{p,\,high}(T,q) = \beta \frac{c_{f,\,high}(T,q)Q(T,q) - P(T,q)}{\left(\ln c_{f,\,high}(T,q)\right)^2}. \tag{9.12}$$

The functions $P(T,\,q)$ and $Q(T,\,q)$ were defined in ▶ Chap. 5: $P(T,q) = -\frac{\nu_r}{q}\exp\left(-\frac{Q_r}{RT} - \frac{B}{T-T_0}\right)$, and $Q(T,q) = -\frac{\nu_r}{q}\exp\left(-\frac{Q_r}{RT}\right)$.

Viscous Flow and Specific Heat of Amorphous Metallic Alloy $Fe_{25}Zr_{75}$ Under Different Constant Heating Rates

The temperature dependence of the nonisothermal viscosity of $Fe_{25}Zr_{75}$ glassy alloy was studied at two heating rates, namely, at 10 K/min by using silica glass dilatometer Heraeus TMA 500 and at 1.25 K/min by using silica glass dilatometer Perkin-Elmer TMS-2. The measurements of the specific heats C_p^g and C_p^{cr} of the as-quenched amorphous and crystallized $Fe_{25}Zr_{75}$ ribbons were carried out by using differential scanning calorimeter Perkin-Elmer DSC 7 at heating rate of 40 K/min, as described in [11, 12]. The difference of $\Delta C_p(T,q)$ between C_p^g and C_p^{cr} was subjected to multiparameter nonlinear regression analysis based on Eqs. (9.10) and (9.11) [11, 12].

Figure 9.5 shows the viscosity temperature dependence of the glassy alloy studied at two different heating rates by using two different silica dilatometers together with the best fit curves resulted from the FVM regression analysis. The calculated temperature dependence (▶ Eq. 5.9) of the quasi-equilibrium viscosity η_{eq} is also shown. The values of FVM regression coefficients are shown in Table 9.3 for heating rates of $q = 1.25$ and $q = 10$ K/min. They result from the *simultaneous fitting* of both viscosity temperature dependencies.

Figure 9.6 shows the experimentally determined deviation $\Delta C_p(T, q)$ from extrapolated to higher-temperature linear temperature dependence of the specific heat $C_p^g(T)$ in the low temperature range (T < 400 K) of as-quenched sample of $Fe_{25}Zr_{75}$ glassy alloy. In this temperature range, $C_p^g(T)$ coincides with the temperature dependence of $C_p^{cr}(T)$ of the crystallized samples of the same glassy alloy. The last one remains linear in the

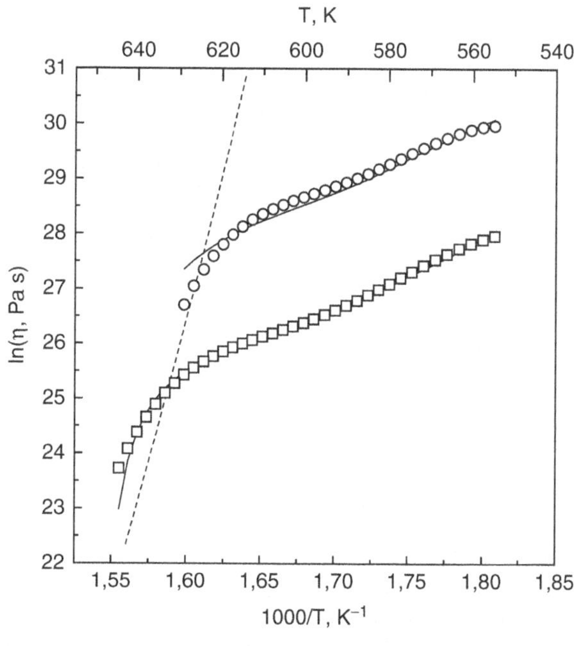

Fig. 9.5 Viscosity temperature dependencies of $Fe_{25}Zr_{75}$ glassy alloy determined at two different heating rates: o, 1.25 K/min, and Δ, 10 K/min, together with the best fit FVM curves. The steeper broken line represents the temperature dependence of the quasi-equilibrium viscosity η_{eq}

Table 9.3 FVM parameters from the regression analysis of the viscosity experimental data $\eta(T, q)$ of the $Fe_{25}Zr_{75}$ glassy alloy (▶ Eqs. 5.8 and ▶ 5.17) and of the specific heat $\Delta C_p(T, q)$ (Eqs. 9.10 and 9.11). For quick comparison purpose, the FVM parameters from the thermal expansion data of the same glassy alloy are presented

FVM parameter, dimension	Viscosity	Thermal expansion	From data $\Delta C_{p,high}$	From data $\Delta C_{p,low}$
q, K/min	1.25, 10	10	40	40
η_o, Pa s/K	5.9×10^{-20}	–	–	–
Q_η, kJ/mol	154	–	–	–
B, K	5749	5500	5730	–
T_o, K	460	458	458	–
ν_r, 1/s	7×110^{19}	6.4×110^{19}	4.7×110^{18}	4.5×110^{18}
Q, kJ/mol	89	88	87	89
$c_{f,o}$	3.2×110^{-14}	2.9×110^{-14}	9.8×110^{-12}	7.6×110^{-11}
β, kJ/mol K	–	–	167	130

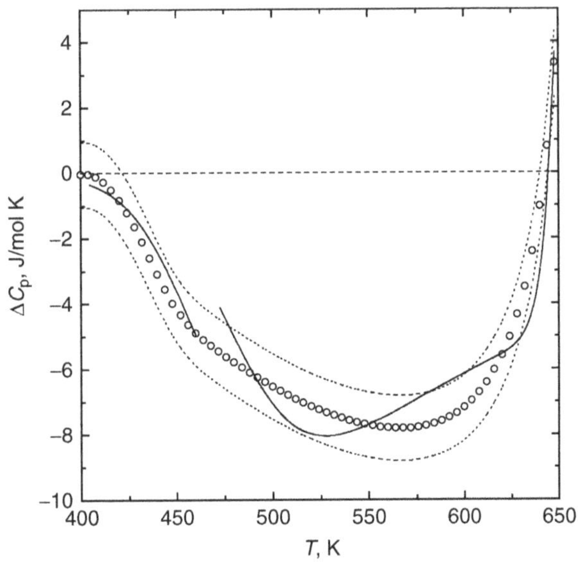

Fig. 9.6 $\Delta C_p(T,q)$ temperature dependence of glassy $Fe_{25}Zr_{75}$ alloy at heating rate of 40 K/min: o, experimental data; *solid lines*, best fit curves, calculated based on Eqs. (9.10) and (9.11). The *dotted lines* indicate the experimental data scattering limits. The *horizontal broken line* indicates the zero level

whole temperature range of investigation. When comparing the temperatures of visible deviation appearance of ΔL_f in thermal expansion experiments with the temperature of visible deviation appearance of $\Delta C_p(T, q)$ in specific heat experiments, it becomes obvious that the specific heat measurements "notice" the beginning of free volume annihilation at substantially lower temperature as compared with the temperature at which this becomes true in the thermal expansion experiments (see section "Viscous Flow and Thermal Expansion of Amorphous Metallic Alloys $Pd_{82}Si_{18}$ and $Fe_{25}Zr_{75}$").

The experimental data processing based on Eq. (9.10) is possible only at temperatures higher than the so-called ideal glass transition temperature T_o. In the case of the glassy alloy studied, this temperature is approximately 460 K (see section "Viscous Flow and Thermal Expansion of Amorphous Metallic Alloys $Pd_{82}Si_{18}$ and $Fe_{25}Zr_{75}$"). The starting temperature of visible appearance of $\Delta C_p(T, q)$ signal is however with 60 K lower. This why in the temperature range 400–450 K, the experimental data were subjected to regression analysis based on Eq. (9.11). At higher temperatures, Eq. (9.10) was applied. The FVM parameter values are presented in Table 9.4. When comparing them with the FVM regression coefficients obtained by the thermal expansion experiments, it becomes obvious that they are in satisfactory agreement.

It could be concluded that:

- The extended free volume model by the authors is able to propound a theoretical description of the nonisothermal viscous flow behavior, thermal expansion, and specific heat features of metallic glasses at different heating rates. These phenomena can be successfully related to the current amount of free volume in the glassy metal structure. The lack of internal inconsistency in the FVM is apparent.

Viscous Flow, Thermal Expansion, and Specific Heat of $Al_{25}La_{50}Ni_{25}$ Amorphous Alloy

The free volume model (*FVM*) describes the atomic mobility in terms of the volume available for the atoms. The structural relaxation is also identified as an approach of the disordered structure to the (quasi-) equilibrium defect concentration $c_{f,e}$. By rapid solidification of liquid alloys, an excess defect concentration $c_{f,o}$ is frozen into the glassy state. Upon isothermal heating, the sample $c_{f,o}$ reduces toward $c_{f,e}$. By subsequent heating of the material to a higher temperature, the defect concentration increases until equilibrium has again been established [3, 13, 14]. These processes of annihilation or creation of free volume can be experimentally detected when studying the viscosity, relaxation of bend stresses both under isothermal and nonisothermal conditions, thermal expansion, and heat capacity of glassy metals.

The first aim of the present book section is to demonstrate the capability of FVM [1, 5, 9, 10, 15–17] to be used for the quantitative representation of the results under constant heating rate conditions. The scope is restricted to viscosity creep around the glass transition temperature T_g and thermal expansion and heat capacity properties on the other hand, using $Al_{25}La_{50}Ni_{25}$ glassy alloy with a very broad temperature interval between the glass transition and the onset temperatures of crystallization [18] as an example.

It should become clear from the exposed till now that the FVM, describing the atomic mobility via the system free volume, available for the atoms, provides the possibility to present the structural relaxation of glassy alloys as a trend of their chaotically ordered structure to approach the temperature-dependent quasi-equilibrium concentration of structural defects. This is realized via annihilation ($T < T_g$) or production ($T > T_g$) of structural defects, depending upon the circumstance on which side of the quasi-equilibrium state the amorphous system (glassy alloy) is situated. It

Table 9.4 Unique combination of FVM regression parameters, describing the experimental data from viscosity determinations, the isothermal and nonisothermal bend stress relaxation, and the peculiarities of thermal expansion and specific heat of $Al_{25}La_{50}Ni_{25}$ glassy alloy. *Measured or estimated values of some $Al_{25}La_{50}Ni_{25}$ glassy alloy specific constants

		FVM parameters				
Kind of experiment	Heating rates, K/min	η_o, Pa s	$c_{f,o}$	ν_r, 1/s	Q_r, kJ/mol	Q_η, kJ/mol
Viscosity	2.5, 10, 20	1.15×10^{-17}	1.79×10^{-7}	8.55×10^{21}	134.4	149.64
Relaxation of bend stresses, isothermal	–	1.15×10^{-17}	1.79×10^{-7}	8.55×10^{21}	134.4	149.64
Relaxation of bend stresses, nonisothermal	2.5, 5, 10, 20, 40	1.15×10^{-17}	1.79×10^{-7}	8.55×10^{21}	134.4	149.64
Peculiarities of specific heat	5, 10, 20, 40	–	1.79×10^{-7}	8.55×10^{21}	134.4	–
Peculiarities of thermal expansion	2.5, 20	–	1.79×10^{-7}	8.55×10^{21}	134.4	–
		FVM parameters				
Kind of experiment	Heating rates, K/min	B, K	T_o, K	β, kJ/mol	E, Pa	
Viscosity	2.5, 10, 20	3772	320	–	–	
Relaxation of bend stresses, isothermal	–	–	–	–	85×110^9	
Relaxation of bend stresses, nonisothermal	2.5, 5, 10, 20, 40	–	–	–	85×110^9	
Peculiarities of specific heat	5, 10, 20, 40	3772	320	43.8	–	
Peculiarities of thermal expansion	2.5, 20	3772	320	–	–	

*$L_o = 2.4 \times 10^{-3}$ m, $\alpha_l = 2 \times 10^{-5}$ 1/K, $V_{mol} = 1.37 \times 10^{-5}$ m^3, $\Omega_o = 2.4 \times 10^{-29}$ m^3, $\gamma = 0.744$

has been shown as well, by using various glassy metallic systems as an example, that the processes of annihilation and/or production of free volume can be successfully monitored experimentally via studying the viscosity temperature dependence, the bend stress relaxation, and the thermal expansion and specific heat peculiarities.

The second aim of the present book section is to demonstrate once again the unambiguous capability of FVM to describe quantitatively the abovementioned relaxation phenomena in glassy metals taking place under isothermal and/or nonisothermal conditions. This is attained via different kinds of experiments at several constant heating rates upon the $Al_{25}La_{50}Ni_{25}$ glassy alloy [19]. The results of the different experiments are interpreted *simultaneously* on the basis of the FVM, aiming to demonstrate the possibility to describe them quantitatively with common and unique combination of FVM parameters.

The amorphous $Al_{25}La_{50}Ni_{25}$ ribbons were produced using melt-spinning method at low He atmosphere pressure (1.3×10^4 Pa) as described in section ▶ "Viscosity and Relaxation Under Nonisothermal Conditions" (Chap. 5) of this book. The resulting ribbon studied was 2.5 mm wide by 0.038 mm thick. Viscosity measurements of the glassy alloys studied were performed with the aid of a Perkin-Elmer TMS-2 silica glass dilatometer using the experimental approach described in detail in section ▶ "Influence of Vanadium Alloying Additions" (Chap. 4) of this book and in [6, 20, 21]. ▶ Figure 8.13 (▶ Chap. 8) shows the viscosity temperature dependence of the alloy studied at heating rates of 2.5, 10, and 20 K/ min. The experimental data were subjected simultaneously on a regression analysis based on (▶ Eqs. 5.8) and (▶ 5.17) (▶ Chap. 5). The values of the FVM parameters obtained are given in Table 9.4. The temperature dependence of the quasi-equilibrium viscosity η_{eq}, calculated by using the FVM parameters, is also shown in ▶ Fig. 8.13 (▶ Chap. 8). The intercept points of η_{eq} – curve with the curves of the dynamic viscosity at the three heating rates – correspond to the reciprocals of the glass transition temperatures at heating rates of 2.5, 10, and 20 K/min. They are $T_g \cong 485$ K, $T_g \cong 492$ K, and $T_g \cong 495$ K/min at heating rates of 2.5 K/min, 10 K/min, and 20 K/min, respectively.

The peculiarities of thermal expansion of the $Al_{25}La_{50}Ni_{25}$ glassy alloy were experimentally determined with a silica glass dilatometer Perkin-Elmer TMS-2 at heating rates of 2.5 and 20 K/min also, as described in detail in [19]. Three separate experiments for each heating rate used were performed. The mean value of $\Delta L_f(T)$ from the experiments at the two heating rates is shown in Fig. 9.7. The experimental data were interpreted according to the FVM (Eq. 9.8). The solid lines are the best fit curves calculated by using the FVM regression parameters. The FVM parameters are given in Table 9.4. Two additional estimated parameters, added as footnote to Table 9.4, are the geometric atomic free volume overlap factor γ and the magnitude of the critical free volume v*, respectively.

Due to the circumstance that at temperatures higher than 520 K the viscosity of the glassy alloy studied becomes lower than 10^9 Pa s, reliable experimental data for the thermal expansion can be obtained only up to this temperature. At higher temperatures, even the negligible load (about 0.1 g) applied upon the coiled ribbon sample, needed to ensure the contact between the silica glass sensor and the sample, is sufficient to cause significant sample flattening deteriorating the reliability of the data.

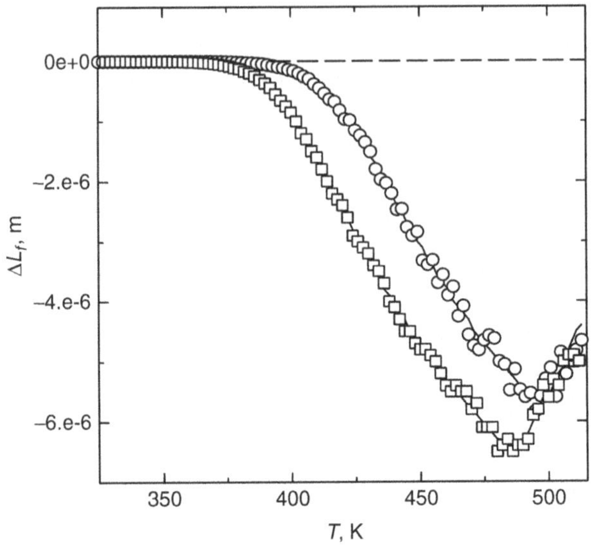

Fig. 9.7 ΔL_f temperature dependence of $Al_{25}La_{50}Ni_{25}$ glassy alloy at heating rates of 2.5 and 20 K/min: □, 2.5 K/min, and o, 20 K/min. The *solid lines* represent the best fit curves from the FVM regression analysis based on Eq. (9.8)

The peculiarities of the specific heat C_p of $Al_{25}La_{50}Ni25$ glassy alloy were determined with the aid of Perkin-Elmer DSC 7 according to a method described in [11].

Figure 9.8 shows the temperature dependencies of ΔC_p – data for heating rates of 5 and 40 K/min, together with the best fit curves resulted from the regression analysis based on Eq. (9.11). Analogous data and curves (not shown graphically) were obtained at heating rates of 10 and 20 K/min also. The FVM regression parameters are shown in Table 9.4.

It becomes clear from the experimental results of the five kinds of experiments for study of the relaxation phenomena in amorphous metallic alloys, and especially $Al_{25}La_{50}Ni_{25}$ glassy alloy, that they could be conveniently described with the aid of FVM basic concepts. This is due to the circumstance the relaxation phenomena in glassy metals really directly reflect the magnitude and changes of the free volume available for the atoms in a definite moment of structural relaxation, both under isothermal and nonisothermal conditions. By the description of this time/temperature dependence of glassy metal free volume, one has to operate with a complex of FVM parameters, namely, ν_r, Q_r, $c_{f,o}$, B, and T_o, respectively, together with some specific for the different kind of experiment parameters and constants, such as the pre-exponential viscosity factor, η_o, the activation energy for viscous flow Q_η, the Young modulus E, the coefficient of linear thermal expansion α_l, and the proportionality constant β by studying the specific heat peculiarities. When studying the relaxation phenomena in one and the same glassy metallic alloy, it should be expected that the values of the FVM parameters will remain one and the same for all kinds of experiments. The FVM parameter values should be physically significant as well.

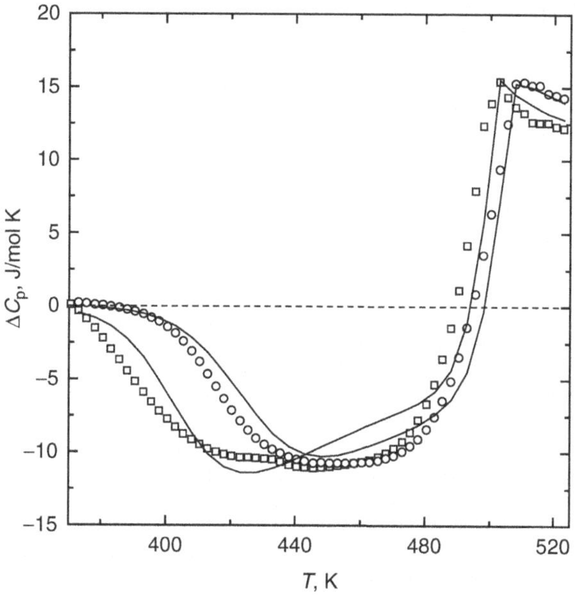

Fig. 9.8 ΔC_p, temperature dependencies of as-quenched $Al_{25}La_{50}Ni_{25}$ glassy alloy samples at heating rates of 5 and 20 K/min: □, 5 K/min, and O, 40 K/min. The *solid lines* represent the best fit curves from the FVM regression analysis based on Eq. (9.11)

Aiming to prove to which extent these expectations are fulfilled, the authors tried on the one hand to ensure maximally well calibration of all devices used, concerning the temperature and heating rate and minimizing the degree of experimental errors. On the other hand, all experimental data under varying conditions from a definite experiment were subjected *simultaneously* to a nonlinear multiparameter regression analysis according to the specific for the different kinds of experiment theoretical FVM equations. The values of the FVM parameters are presented in Table 9.4.

The Specific Heat Proportionality Constant β. The value of this model parameter was determined to be 43.8 kJ/mol. Taking into account that the enthalpies of monovacancy formation β_{cr} for the pure components [22] are Al, 63 kJ 1/mol; La, 103 kJ/ mol; and Ni, 135 kJ/mol, the enthalpy of monovacancy formation of an average atomic volume Ω_0 for the amorphous alloy studied can be estimated according to Miedema [23] to be about 100 kJ/mol. From the ratio β/β_{cr}, the conclusion can be drawn that the value of the critical free volume υ^* should be about 0.438 Ω_0, which is physically meaningful and also in a good agreement with the estimation of υ^*/Ω_0 in this book (▶ Chap. 8).

Conclusions

The free volume theory describing the atomic mobility during the relaxation of glassy alloys can be successfully applied for quantitative presentation of experimental results obtained by studying the viscosity, bend stress relaxation, thermal expansion, and specific heat peculiarities with a single and unique combination of FVM parameters.

References

1. Russew K, Stojanova L, Sommer F (1995) Nonisothermal viscous flow and thermal expansion of $Pd_{82}Si_{18}$ glassy alloy considered as Free Volume related phenomena. Int J Rapid Solidif 8(4):267
2. Cohen MH, Turnbull D (1959) J Chem Phys 31:1164
3. Turnbull D, Cohen MH (1970) J Chem Phys 52:3038
4. Koebrugge GW, Sietsma J, van den Beukel A (1992) Acta Metall Mater 40:753
5. Duine PA, Sietsma J, van den Beukel A (1992) Acta Metall Mater 40:743
6. Russew K, Stojanova L (1995) Viscosity and structural relaxation in $Pd_{82}Si_{18}$ metallic glass. J Mater Sci Technol 3:40
7. Russew K (1997) Free volume related viscous flow, relaxation of bend stresses, thermal expansion and heat capacity of glassy metals: theory and experiment. In: 9th international conference on rapidly quenched and metastable materials RQ9, 1996, Bratislava. Mat Sci Eng A226–228, p 779
8. Russew K, Stojanova L, Sommer F (1997) Viscous flow, thermal expansion and heat capacity of $Fe_{25}Zr_{75}$ glassy alloy under nonisothermal conditions at different heating rates. In: 9th international conference on rapidly quenched and metastable materials RQ9, 1996, Bratislava. Mat Sci Eng A226–228, p 344
9. Russew K, Stojanova L (1996) Viscous flow, thermal expansion and relaxation of amorphous metallic alloys: theory and experiment. In: Markov KZ (ed) Proceedings of 8th international symposium continuum models and discrete systems, Varna, 1995. World Scientific Publishing Co, London, p 73
10. Van den Beukel A, Sietsma J (1990) Acta Metall Mater 38:383
11. Zappel BJ, Sommer F (1994) Mater Sci Eng A179–180:283
12. Zappel BJ (1995) Untersuchung unterkühlter Legierungsschmelzen im Glasübergangsbereich. In: Fortschrittsberichte VDI, Reihe 5, Nr. 407. VDI-Verlag, Dusseldorf, PhD Dissertation (in German)
13. Tsao SS, Spaepen F (1985) Acta Metall 33:881
14. Volkert CA, Spaepen F (1990) Acta Metall 37:1355
15. Russew K, Zappel BJ, Sommer F (1995) Nonisothermal viscous flow behaviour of $Pd_{40}Ni_{40}P_{20}$ glassy alloy considered as a free volume related phenomenon. Scr Metall Mater 32:271
16. Russew K, de Hey P, Sietsma J, van den Beukel A (1997) Viscous flow of amorphous $Fe_{40}Ni_{40}Si_6B_{14}$ glassy alloy studied by direct creep measurements and relaxation of bend stresses under nonisothermal conditions. Acta Mater 45(5):2129
17. Tuinstra P, Duine PA, Sietsma J (1993) J Non-Cryst Solids 156–158:519
18. Inoue A, Zhang T, Masumoto T (1993) J Non-Cryst Solids 156–158:473
19. Russew K, Sommer F (1998) Determination of the viscosity, thermal expansion, heat capacity and bend stress relaxation temperature dependence of glassy alloys - a tool for understanding of relaxation phenomena. Int J Non-Equilibr Process 11:3
20. Russew K, Sommer F, Stojanova L (1993) Influence of Cr-alloying additions on the viscous flow behaviour of $Fe_{82-x}Cr_xB_{18}$ amorphous alloy. In: Duhaj P, Mrafko P, Svec P (eds) Proceedings of conference on amorphous metallic materials AMM III. Trans Tech Publications, Slovakia, p 625
21. Russew K, Sommer F, Duhaj P, Bakonyi I (1992) Viscous flow behaviour of Ni_xZr_{100-x} metallic glasses from $Ni_{30}Zr_{70}$ to $Ni_{64}Zr_{36}$. J Mater Sci 27:3565
22. Seeger A, Schumacher D, Schilling W, Diehl J (eds) (1970) Vacancies and interstitials in metals. North Holland, Amsterdam
23. Miedema AR (1979) Zs Metallkde 70:345

Relation Between the Density Changes and Thermal Expansion Anomalies in the Amorphous Metallic Alloys

10

Abstract

The structural relaxation of bulk samples of the amorphous metals with $Pd_{40}Cu_{30}Ni_{10}P_{20}$ bulk glassy alloy as an example, by the determination of the temperature dependences of the thermal expansion at a constant heating rate of 10 K/min and isothermal measurement of the density changes, has been studied. An expression for the theoretical description of density changes of amorphous metallic alloys as a function of the heat treatment conditions in the framework of the free volume model (FVM) has been derived. Experimental determination of the density changes of a $Pd_{40}Cu_{30}Ni_{10}P_{20}$ glassy alloy is carried out. A good agreement between the experimental data and the FVM theoretical expectations is established. Calculated using the derived theoretical description via inserting in it the experimentally obtained thermal expansion, experimental data have been carried out. They are in good agreement with the experimental data for the density changes of the same system.

Keywords

Glassy metals • Amorphous metallic alloys • $Pd_{40}Cu_{30}Ni_{10}P_{20}$ bulk glassy alloy free volume model • Relaxation • Density • Viscosity • Thermal expansion

The free volume model describes the atomic mobility in terms of the volume available for the atoms. The structural relaxation is also identified as an approach of the disordered structure to the (quasi-) equilibrium defect concentration $c_{f,eq}$. By rapid solidification of liquid alloys, an excess defect concentration $c_{f,o}$ is frozen into the glassy state. Upon isothermal heating the sample $c_{f,o}$ reduces toward $c_{f,eq}$. By subsequent heating of the material to a higher temperature, the defect concentration increases until equilibrium has again been established. These processes of annihilation or creation of free volume can be experimentally detected when studying the viscosity, thermal expansion, and density changes of glassy metals.

K. Russew, L. Stojanova, *Glassy Metals*,
DOI 10.1007/978-3-662-47882-0_10

In ▶ Chap. 9 of the present book, an FVM analysis was carried out about the "anomalous" deviations of the thermal expansion of glassy metals from the simple low-temperature linear temperature dependence. It was shown that linear size $L_f(T,q)$ of the glassy metallic alloy depends upon the reached temperature T at a constant heating rate q as follows:

a/for $T < T_o$

$$L_f(T,q) = L_0(T) = L_0(T_B)\left[1 + \alpha_l^0(T - T_B)\right]; \tag{10.1}$$

b/for $T > T_o$

$$L_f(T,q) = L_o(T_B)\left[1 + \alpha_l^o(T - T_B)\right] + \frac{\gamma \upsilon^* N_A L_o(T_B)}{3V_{mol}\left[1 + \alpha_l^o(T - T_B)\right]^2}\left[\frac{1}{\ln c_{f,o}} + \frac{1}{\ln c_f(T,q)}\right], \tag{10.2}$$

where $L_o(T_B)$ is the linear size of the amorphous metallic sample at the starting temperature of heating (e.g., $T_B = 333$ K), α_l^o is the coefficient of linear thermal expansion in the low temperature range $T < T_o$, T_o is the ideal glass temperature, V_{mol} is the molar volume, υ^* is the so-called critical atomic free volume, γ is a geometric overlap factor, N_A is the Avogadro number, and $c_f(T,q)$ is the concentration of flow defects as defined by (▶ Eq. 5.17) (▶ Chap. 5). The coefficient of linear thermal expansion α_l^o can be considered as a temperature independent.

The "anomalous" deviation $\Delta L_f(T,q)$ from the linear sample length temperature dependence is presented as

$$\Delta L(T,q) = \frac{\gamma \upsilon^* N_A L_o(T_B)}{3V_{mol}\left[1 + \alpha_l^0(T - T_B)\right]^2} \mathrm{x} \\ \mathrm{x}\left[\frac{1}{\ln c_{f,o}} + \frac{1}{\ln\left(\frac{1}{c_{f,0}} - \int_{T_0}^{T} Q(\theta,q)e^{-\int_{T_0}^{T} P(\theta',q)d\theta'} d\theta\right) + \int_{T_0}^{T} P(\theta',q)d\theta'}\right], \tag{10.3}$$

where $P(T,q) = -\frac{\nu_r}{q}\exp\left(-\frac{Q_r}{RT} - \frac{B}{T-T_0}\right)$ and $Q(T,q) = -\frac{\nu_r}{q}\exp\left(-\frac{Q_r}{RT}\right)$, as shown in ▶ Chap. 5 of the present book (see also the book Appendix). Here ν_r is the attempt frequency, Q_r is the activation energy of relaxation, $c_{f,0}$ is the initial defect concentration, B and T_o are two model parameters which can be related to the empirical constants of the Vogel-Fulcher-Tammann equation, and R is the universal gas constant.

Under isothermal relaxation conditions, the change of the sample linear size can be presented as

$$\Delta L_f(T,t) = \frac{\gamma v * N_A L_o}{3V_{mol}} \left(\frac{1}{\ln c_{f,o}} - \frac{1}{\ln c_f(T,t)} \right), \tag{10.4}$$

where T and t are the temperature and time of isothermal annealing, respectively.

It was shown in ▶ Chap. 5 that the most probable differential equation describing the change of structural defect concentration $c_f(T,t)$ along with increasing the time of annealing at constant temperature T is

$$\frac{dc_f(T,t)}{dt} = -k_r c_f \left(c_f - c_{f,e} \right), \tag{10.5}$$

where $k_r = \nu_r \exp\left(-\frac{Q_r}{RT}\right)$ is the rate constant of relaxation, $c_{f,e}(T)$ is the equilibrium defect concentration at the temperature of isothermal annealing, and R is the universal gas constant. The solution of Eq. (10.5) is

$$c_f(T,t)^{-1} = c^{-1}{}_{f,e} + \left(c^{-1}{}_{f,o} - c^{-1}{}_{f,e} \right) \exp\left(-k_r c_{f,e} t\right), \quad c_f(t = 0s) = c_{f,0}. \tag{10.6}$$

$c_{f,o}$ is here the defect concentration at the start of the isothermal heat treatment. The temperature dependence of $c_{f,e}(T)$ is given by an empirical Vogel-Fulcher-Tammann-type equation [1, 2]:

$$c_{f,e}(T) = \exp\left(-\frac{B}{T - T_o} \right), \tag{10.7}$$

where B and T_o are constants.

Relaxation Changes of Density of Amorphous Metallic Alloys: Free Volume Model Description

The density is one of the basic properties of condensed matter bodies. This is also valid for the amorphous metallic alloys, which do not possess atomic long-range order, on the contrary to the crystalline bodies. The validity of any new structural model is usually proven via its capability to describe properly the changes in density caused by a structural relaxation. The definition of the density from the macroscopic point of view is quite simple, while its interpretation on an atomic scale is much more complicated. The free volume model extended by the authors of the present book provides the possibility for such interpretation.

The initial density of an amorphous metallic sample ρ_o is obviously equal to $\rho_o = \frac{W}{V_o} = \frac{W}{L_o^3}$, where W and V_o are the initial sample weight and volume, respectively. The density change of a cube of initial edge length L_o, caused by nonisothermal or isothermal structural relaxation, can be presented as follows:

$$\Delta\rho(T,q) = \rho(T,q) - \rho_o = \frac{W}{\left(L_o + \Delta L_f(T,q)\right)^3} - \frac{W}{L_o^3} = -\frac{3W\left(\Delta L_f(T,q)/L_o\right)}{L_o^3\left(1 + 3\Delta L_f(T,q)/L_o\right)}, \qquad (10.8)$$

$$\frac{\Delta\rho(T,q)}{\rho_o} = -\frac{3\left(\Delta L_f(T,q)/L_o\right)}{1 + 3\left(\Delta L_f(T,q)/L_o\right)}, \qquad (10.9)$$

under nonisothermal relaxation conditions, and

$$\frac{\Delta\rho(t)}{\rho_o} = -\frac{3\left(\Delta L_f(T,t)/L_o\right)}{1 + 3\left(\Delta L_f(T,t)/L_o\right)}, \qquad (10.10)$$

under isothermal relaxation conditions. Here $\Delta L_f(T,q)$ и $\Delta L_f(T,t)$ are described by Eqs. (10.3) and (10.4), respectively. It should be clear here that the denominator in the middle part of Eq. (10.8) was expanded in a series, and the terms ΔL_f^3 и ΔL_f^2 were neglected as extremely small.

A quite limited number of critical surveys can be found in the scientific bibliography concerning with the problem of density measurements of ribbonlike amorphous metallic alloys [3–5]. Regardless of the fact that the density of solids could be measured with a high precision (0.1 % or even better), its determination by ribbonlike glassy metals remains a big problem, which is practically impossible to be solved in gas environment. This is due to the very big ratio sample surface/sample weight. Air microbubbles adhere to sample surface in an uncontrollable amount. This has a strong and negative effect upon the measurement reproducibility of density. This problem is extensively discussed by several researchers, e.g., G. Konczos [6], H.-R Sinning, et al. [7], etc. The last researchers even regarded that the dilatometry is the most sensitive method for the study of the volume changes of glassy alloys caused by structural relaxation during heat treatment. This method used under isothermal experimental conditions is however connected with series of specific problems when dealing with ribbonlike samples. These problems are discussed in Refs. [8–12]. This is why the best solution of the problem for checking the validity of above-derived FVM equations describing the density changes of glassy metals caused by heat treatment is to use bulk glassy samples. Suitable bulk glassy alloy for this purpose is $Pd_{40}Cu_{30}Ni_{10}P_{20}$.

Relaxation Changes of Density of Amorphous Metallic Alloys: Experimental Proof

The $Pd_{40}Cu_{30}Ni_{10}P_{20}$ bulk glassy alloy was produced via melting of pure copper and palladium in suitable mixture of intermetallic compounds Ni_2P and Pd_7P_3. More details are presented in ▶ Chap. 1 of the present book, as well as in Ref. [13]. Ribbonlike samples of the glassy alloy of cross section 2.5 × 0.04 mm were produced by using the CBMS method. The amorphous structure of the as-quenched samples was checked with the aid of X-ray diffractometer by using Cu K_α radiation. The density of these samples was measured by using an electronic balance, Mettler AE163 of 10^{-5} g sensitivity, specially designed for density measurements, applying the buoyancy method. Liquid CCl_4, with a density of 1.5925 g/cm^3 at 20 °C, was used as the buoyancy fluid. The temperature of CCl_4 was kept constant during the density measurements at 20 ± 0.01 °C using a thermostat. The length change of the amorphous $Pd_{40}Cu_{30}Ni_{10}P_{20}$ ribbons at a constant heating rate of 10 K/min was monitored up to the crystallization temperature using a Perkin-Elmer TMS-2 silica glass dilatometer with noise level of 5 × 10^{-5} mm. The ribbonlike sample was coiled into a silica glass ring of a height smaller than the width of the ribbon itself. A very thin and light silica glass plate was placed onto the upper edge of the ribbon in order to ensure contact between the sample and the silica glass rod of the probe. A power-compensated differential scanning calorimeter (DSC, Pyris-1, Perkin-Elmer) was used to determine the heating rate dependence of the glass transition temperature T_g of bulk amorphous $Pd_{40}Cu_{30}Ni_{10}P_{20}$. The reproducibility of temperature measurements was better than ±0.3 K.

The inflection point in the DSC curve in the glass transition region was assigned to T_g. Prior to each measurement of each heating curve, the sample was given a specified thermal history by cooling it through the glass transition region at a rate equal to the subsequent heating rate. The density changes of bulk amorphous $Pd_{40}Cu_{30}Ni_{10}P_{20}$ samples were measured under five different isothermal heat treatment conditions. The cylindrical samples 6 mm in diameter and 5 mm long were all cut from the water-quenched sample using a lathe. The samples were sealed in silica glass ampoules under vacuum of 10^{-3} Pa and were annealed at a constant temperature for a chosen time. The annealing was interrupted by quenching the samples into water at room temperature. The measured density ρ_o, obtained as the mean value of ten consecutive measurements, of the as-quenched amorphous samples varied between 9.2472 ± 0.0007 and 9.2886 ± 0.0007 g/cm^3. The value of 9.275 g/cm^3 reported by Inoue [14] of the same amorphous alloy lies within these limits. The scatter in the absolute value of the density is obviously due to the existence of small casting voids of varying volume fraction in different samples. Such casting voids were also observed using metallography in the cross section of the samples, and some of them could be seen even without any optical means – see Fig. 10.1. However, the scattering of the initial density does not influence the relative density change $\Delta\rho/\rho_o$. The accuracy of the measured $\Delta\rho(t)/\rho_o$ data was better than 0.01 %.

Figure 10.2 shows the relative density changes $\Delta\rho(t)/\rho_o$ obtained at 500, 520, 550, 573, and 593 K as a function of isothermal annealing time. The fitting curves (solid lines) are calculated using Eq. (10.10) and the FVM parameters given in Table 10.1. The maximum value of $(\Delta\rho(t)/\rho_o) = 0.11$ %, determined experimentally by Inoue [14] after 900 s isothermal annealing at 573 K, is in accordance with the result given in Fig. 10.2.

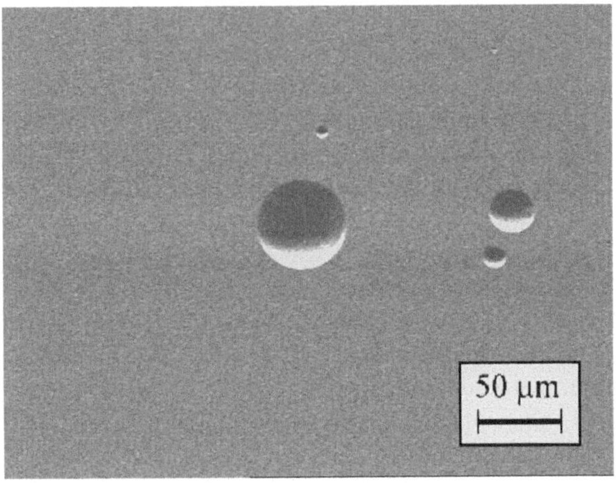

Fig. 10.1 Casting voids in the volume of $Pd_{40}Cu_{30}Ni_{10}P_{20}$ bulk glassy alloy – SEM image

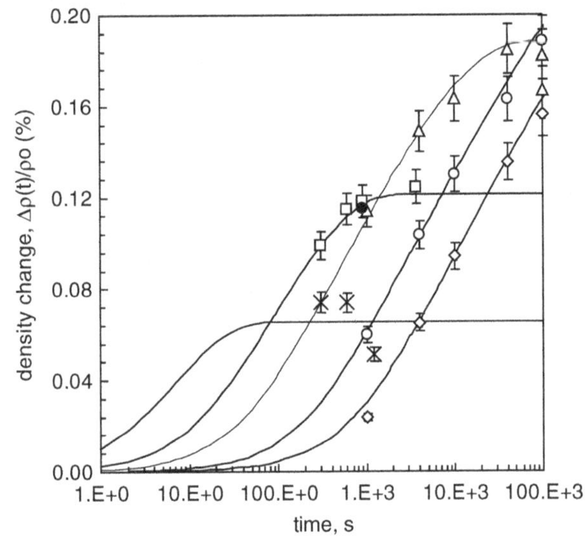

Fig. 10.2 Relative density changes $\frac{\Delta\rho(t)}{\rho_o}$ of amorphous $Pd_{40}Cu_{30}Ni_{10}P_{20}$ bulk glassy alloy samples measured at 500 K (◇), 520 K (o), 550 K (Δ), 573 K (□), and 593 K (×), as a function of isothermal annealing time. The graphic symbol (•) denotes the value $\frac{\Delta\rho(t)}{\rho_o} = 0, 11\%$ obtained by Inoue for the same bulk amorphous alloy after annealing at 573 K for 900 s

Table 10.1 FVM parameters obtained by the regression analysis of the time dependence of $\frac{\Delta\rho(t)}{\rho_o}$ experimental data, from the different temperatures of isothermal annealing in the temperature range 500–593 K. $Pd_{40}Cu_{30}Ni_{10}P_{20}$ bulk glassy alloy

FVM parameter, dimension	Value	FVM parameter, dimension	Value
ν_r, 1/s	3.12×10^{19}	a_l^o, 1/K	1.86×10^{-5}
Q_r, кJ/mol	130.45	V_{mol}, m^3	7.96×10^{-6}
$c_{f,o}$	1.38×10^{-9}	υ^*, m^3	1.2×10^{-29}
T_o, K	307	γ	0.8
B, K	6324	N_A, 1/mol	6.02×10^{23}

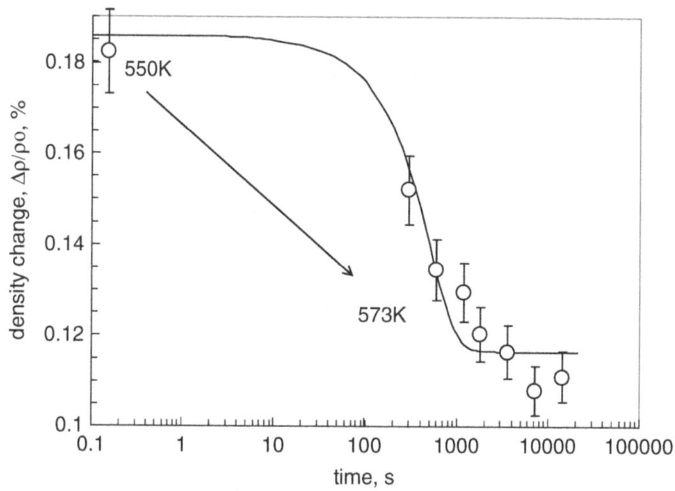

Fig. 10.3 Reversible relative density changes $\Delta\rho/\rho_o$ (in %) of a bulk amorphous $Pd_{40}Cu_{30}Ni_{10}P_{20}$ sample firstly annealed at 550 K, and then the temperature was changed to 573 K. The *solid line* represents the FVM model fitting curve

Table 10.2 FVM parameters obtained by the regression analysis of the time dependence of $\frac{\Delta\rho(t)}{\rho_o}$ experimental data from the isothermal annealing at 573 K, after the preliminary annealing at 550 K. $Pd_{40}Cu_{30}Ni_{10}P_{20}$ bulk glassy alloy

FVM parameter, dimension	Value	FVM parameter, dimension	Value
ν_r, 1/s	3.12×10^{19}	a_l^o, 1/K	1.86×10^{-5}
Q_r, kJ/mol	130.45	V_{mol}, m^3	7.96×10^{-6}
$c_{f,o}$	2.57×10^{-12}	υ^*, m^3	1.2×10^{-29}
B, K	6324	N_A, 1/mol	6.02×10^{23}
T_o, K	307	γ	0.8

At 550 K the density increases and saturates after approximately 4×10^4 s, whereas at the investigated two lower temperatures, the density change increases monotonously and exhibits no tendency for saturation within the given annealing time. At annealing temperatures of 573 and 593 K, a plateau value is reached for annealing times of about 2.5×10^3 and 6×10^2 s. The plateau data could be assigned to the quasi-equilibrium density value at these temperatures. In order to prove this assumption and the reversibility of the density changes at these temperatures, the annealing temperature of the sample used for density measurements at 550 K was changed from 550 to 573 K. The resulting density changes as a function of the annealing time at 573 K are shown in Fig. 10.3. The solid line represents the FVM model fitting curve (see Table 10.2), obtained by fitting the experimental data based on Eq. (10.10) under the new annealing conditions. The starting $c_{f,o}$ value of 1.38×10^{-9} of the as-quenched samples (see Table 10.1) has changed to 2.57×10^{-12} and almost coincides with the final value of the defect concentration

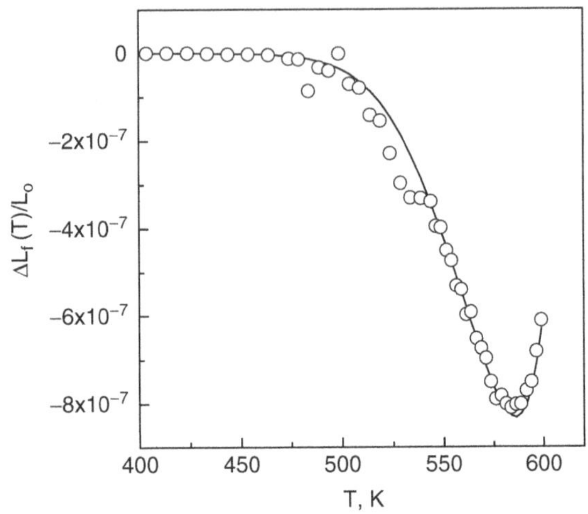

Fig. 10.4 Temperature dependence of the $\Delta L_f(T)/L_o$ deviations at a heating rate of 10 K/min. The experimental data are presented with symbols (o), whereas the regression curve, describing them, is presented with *solid line*

reached after the last anneal of the sample at 550 K, namely, 1.58×10^{-12}. The coincidence between the experimentally measured and the calculated density changes under these conditions is relatively good. The quasi-equilibrium defect concentration at 573 K is reached after an annealing time of approximately 1×10^4 s.

Figure 10.4 shows the temperature dependence of the deviations of thermal expansion $\Delta L_f(T)/L_o$ with respect to the extrapolated low-temperature linear temperature dependence $L_O(T)$ at a heating rate of 10 K/min (Table 10.3). The thermal elongation experimental data are presented with symbols, whereas the regression curve, describing them, is presented with solid line and was calculated by using regression analysis on the basis of Eq. (10.3). The parameters as obtained by the fitting are given in Table 10.3.

Free Volume Model Calculations

The experimental $\Delta\rho(t)/\rho_o$ data (see Fig. 10.2) were subjected to regression analysis based on Eq. (10.10). The FVM fitting model parameters are given in Table 10.1. The model parameters were obtained by the simultaneous regression analysis of all experimental data for the different annealing temperatures. The calculated results describe the given data almost perfectly, with the exception of the data obtained during the annealing at 593 K.

The reversible relative density changes, shown in Fig. 10.3, were fitted based on Eq. (10.10) by varying only the value of c_f. As a starting $c_{f,o}$ value, 1.58×10^{-12} was used, which represents the defect concentration reached after the preceding anneal of the sample at 550 K. Except the $c_{f,o}$ all other parameters were kept constant and equal to the fitting model parameters obtained from fitting of $\Delta\rho(T)/\rho_o$ set of data shown in Fig. 10.2. By plotting the data in Fig. 10.3, it is taken into account that instead of zero, the initial $\Delta\rho(t)/\rho_o$ value (0.183 %) should correspond to the final density change value reached after the final annealing at

Table 10.3 FVM parameters obtained by the regression analysis of the temperature dependence of $\Delta L_f\,(T)$ of $Pd_{40}Cu_{30}Ni_{10}P_{20}$ bulk glassy alloy at a heating rate of 10 K/min

FVM parameter, dimension	Value	FVM parameter, dimension	Value
ν_r, 1/s	3.12×10^{19}	a_l^o, 1/K	1.86×10^{-5}
Q_r, kJ/mol	130.45	V_{mol}, m^3	7.96×10^{-6}
$c_{f,o}$	2.5×10^{-9}	υ^*, m^3	1.2×10^{-29}
T_o, K	307	γ	0.8
B, K	6324	N_A, 1/mol	6.02×10^{23}
L_o, m	2.5×10^{-3}	T_B, K	323

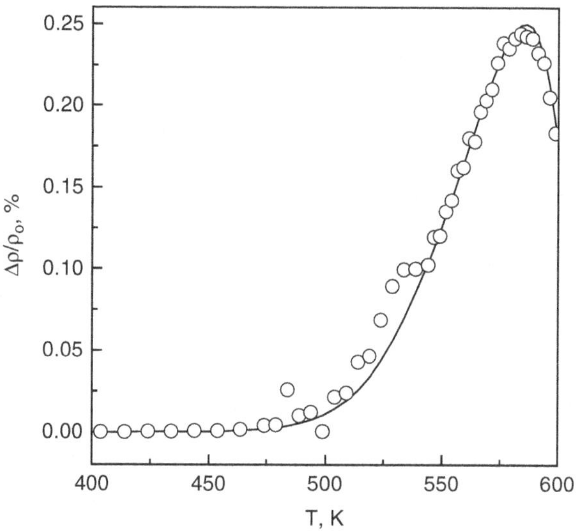

Fig. 10.5 Temperature dependence of the estimated density changes $\Delta\rho/\rho_o$ of $Pd_{40}Cu_{30}Ni_{10}P_{20}$ glassy alloy at a heating rate of 10 K/min, calculated using Eq. (10.9), inserting the experimentally obtained ΔL_f data from the thermal expansion experiments (o – symbols) and the fitting values from Table 10.3 (*solid line*)

550 K. The agreement between measured and calculated density changes is relatively good. The solid line in Fig. 10.4 reflects the fitting obtained from the nonlinear multiparameter regression analysis of the experimental $\Delta L_f(T)/L_o$ data based on Eq. (10.3). The parameters are given in Table 10.3.

Figure 10.5 shows the temperature dependence of the expected density changes $\Delta\rho(T)/\rho_o$ of amorphous $Pd_{40}Cu_{30}Ni_{10}P_{20}$ at a heating rate of 10 K/min. The "experimental" data are calculated from Eq. (10.9) inserting the ΔL_f data from the thermal expansion experiments.

The temperature dependence of the quasi-equilibrium defect concentration of the bulk amorphous alloy studied can be calculated based on (▶ Eq. 5.3) (▶ Chap. 5) with the values for B and T_o taken from Table 10.1. The value for the initial defect concentration (1.38×10^{-9}) (see Table 10.1) corresponds to a quasi-equilibrium defect concentration at approximately 610 K (see Fig. 10.6). It means that during the rapid quenching process, the undercooled melt behaves ergodic down to this temperature and changes then to a non-ergodic amorphous phase.

The heating rate dependence of the glass transition temperature T_g of the bulk amorphous $Pd_{40}Cu_{30}Ni_{10}P_{20}$ alloy studied was obtained using data from DSC,

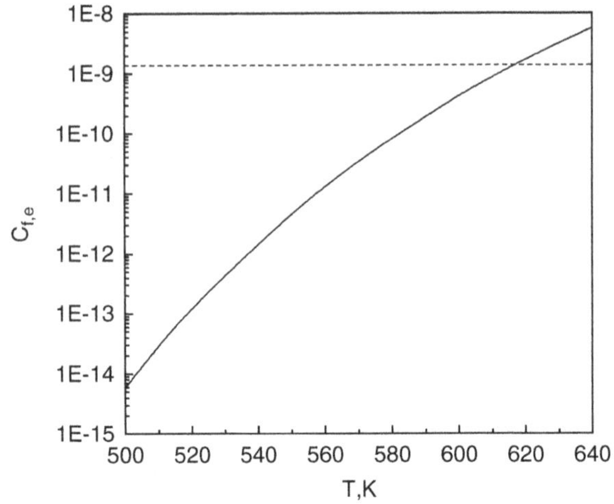

Fig. 10.6 Temperature dependence of the quasi-equilibrium defect concentration $c_{f,e}$ of the bulk amorphous $Pd_{40}Cu_{30}Ni_{10}P_{20}$ alloy studied. As is seen, the initial defect concentration $c_{f,o}$ of 1.38×10^{-9} (*dashed horizontal line*) corresponds to the quasi-equilibrium defect concentration at approximately 610 K

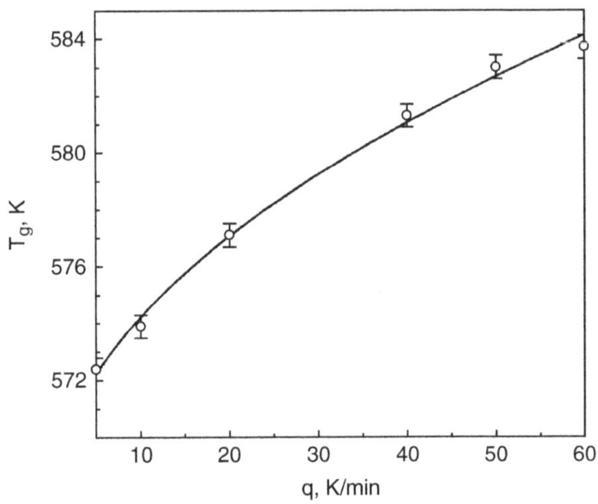

Fig. 10.7 Heating rate dependence of the glass transition temperature T_g of the bulk amorphous $Pd_{40}Cu_{30}Ni_{10}P_{20}$ alloy studied

Pyris-1, Perkin-Elmer measurements and is shown in Fig. 10.7. This dependence can be described empirically by the equation

$$T_g(q) = 567.42 + 2.157q^{1/2}, \tag{10.11}$$

with a correlation coefficient of 0.996. The cooling rate at which 610 K appears to be the glass transition temperature was calculated to be 6.4 K/s, using Eq. (10.11). This is a quite realistic cooling rate value which is also in acceptable agreement with the critical cooling rate of 2 K/s for bulk amorphous $Pd_{40}Cu_{30}Ni_{10}P_{20}$ alloy, as experimentally determined by Inoue [14].

Except for the atomic attempt frequency $\nu_r(3.12 \times 1019\ s^{-1})$, which should be of Debye frequency ($\approx 10^{13}\ s^{-1}$) order of magnitude, all other FVM parameters, determined in this study, are physically significant and reasonable. It should be mentioned that the attempt frequency is very near and in agreement with the attempt frequency determined by Tsao and Spaepen [15], Duine et al. [16], and Berry [17]. A possible explanation for the very high value of ν_r is given in ▶ Chap. 8 of the present book. Additionally, one has to study the considerations of R. L. McGreevy et al. [18], P. A. Duine et al. [19], I. R. Lu et al. [20], and Seeger [21], discussing this problem.

Conclusions

- The determination of length and density changes, due to relaxation annealing under nonisothermal and isothermal conditions, is a useful method for studying the structural relaxation of amorphous solids.
- The free volume model provides a simple and predictive model description for structural relaxation phenomena of amorphous solids below and close to the glass transition temperature.

References

1. Cohen MH, Turnbull D (1959) J Chem Phys 31:1164
2. Turnbull D, Cohen MH (1970) J Chem Phys 52:3038
3. Cargill GS III (1975) In: Ehrenreich H, Seitz F, Turnbull D (eds) Solid state physics, vol 30. Academic, New York, p 227
4. Turnbull D (1977) Scr Metall 11:1131
5. Gaskell PH (1981) Acta Metall 29:1203
6. Konczos G, Sas B (1987) Key Eng Mater 13–15:19
7. Sinning HR, Leonardsson L, Cahn RW (1984/1985). Int J Rapid Solidif 1:175
8. Pratten NA (1981) J Mater Sci 16:1737
9. Cahn RW et al (1984) In: Kear BH (ed) Rapidly solidified metastable materials. Amsterdam, North-Holland, p 241
10. Girt E, Kursumovic A, Mihac T (1980) J Phys E 13:898
11. Dietz G, Hueller K (1982) J Non-Cryst Solids 47:377
12. Russew K, Sommer F (2003) Length and density changes of amorphous $Pd_{40}Cu_{30}Ni_{10}P_{20}$ alloys due to structural relaxation. J Non-Cryst Solids 319(3):289
13. Russew K, Stojanova L, Sommer F (2000) Preparation of bulk and ribbon-like amorphous metallic alloys and study of their structure and rheological properties. PART I: preparation and structural study. J Mater Sci Technol 8(1):25
14. Inoue A (1998) Bulk amorphous alloys, preparation and fundamental characteristics, vol 4, Materials science foundations. Trans Tech Publications, Uetikon-Zuerich
15. Tsao SS, Spaepen F (1985) Acta Metall 33:881
16. Duine PA, Sietsma J, van den Beukel A (1992) Acta Metall Mater 40:743
17. Berry BS (1982) Scr Metall 16:1407
18. McGreevy RL, Pusztai L (1988) Mol Simul 1:359
19. Duine PA, Sietsma J, van den Beukel A (1995). In: Fisher DJ (ed) Defect and diffusion forum 117–118, p 91
20. Lu I-R, Goerler P, Fecht H-J, Willnecker R (2000) J Non-Cryst Solids 274:294
21. Seeger A, Schumacher D, Schilling W, Diehl J (eds) (1970) Vacancies and interstitials in metals. Amsterdam, North Holland

Crystallization Kinetics of Bulk Amorphous Metallic Alloys and Its Relation to the Forgoing Relaxation Processes

11

Abstract

The crystallization behavior of undercooled liquid $Pd_{40}Cu_{30}Ni_{10}P_{20}$ metallic alloy has been studied with the aid of differential scanning calorimetry, optical metallography, and electron microscopy methods. The crystallization occurs as a single-stage process starting preferably on the surface of casting voids in the sample volume and on the outer sample cutting planes. The linear crystal growth velocity of crystallizing cells was measured at several constant temperatures in the range of 590–620 K. It is constant at constant temperatures and its temperature dependence is in good agreement with the theoretical concepts of Uhlmann calculated using the temperature dependence of the experimentally determined viscosity temperature dependence of the alloy studied. Differential scanning calorimetry study of isothermal overall volume transformation kinetics of samples pre-annealed at 600 K for 600 s has been also carried out at several constant temperatures. The experimental data are interpreted on the basis of the equation of Kolmogorov–Avrami–Johnson–Mehl for the case of 3D growth of fixed number of preexisting nuclei. The possibility to re-vitrify the crystallized samples by their melting in differential scanning calorimeter, followed by quick cooling with cooling rate of 5 K s^{-1} is shown. The volume transformation kinetics of re-vitrified samples has been also studied.

Keywords

Amorphous • Bulk glassy alloys • $Pd_{40}Cu_{30}Ni_{10}P_{20}$ • Isothermal crystallization • Viscosity • Re-vitrification • Free Volume Model • Overall volume transformation kinetics

As it was discussed in ▶ Chap. 3 of the present book, amorphous metallic alloys are most frequently produced via rapid quenching from the melt in a way avoiding the beginning of their crystallization. This is schematically shown as a

K. Russew, L. Stojanova, *Glassy Metals*,
DOI 10.1007/978-3-662-47882-0_11

time–temperature–transformation (TTT) curve in ▶ Fig. 3.1, ▶ Chap. 3. This TTT diagram, however, does not provide any information about the kind, the number, and the kinetics of the arising of the crystalline phases. It is known also that practically all glassy alloys crystallize during heat treatment, forming one or more metastable crystalline phases. By prolonged isothermal or nonisothermal annealing, these crystalline phases additionally endured transformation into stable phases according to the respective equilibrium phase diagrams. The crystallization of metallic glasses is most frequently traced out with the aid of differential scanning calorimeter. The observed DSC crystallization traces are as a rule exothermic and consist of a single or several overlapping peaks. The released heat of crystallization is about one half of the enthalpy of melting of respective crystalline counterpart of the glassy alloy. The calorimetric research methods for crystallization investigation are convenient for quantitative tracing of the overall crystallization kinetics. It deals with the determination of the crystallized volume fraction as a function of the time and temperature of heating under isothermal or nonisothermal conditions, respectively. Often this is considerably complicated by the overlapping of the crystallization peaks. For obtaining of a reliable information about the morphology and nature of crystallization products, one needs the application of metallographic and X-ray methods. The application of metallography is complicated by the circumstance that the samples are ribbonlike, with thickness of several micrometers. These hindrances can be easily overcome by using of bulk amorphous metallic samples of high thermal stability.

The crystallization mechanism of glassy alloys includes nucleation of crystalline particles and their subsequent growth, fully analogous to the crystallization of undercooled metallic melts below and close to the melting point of the glass-forming alloy studied. These processes proceed so fast that it is extremely difficult to follow their kinetics quantitatively. On the contrary, due to the much bigger undercooling realized by the production of glassy metals by their rapid solidification, the nucleation and growth of crystalline particles are orders of magnitude slower as compared to the lightly undercooled metallic melts. This circumstance provides the unique possibility for quantitative study of crystallization processes in an isotropic amorphous phase. The crystallization of amorphous materials, including metallic glasses, has been an extensive field of theoretical and experimental studies and is an object of numerous scientific articles, books, and scientific medleys, for example, [1, 2]. This is why in this book section, some authors' own results on bulk glassy samples will be represented and discussed. These studies deal with the overall crystallization kinetics of thin platelike, and the kinetics of crystal growth in bulk glassy alloys, respectively. Calorimetric and metallographic methods were used.

The aim of this book chapter is to present the study on the crystal growth and overall volume crystallization behavior of bulk $Pd_{40}Cu_{30}Ni_{10}P_{20}$ amorphous metallic alloy during different isothermal annealing treatments by using differential scanning calorimetry (DSC), optical metallography, and electron microscopy methods.

Experimental Approach

The bulk amorphous $Pd_{40}Cu_{30}Ni_{10}P_{20}$ alloy was obtained in the form of rods of 5 mm diameter by rapid water quenching of its melt as described in [3]. Approximately 1.5 and 0.15 mm thick plates for measurements of crystal growth rate and for study of overall volume crystallization kinetics, respectively, were cut from these rods. The amorphous structure of the as-quenched samples was proved by X-ray diffraction measurements. All as-quenched samples used in this study were X-ray amorphous – see Fig. 11.1.

The isothermal crystal growth rate of bulk amorphous $Pd_{40}Cu_{30}Ni_{10}P_{20}$ alloy was studied by using differential scanning calorimeter Perkin-Elmer DSC 2C as a very precise annealing furnace at 590, 600, 605, 610, 615, and 620 K, respectively. The DSC studies of as-quenched samples at a heating rate of 20 K/min have shown that a single-stage crystallization process takes place – Fig. 11.2. The isothermal annealing times varied from 10 to 540 min, depending on the temperature of annealing, respectively. After every annealing the heat-treated samples were rapidly cooled down to room temperature with a cooling rate of 320 K/min and subjected to metallographic observations and measurements. The microstructure of the as-quenched and heat-treated samples was developed with the aid of *Aqua regia* at room temperature [4].

The overall volume crystallization kinetics was studied with the aid of a Perkin-Elmer DSC 2C at 610, 615, 620, and 630 K, respectively. Two series of experiments were carried out. The first one included as-quenched samples, pre-annealed at 600 K for 600 s, followed by isothermal annealing at the temperatures stated above. For the second series of experiments, the samples from the first series were heated up to the temperature of 900 K (approximately 100 K higher than the melting point, $T_m \approx 804$ K, of the alloy studied) and tempered in molten state for 300 s in order to ensure better homogeneity of the melt. After that, using cooling rate of 320 K/min (≈5 K/s), the samples were cooled down in order to obtain them in an amorphous state again. Figs. 11.3, 11.4 shows a typical DSC peak of melting of a sample after its crystallization.

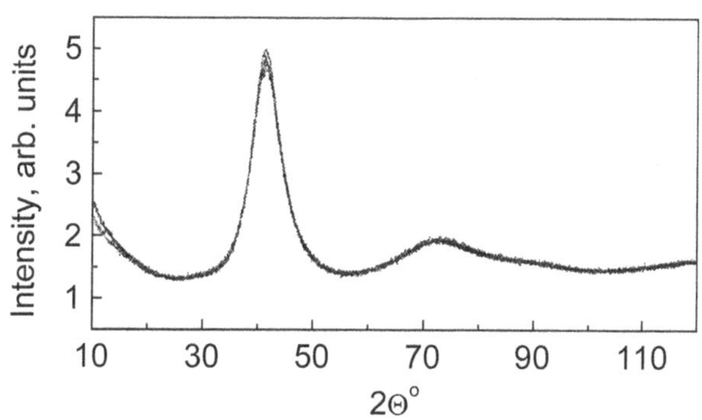

Fig. 11.1 All as-quenched samples used in this study were X-ray amorphous without any traces of crystallinity

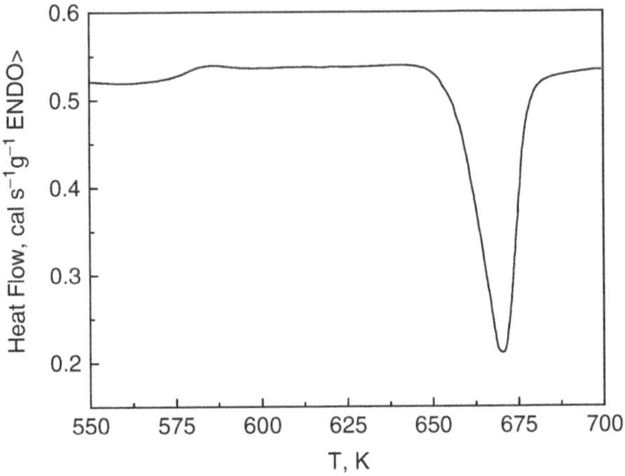

Fig. 11.2 DSC crystallization trace of bulk amorphous $Pd_{40}Cu_{30}Ni_{10}P_{20}$ at a heating rate of 20 K/min. Both broad glass transition region and the single-stage crystallization peak are clearly seen

Figure 11.4 presents a DSC crystallization trace of crystallized, remelted and re-vitrified again in the DSC unit bulk amorphous $Pd_{40}Cu_{30}Ni_{10}P_{20}$ alloy. The similarity of DSC traces in Figs. 11.2 and 11.4 is obvious with the exception that the DSC trace of the re-vitrified sample is shifted to lower temperatures. The possible reasons of the differences in both DSC traces are discussed in [3]. It can be concluded that the samples treated in the above-described way were really successfully vitrified, at least to a great extent, at a cooling rate not higher than approximately 5 K/s. We estimated the critical cooling rate for a $Pd_{40}Cu_{30}Ni_{10}P_{20}$ bulk amorphous alloy as 6.4 K/s [5], while the experimentally estimated critical cooling rate by A. Inoue [6], has been reported to be 2 K/s. In our opinion the range of the estimated critical cooling rate (2–6 K/s) is quite narrow and its accuracy is acceptable.

Crystal Growth and Viscous Flow Behavior of Amorphous $Pd_{40}Cu_{30}Ni_{10}P_{20}$ Metallic Alloy

Crystal Growth Rate Kinetics

Figure 11.5 shows a scanning electron microscope (SEM) image of the microstructure of an as-quenched amorphous sample. The only visible structural elements are several casting voids. Figure 11.6a–d shows the microstructure development in the alloy studied after annealing at 610 K for 30, 60, 75, and 152 min, respectively. The most active crystallization sites are the surfaces of casting voids as well as the planes of diamond saw sample cutting – see Fig. 11.6a, the SEM image in Fig. 11.7a, as well as the optical microscope image in Fig. 11.7b.

This is obviously due to mechanical stresses introduced by cutting and to the increased surface free energy due to the curvature of the casting void surfaces. Although this could be theoretically expected, it was observed experimentally by us for the first time for this bulk amorphous metallic alloy.

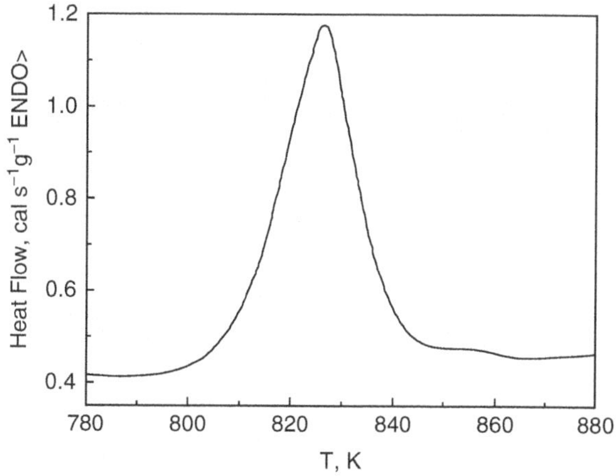

Fig. 11.3 DSC trace of melting of crystalline $Pd_{40}Cu_{30}Ni_{10}P_{20}$ alloy at a heating rate of 40 K/min

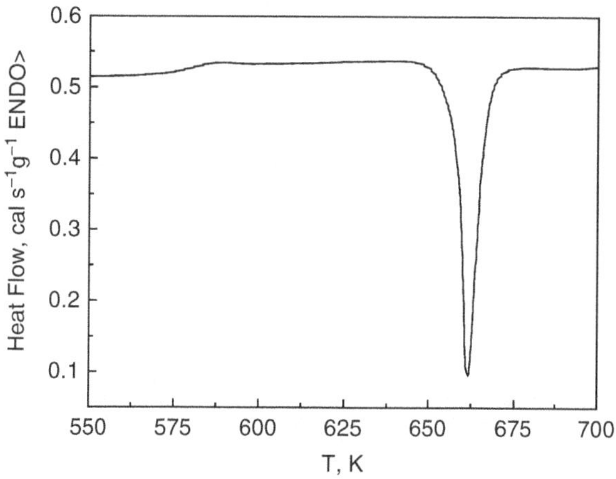

Fig. 11.4 DSC crystallization trace of crystallized, remelted, and re-vitrified again in the DSC unit bulk amorphous $Pd_{40}Cu_{30}Ni_{10}P_{20}$ alloy at a heating rate of 20 K/min

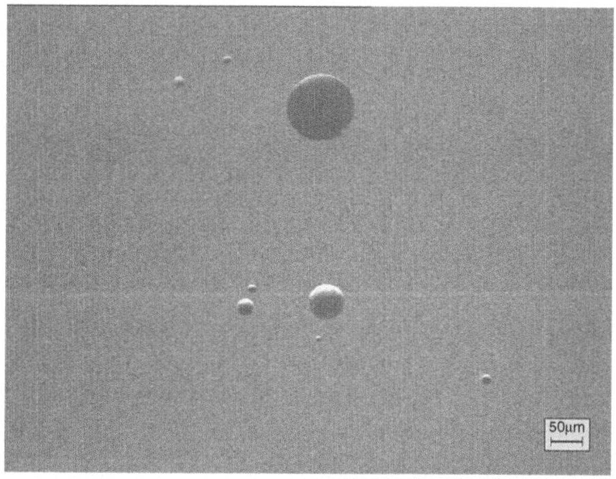

Fig. 11.5 Scanning electron microscope (SEM) image of the microstructure of an as-quenched bulk amorphous $Pd_{40}Cu_{30}Ni_{10}P_{20}$ alloy sample. The only visible structural elements are several casting voids

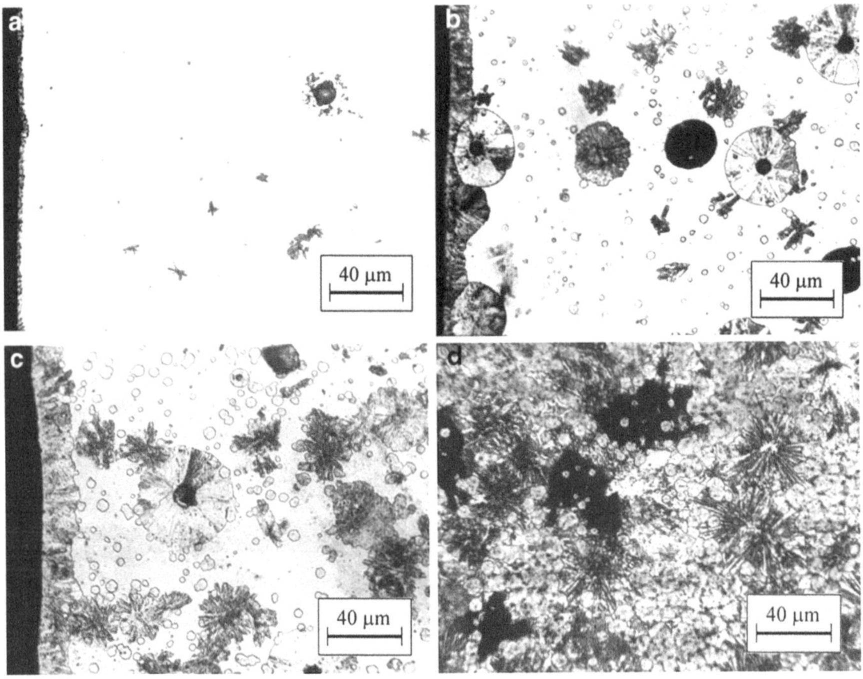

Fig. 11.6 (**a**–**d**): Microstructure development by the crystallization of bulk amorphous $Pd_{40}Cu_{30}Ni_{10}P_{20}$ alloy after annealing at 610 K for 30 min (**a**), 60 min (**b**), 75 min (**c**), and 152 min (**d**), respectively

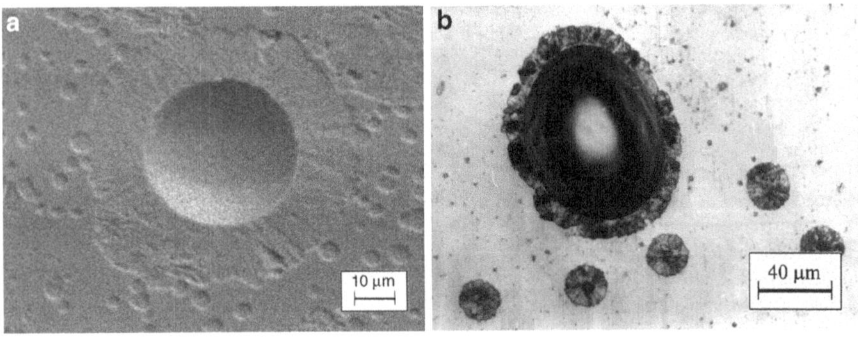

Fig. 11.7 SEM image (**a**) and optical microscope image (**b**) of casting voids acting as preferential sites of crystallization start in bulk amorphous $Pd_{40}Cu_{30}Ni_{10}P_{20}$ alloy. Crystallization at 610 K for 60 min

The fine microstructure of crystalline regions, according to Gutzow and Schmelzer [2, 7] and Kim et al. [7], is composed typically from a mixture of Pd_2Ni_2P, $Pd_{15}P_2$, Ni_3P, and Cu_3P in the form of a cellular dendritic precipitation with a eutectic-like background – see Fig. 11.8. In order to determine

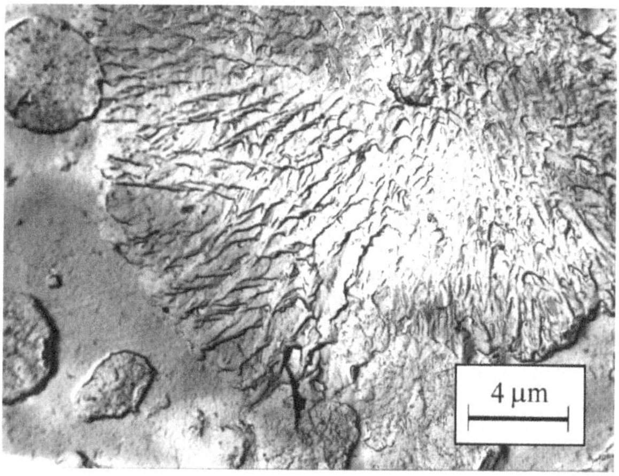

Fig. 11.8 Electron microscope image of the fine eutectic-like microstructure of crystalline regions of bulk amorphous $Pd_{40}Cu_{30}Ni_{10}P_{20}$ alloy at annealing temperature of 610 K. Carbon replica techniques

experimentally the growth rate of crystalline regions, the maximum crystalline layer thickness at the planes of cutting, as well as of the crystalline layers around the casting voids as a function of annealing time, was measured with the aid of metallographic observations at the different constant temperatures of annealing. In all cases three different samples were evaluated in order to obtain a better statistics of growth rates determined. The dependence of the maximal thickness of crystalline layers on the annealing time in bulk amorphous $Pd_{40}Cu_{30}Ni_{10}P_{20}$ alloy samples is linear. The crystallization growth rate for the different constant temperatures of annealing was determined from the slopes of the straight lines obtained. Their temperature dependence is shown in Fig. 11.9, together with the temperature dependence of previously obtained at much lower undercooling data of J.-H. Kim et al. [7]. These authors have used an in situ observation of a confocal scanning laser microscope combined with an infrared image furnace and high-speed video camera.

In order to reach nucleation site saturation, the samples were firstly annealed for 10 min at 600 K. The metallographic observations have shown that practically no visible crystallization could be detected after this heat pretreatment. The structure of a pre-annealed for 10 min at 600 K $Pd_{40}Cu_{30}Ni_{10}P_{20}$ bulk amorphous alloy sample, followed by isothermal crystallization at 610 K for 60 min, is shown in Fig. 11.10 as a SEM image. Almost perfect, impinging-each-other spherical layers are 3D growing into the amorphous matrix.

Relation Between Crystal Growth and Viscous Flow Behavior

Figure 11.9 shows the temperature dependence of the experimentally determined linear crystal growth rates in bulk amorphous $Pd_{40}Cu_{30}Ni_{10}P_{20}$ alloy together with the temperature dependence of crystal growth rates G calculated according to the Uhlmann's approximation [8]:

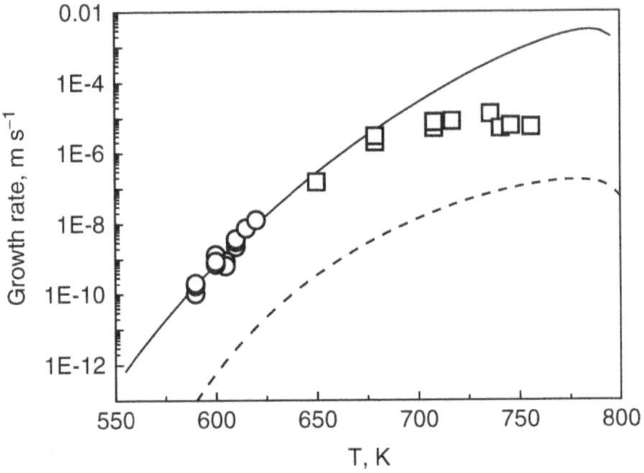

Fig. 11.9 Temperature dependence of linear crystallization growth rate of bulk amorphous $Pd_{40}Cu_{30}Ni_{10}P_{20}$ alloy; o, our data; □, data of J.- H. Kim et al. [7]. The *solid line* presents the fitting of our data according to the Free Volume Model – see [10–12] – by using the approximation of Uhlmann [8], inserting in it the temperature dependence of the quasiequilibrium viscosity η, experimentally determined by K. Russew et al. [13]. The *broken line* presents the calculated temperature dependence of crystallization growth rate according to Uhlmann, inserting in it the viscosity temperature dependence proposed by N. Nishiyama et al. [14]

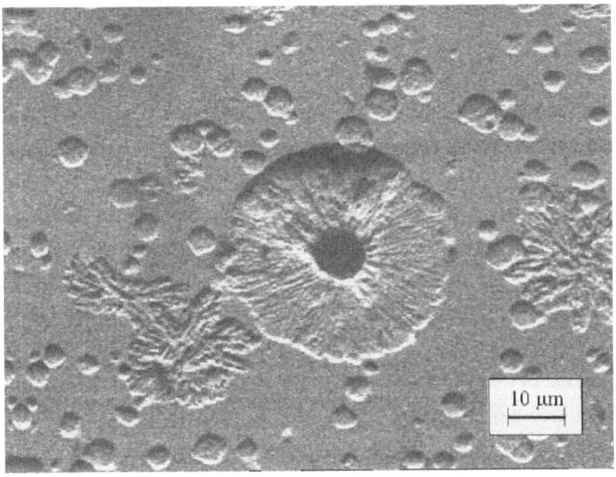

Fig. 11.10 SEM image of the of structure of bulk amorphous $Pd_{40}Cu_{30}Ni_{10}P_{20}$ alloy pre-annealed 600 s at 600 K and after that subjected to crystallization for 60 min at 610 K

$$G = \frac{fkT}{3\pi a_o \eta}\left[1 - \exp\left(\frac{-\Delta G_m}{RT}\right)\right], \tag{11.1}$$

where η denotes the coefficient of viscous flow responsible for atomic motion required for growth, a_o is the mean atomic diameter, ΔG_m is the molar free energy change, and f is the fraction of sites at the interface where crystal growth occurs. For rough interfaces $f = 1$ was chosen, while ΔG_m was accepted [8], to be equal to $\frac{\Delta H_m (T_m - T)T}{T_m^2}$. Here, ΔH_m, T_m, and R denote the molar heat of fusion, the melting temperature, and the ideal gas constant, respectively. The thermodynamic parameters in Eq. (11.1), similarly to J.-H. Kim et al. [7], were taken to be equal to the data of Nishiyama et al. [9], namely, $a_o = 2.55\times10^{-10}$ m, $\Delta H_m = 4{,}840$ J/mol, and $T_m = 804$ K, while for the temperature dependence of the coefficient of viscous flow η, the Free Volume Model approximation [10–12], for previously experimentally obtained by us [13], viscosity data was used.

Here the authors have to remind to the studious reader that the temperature dependence of viscosity η both under isothermal and nonisothermal constant heating rate conditions can be presented as [12]:

$$\eta = \eta_0 T \exp\left(\frac{Q_\eta}{RT}\right)\left(\frac{1}{c_f}\right), \qquad (11.2)$$

with η_o as a pre-exponential factor, Q_η the activation energy of viscous flow, and c_f the concentration of flow defects. The parameter c_f presents the probability of an atom to be encircled with free volume υ^* necessary for an atomic transport to take place at its close proximity. It relates to the so-called reduced free volume $x = \gamma\upsilon^*/\upsilon_f$ according to

$$c_f = \exp\left(-\frac{1}{x}\right), \qquad (11.3)$$

where γ is an overlap factor between 0.5 and 1, and υ_f is the mean free volume per atom. The equilibrium concentration of flow defects is given by

$$c_{f,eq}(T) = \exp\left(-\frac{B}{T - T_0}\right), \qquad (11.4)$$

where B and T_o are two model parameters. The temperature dependence of the quasiequilibrium viscosity η_{eq} arises from the combination of Eqs. (11.2 and 11.4):

$$\eta_{eq} = \eta_o T \exp\left(\frac{Q_\eta}{RT}\right)\exp\left(\frac{B}{T - T_o}\right). \qquad (11.5)$$

In the case that the glassy alloy has not yet reached the (quasi)equilibrium state, c_f changes with temperature following bimolecular kinetics [12]. The proper equation describing the change of c_f with temperature at a constant heating rate q in the temperature range under and around the glass transition temperature T_g is given by

$$c_{f,high}^{-1}(T,q) = \left(c_{f,0}^{-1} - \int_{T_0}^{T} Q(T')\exp\left(-\int_{T_0}^{T'} P(T'')dT''\right)dT'\right) \times \exp\left(\int_{T_0}^{T} P(T')dT'\right) \qquad (11.6)$$

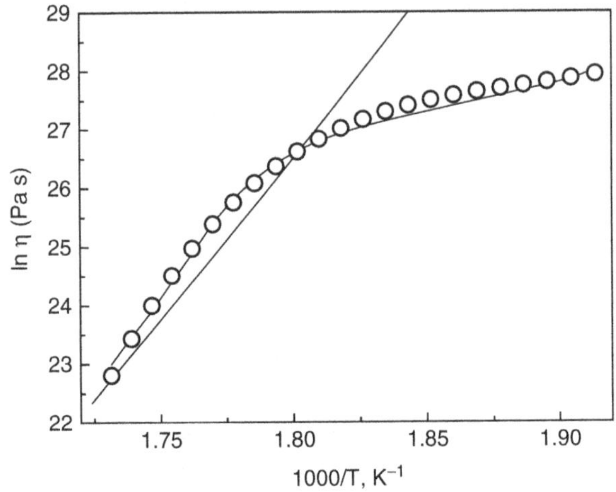

Fig. 11.11 Temperature dependence of the viscosity η of ribbonlike amorphous $Pd_{40}Cu_{30}Ni_{10}P_{20}$ alloy as obtained by K. Russew et al. [13]. The steeper line represents the temperature dependence of the quasiequilibrium viscosity

with $P(T) = -\frac{\nu_r}{q}\exp\left(-\frac{Q_r}{RT} - \frac{B}{T-T_0}\right)$, and $Q(T) = -\frac{\nu_r}{q}\exp\left(-\frac{Q_r}{RT}\right)$, and ν_r is the attempt frequency, and Q_r is the activation energy of relaxation.

In a previous study on the rheological properties of ribbonlike and bulk amorphous $Pd_{40}Cu_{30}Ni_{10}P_{20}$ alloy [13], we have studied experimentally the viscous flow behavior of ribbonlike samples of amorphous $Pd_{40}Cu_{30}Ni_{10}P_{20}$ alloy. The temperature dependence of η is shown in Fig. 11.11.

The fitting parameters of interest, determined by fitting the experimental viscosity data shown in Fig. 11.11 according to the combination of Eqs. (11.2 and 11.6), were $\eta_o = 2.7\times10^{-22}$ Pa s/K, $Q_\eta = 208.5$ kJ/mol, $T_o = 307$ K, and $B = 6{,}153$ K. The curve of model calculations according to the combination of Eqs. (11.1 and 11.5) gives a best fit of measured by the authors' crystal growth rates by using the following values of FVM model parameters: $\eta_o = 2.6\times10^{-22}$ Pa s/K, $Q_\eta = 190$ kJ/mol, $T_o = 307$ K, and $B = 6{,}100$ K. With the exception of the parameters Q_η and B, all other parameters coincide practically perfectly. This is not the case however, when one uses the temperature dependence of viscosity proposed by N. Nishiyama et al. [14], namely,

$$\eta = 9.34 \times 10^{-3}\exp\left(\frac{4135}{T - 447}\right). \tag{11.7}$$

This is shown in Fig. 11.9 where the model calculations on the basis of Uhlmann's approximation [8], performed by J.-H. Kim et al. [7], are shown with a broken line. The experimental crystal growth rates are at least two orders of magnitude higher than the model predictions by using the viscosity temperature dependence of N. Nishiyama et al. [14].

Overall Volume Transformation Kinetics of Bulk Amorphous $Pd_{40}Cu_{30}Ni_{10}P_{20}$ Alloy

It was shown (see Figs. 11.6 and 11.7) that almost perfect, impinging-each-other spherical layers are 3D growing into the amorphous matrix. The volume fraction of the flat crystalline layers growing from the sample planes of cutting is negligible as compared to the volume fraction of spherical crystalline agglomerates growing in the sample volume and could be not taken into account. Such kind of overall volume transformation kinetics can be very well described by the well-known equation of Kolmogorov [15], Avrami [16], and Johnson–Mehl [17] (KAJM):

$$\zeta = 1 - \exp(-V_e), \tag{11.8}$$

where $\xi = V_{cr}/V_o$ is the volume fraction of crystallized regions, V_{cr} is the overall volume of crystallized regions, V_o is the total volume of the sample, and V_e is the so-called extended volume of growing crystalline particles per unit volume of the matrix. By the calculation of the extended volume, the impingement of crystalline particles is not taken into account, imaging that they are able to grow through each other, similarly to the wave circles caused by the water droplets onto the lake surface during rain. For the case of spherical crystalline particles which number per unit volume is $N(t)$:

$$V_e = \frac{4}{3}\pi N(t)\, G^3(T)\, t^3, \tag{11.9}$$

where t is the time of crystallization and $G(T)$ is the crystal growth rate at a temperature T. At constant temperatures $G(T)$ is constant also. The temperature dependence of $G(T)$ has a complicated nature – see Eq. (11.1), but at very high undercooling it can be empirically presented by an Arrhenius-type equation also [18], namely,

$$G(T) = G_o \exp\left(-\frac{Q_{eff}}{RT}\right), \tag{11.10}$$

where G_o is a pre-exponential factor and Q_{eff} is the effective activation energy of crystallization. By nucleation site saturation $N(t)$ has a constant value N and Eq. (11.8) can be rewritten as:

$$\xi = 1 - \exp\left(-K(T)t^3\right), \tag{11.11}$$

where $K(T) = NG^3(T)$. Its temperature dependence can be presented as:

$$\ln(K(T)) = \ln\left(NG_o^3\right) - 3\frac{Q_{eff}}{RT}. \tag{11.12}$$

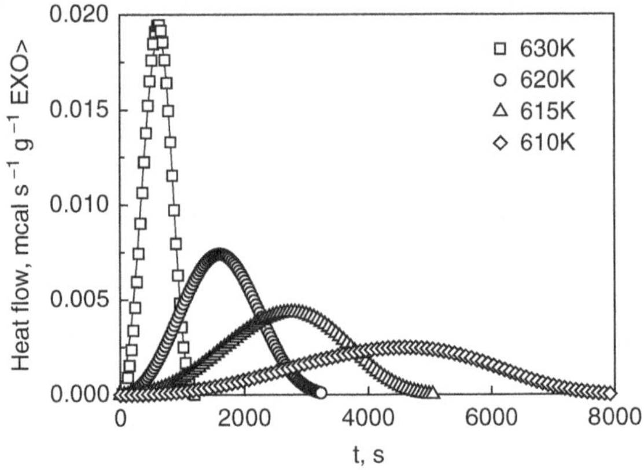

Fig. 11.12 The rate of enthalpy uptake due to crystallization at 610, 615, 620, and 630 K, respectively, of as-quenched amorphous Pd40Cu30Ni10P20 alloy samples, pre-annealed for 600 s at 600 K. The symbols denote the enthalpy uptake as measured and the solid lines as fitted according to the combination of the Eq. (11.8) and Eq. (11.9)

Figure 11.12 presents the rate of enthalpy uptake due to crystallization at 610, 615, 620, and 630 K, respectively, of as-quenched samples, pre-annealed for 600 s at 600 K for ensuring a nucleation site saturation. The symbols denote the enthalpy uptake as measured and the solid lines as fitted according to the combination of Eqs. (11.8 and 11.9). The fitting procedure in this case actually proceeds as one which multiplies the experimentally obtained heat release $\Delta H(t)$ to the ratio $\xi_{fit}(t)/\xi(t)_{exp}$, where $\xi_{fit}(t)$ is the crystallized volume fraction as fitted, and $\xi(t)_{exp}$ is the as experimentally determined crystallized volume fraction. The crystallized volume fraction $\xi(t)$ as a function of annealing time t of as-quenched samples, pre-annealed for 600 s at 600 K, and of samples, which were remelted in DSC and vitrified again, is shown in Figs. 11.14 and 11.15, respectively. The $\xi(t)$ values were calculated as the ratio $\Delta H(t)/\Delta H_{tot}$, where $\Delta H(t)$ is the heat release from the crystallization beginning up to crystallization time t, and ΔH_{tot} is the total enthalpy (heat release) of crystallization. Figure 11.13 presents the same as in Fig. 11.12 for samples, which were remelted in DSC and vitrified again as described above.

The experimentally obtained $\xi(t)$ values for both kinds of samples are presented with symbols in Figs. 11.14 and 11.15, respectively. The solid lines present the fitting results according to Eq. (11.11).

The fitting parameters as well as other important experimental features of the alloy studied are presented in Table 11.1.

The temperature dependence of $K(T)$, see Eq. (11.12), for both kinds of samples is shown in Fig. 11.16. The effective crystallization energies calculated from the slopes of the straight lines in Fig. 11.16 for overall crystallization are 320 and 395 kJ/mol for as-quenched and pre-annealed samples and remelted and re-vitrified samples, respectively. The number of fixed crystallization sites in the first case – approximately $2\times10^{14}/m^3$ – is by an order of magnitude lower than in the case of remelted and again re-vitrified samples. This is not surprising and has to be expected, taking into account that during the remelting and repeated

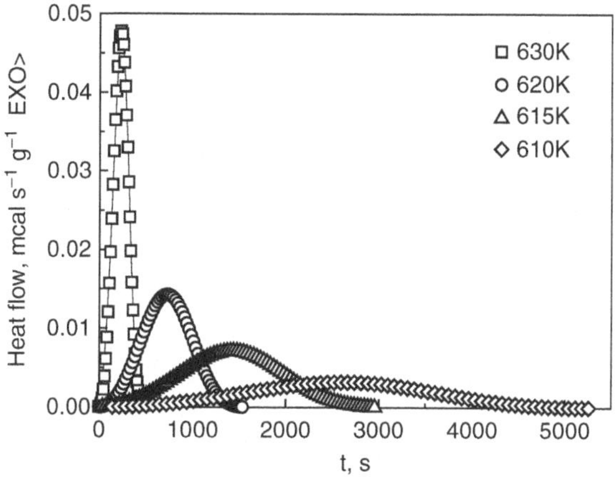

Fig. 11.13 The same as in Fig. 11.12 but for remelted and re-vitrified bulk amorphous $Pd_{40}Cu_{30}Ni_{10}P_{20}$ alloy samples pre-annealed 600 s at 600 K

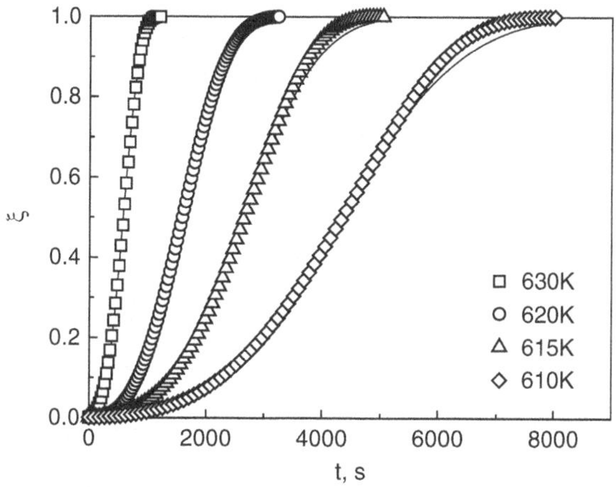

Fig. 11.14 The volume fraction $\xi(t)$ crystallized as a function of annealing time t of as-quenched bulk amorphous $Pd_{40}Cu_{30}Ni_{10}P_{20}$ alloy pre-annealed 600 s at 600 K and after that subjected to crystallization at 610, 615, 620, and 630 K, respectively

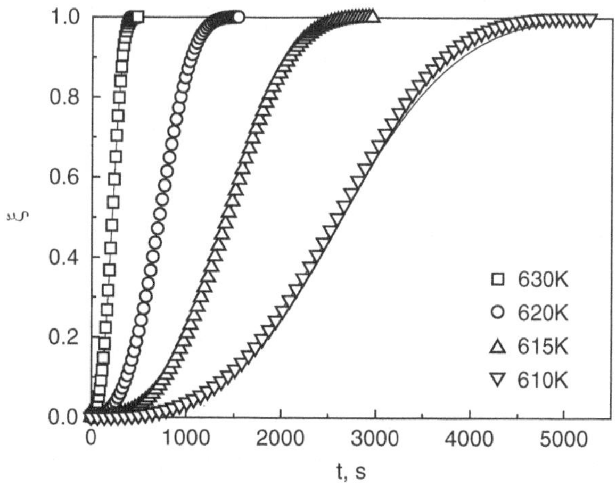

Fig. 11.15 The same as in Fig. 11.14 but for remelted and re-vitrified bulk amorphous $Pd_{40}Cu_{30}Ni_{10}P_{20}$ alloy samples pre-annealed 600 s at 600 K

Table 11.1 Characteristic overall crystallization features of the samples studied

T (K)	ΔH_{cr} (as q), kJ/mol	ΔH_{cr} (re-vitr.), kJ/mol	$K(T)$ (as q), $1/s^3$	$K(T)$ (re-vitr.), $1/s^3$	G(as fit.), m/s	N (as q.) ~ $K\,G^{-3}$, $1/m^3$	N (re-vitr.) ~ $K\,G^{-3}$, $1/m^3$
610	3.13	2.3	8.69×10^{-12}	3.9×10^{-11}	2.73×10^{-9}	4.27×10^{14}	1.92×10^{15}
615	3.14	2.9	3.76×10^{-11}	2.35×10^{-10}	5.14×10^{-9}	2.77×10^{14}	1.73×10^{15}
620	3.43	2.93	1.75×10^{-10}	1.91×10^{-9}	9.52×10^{-9}	2.03×10^{14}	2.21×10^{15}
630	3.3	2.99	3.7×10^{-9}	6.66×10^{-8}	3.11×10^{-8}	1.23×10^{14}	2.21×10^{15}

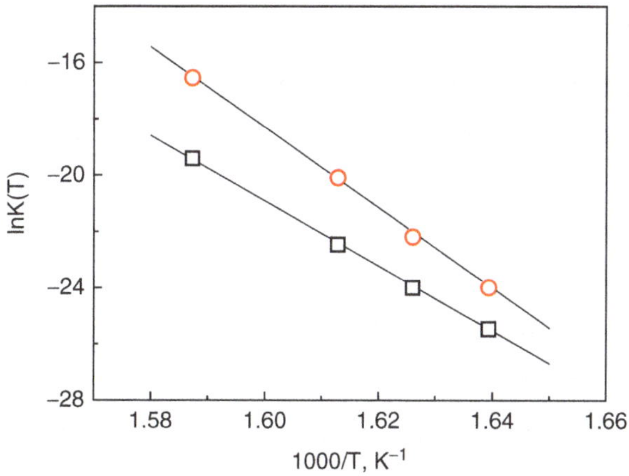

Fig. 11.16 The temperature dependence of the parameter *K(T)* for both kinds of amorphous $Pd_{40}Cu_{30}Ni_{10}P_{20}$ alloy as defined by Eq. (11.12)

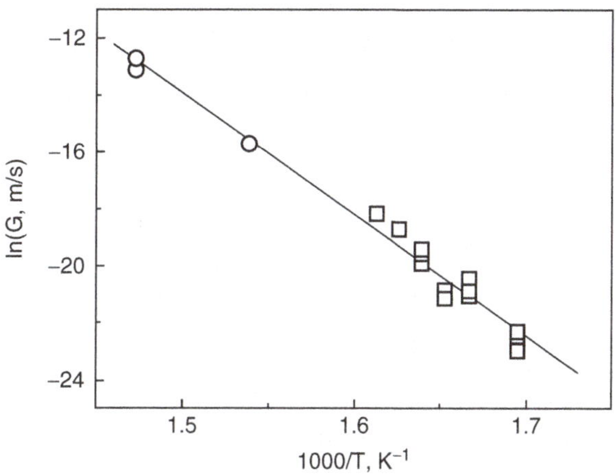

Fig. 11.17 The empirical Arrhenius temperature dependence of linear crystal growth rate in bulk amorphous $Pd_{40}Cu_{30}Ni_{10}P_{20}$ alloy measured in this study (□), together with some data of J.-H. Kim et al (○) [7], obtained by these authors at lower u

vitrification of the sample in the calorimeter, inevitable additional infection with nucleation sites of the melt takes place, due to inclusion of newly created oxide particles.

Figure 11.17 presents the empirical Arrhenius-type temperature dependence of crystal growth rate measured in this study together with some data of J.-H. Kim et al. [7], obtained by these authors at lower undercooling values and close to the temperature range of our measurements. The solid line represents the fitting straight line according to the logarithmic form of Eq. (11.10). The effective activation energy Q_{eff}, obtained from the slope of the straight line in Fig. 11.17, is 355 kJ/mol, e.g., in fairly good agreement with activation energies for overall volume crystallization.

Conclusions

Crystal Growth Rate Kinetics

- The most active sites for nucleation and crystal growth of $Pd_{40}Cu_{30}Ni_{10}P_{20}$ bulk amorphous alloy are the surfaces of those embedded into the volume of the alloy casting voids and the planes of diamond saw cutting of the samples.
- The crystal growth rates of $Pd_{40}Cu_{30}Ni_{10}P_{20}$ bulk amorphous alloy in the temperature range 590–620 K varied from 1×10^{-10} to 1×10^{-8} m s^{-1}. Their temperature dependence can be joined in a natural way to the previously obtained data of J.-H. Kim et al. [7], at much lower undercooling values.
- The temperature dependence of crystal growth rate of $Pd_{40}Cu_{30}Ni_{10}P_{20}$ bulk amorphous alloy can be successfully described by using Uhlmann's approximation on the basis of viscosity temperature dependence derived experimentally by K. Russew et al. [13].

Overall Volume Crystallization Kinetics

- The isothermal overall volume crystallization kinetics of pre-annealed at 600 K for 600 s samples of $Pd_{40}Cu_{30}Ni_{10}P_{20}$ bulk amorphous alloy takes place via 3D crystal growth of fixed constant number of crystallization sites with a constant growth rate.
- The equation of Kolmogorov–Avrami–Johnson–Mehl describes very well the isothermal overall volume crystallization of $Pd_{40}Cu_{30}Ni_{10}P_{20}$ bulk amorphous alloy after saturation of nucleation sites via pre-annealing at 600 K for 600 s. The number of fixed sites is about 10^{14} 1/m^3 for as-quenched, pre-annealed samples and 10^{15} 1/m^3 for remelted, re-vitrified, pre-annealed samples, respectively.
- The effective activation energy for isothermal overall crystallization is about 320 and 390 kJ/mol for as-quenched pre-annealed and remelted, vitrified, and pre-annealed samples, respectively. These values are in agreement with the effective activation energy for crystal growth – about 350 kJ/mol.
- The combination of optical and EM studies with DSC measurements of the enthalpy uptake during isothermal crystallization of amorphous alloys provides a very suitable tool for understanding the crystallization phenomena.

References

1. Scott MG (1983) In: Luborsky FE (ed) Amorphous metallic alloys. Butterworth and Co (Publishers) LTD, London, p 144
2. Gutzow I, Schmelzer J (1995) The vitreous state/thermodynamics. Structure, Rheology and Crystallization. Springer Verlag, Berlin, Heidelberg
3. Russew K, Stojanova L, Kovacheva R (2006) Isothermal crystallization behavior of undercooled liquid Pd40Cu30Ni10P20 in terms of crystal growth, overall volume

transformation kinetics and their relation to the viscosity temperature dependence. Int J Mater Res 97:409–417
4. Beckert M, Klemm H (1966) Handbuch der metallographischen Ätzverfahren. Deutscher Verlag für Grundstoffindustrie, Leipzig
5. Russew K, Sommer F (2003) Length and density changes of amorphous $Pd_{40}Cu_{30}Ni_{10}P_{20}$ alloy due to structural relaxation. J Non-Cryst Solids 319:289–296
6. Inoue A (1998) Bulk amorphous alloys, preparation and fundamental characteristics, materials science foundations, vol 4. Trans Tech Publications, Zürich
7. Kim J-H, Kim S-G, Inoue A (2001) Acta Mater 49:615
8. Uhlmann DR (1972) J Non-Cryst Solids 22:357
9. Nishiyama N, Inoue A (1995) Mater Trans JIM 47(5):1487
10. Chen HS, Turnbull D (1968) J Chem Phys 48:2560
11. Tsao SS, Spaepen F (1985) Acta Metall 33:881
12. Russew K, Zappel BJ, Sommer F (1995) Nonisothermal viscous flow behaviour of $Pd_{40}Ni_{40}P_{20}$ glassy alloy considered as a free volume related phenomenon. Scr Metall Mater 32:271–276
13. Russew K, Stojanova L, Sommer F (2000) Preparation of bulk and ribbon-like amorphous metallic alloys and study of their structure and rheological properties. PART II: study of rheological properties. J Mater Sci Technol 8(1):34–42
14. Nishiyama N, Inoue A (1999) Mater Trans JIM 40(1):64
15. Kolmogorov AN (1937) Izv. Acad. Nauk SSSR. Ser. Fiz 3:355. (in Russian)
16. Avrami M (1939) J Chem Phys 7:1109
17. Johnson WA, Mehl RF (1939) Trans Am Inst Min Metall Eng 135:1
18. Köster U, Herold U (1981) Glassy metals I. In: Günterodt H-J, Beck H (eds) Topics in applied physics, vol 46. Springer Verlag, New York

12 Mechanical Properties of Amorphous Metallic Alloys and Their Relation to the Viscosity Characteristic Parameters

Abstract

The unified theoretical equation of B. Yang et al. on the relation between the mechanical strength and viscous flow behavior of glassy metals is presented and discussed; it is shown that the equation of Yang can be successfully used for estimation of fraction strength of various ribbonlike glassy metals, provided that their viscous flow behavior is precisely determined under nonisothermal experimental conditions. The viscous flow and mechanical properties of Co_{100-x} Zr_x and $Cu_{100-x}Zr_x$ glassy alloys are presented and discussed, together with their comparison with the data existing in the bibliography. A relation between the estimated fraction strength of glassy alloys and their Vickers microhardness is established.

Keywords

Glassy metals • Amorphous metallic alloys • Mechanical properties • Viscous flow • Equation of Yang • Tensile fracture strength • Vickers hardness

The present book deals mostly with the problems concerning the ribbonlike glassy metals produced via rapid solidification from the melt by CBMS and/or PFC methods. Due to this circumstance, the book is scarce of more extended information about the principles of development and selection of suitable alloy compositions for the production of bulk amorphous metallic alloys with sufficiently low critical melt cooling rate, allowing to produce samples of thickness of several centimeters. Such kind of information could be found in the book of A. Inoue [1]. Bulk amorphous alloys can be successfully forged to machine parts of arbitrary complicated form via plastic deformation of their deeply undercooled melts [2]. Their mechanical properties are very high [1, 3], close to the theoretically expected fraction strength of solid bodies with defect-free structure. Examples of such bulk amorphous alloys are the alloys on the basis of Fe, Co, Cu, Ce, etc., elements [3–6]. The mechanical strength of bulk amorphous alloys varies within 300–5000 MPa.

K. Russew, L. Stojanova, *Glassy Metals*,
DOI 10.1007/978-3-662-47882-0_12

Unified Equation of Yang

While the mechanical strength of crystalline materials is determined by the movement of dislocations, in the case of bulk glassy metals, it is directly related to the physical properties determined by the cohesive interaction energy of atoms, i.e., glass transition temperature T_g, the Young modulus, and the coefficient of thermal expansion. B. Yang et al. [7] have carried out a thorough analysis on the available experimental data about the fraction strength σ_f at ambient temperature of a series of bulk amorphous alloys as function of their glass transition temperature. The result is shown in Fig. 12.1. As is seen, regardless of data scattering, an obvious trend of increase of fraction strength σ_f along with increase of the glass transition temperature exists. Proceeding from the assumption that the fraction strength of amorphous metallic alloys originates from the appearance of shear bands and their quick propagation, F. Spaepen was the first [8] who has used the free volume model (FVM) for modeling the plastic deformation of glassy alloys. The shear band deformation can be presented as is shown in Fig. 12.2.

The shear band of transformational cutting is presented as a thin disk of two atomic layers in contact around a site of free volume. By sliding the upper atomic layer, the arising plastic deformation is presented as $\Upsilon_o \approx 1$ having in mind that all atoms in the upper layer cover a distance of one atomic diameter. A lot of similarity exists between the physical processes of plastic deformation and glass transition in bodies of amorphous structure. Both processes are enhanced by collective atomic movements that require considerable energy consumption to overcome the cohesion forces binding together the atoms and result in an increase of the free volume. In a way similar to the glass transition process, the plastic deformation "scarifies" the structure and makes it to "forget" the structural changes taking place due to the relaxation processes. As a result the structure recovers into its as-quenched state. Both processes are characterized with an existence of a threshold value of external impact under which the atomic mobility suddenly increases to macroscopic dimensions. It is established that the packing density for reaching the threshold value of free volume at the glass transition temperature is 85 % calculated on a basis of FVM for dense random packing. This threshold value coincides with the situation observed by the deformation shear band shown in Fig. 12.2, consisting of approximately five atoms and one vacation site of free volume of monatomic size, i.e., $5/(5 + 1) \sim 85$ %. While the thermal energy needed for the glass transition process is uniformly distributed in the sample volume subjected to heating, the mechanical energy of cutting deformation is fixed exceptionally into the shear bands. Taking into account, however, that both processes have to overcome one and the same binding force between the atoms and to generate one and the same product – looser atomic structure – with similar amount of structural defects (free volume), the energy density per unit volume should depend only on the initial and final states of the system and should be equal for both processes. It leads to [7]

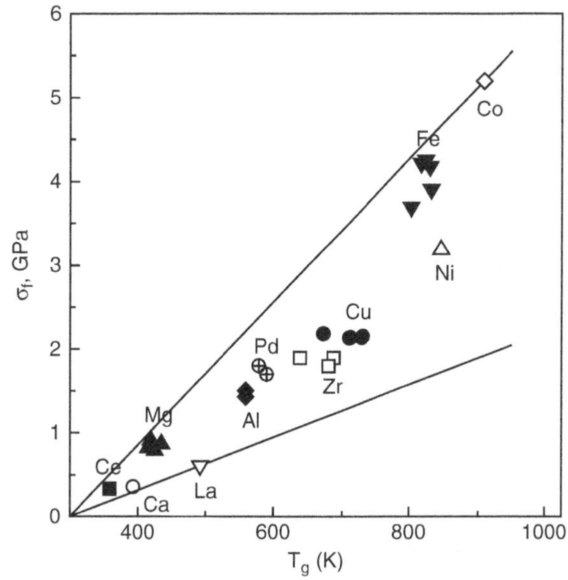

Fig. 12.1 Dependence of fraction strength σ_f on the glass transition temperature T_g of a series of bulk glassy alloys according to Yang et al. [7]

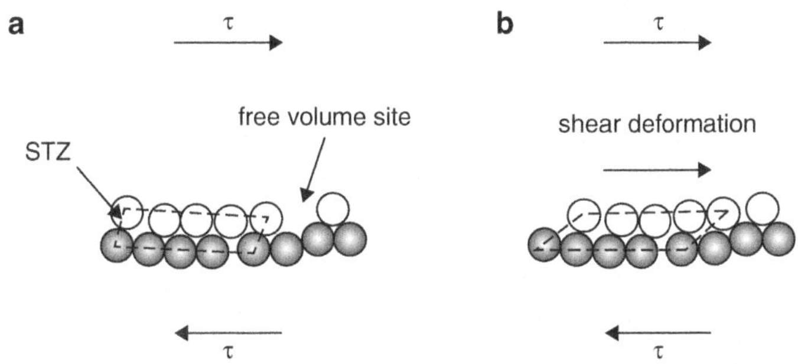

Fig. 12.2 Scheme of deformation shear band according to the FVM as presented by Yang et al. [7]

$$\tau_\gamma \gamma_o = \int_{T_o}^{T_g} \rho C_p dT, \qquad (12.1)$$

where τ_γ is the maximal stress of shear deformation yielding $\gamma_o = 1$ of the basic share unit, ρ is the density of the material, C_p is the molar specific heat, T_g is the glass transition temperature, and T_o is the ambient temperature. It is accepted that the processes are taking place under constant pressure.

Under monotonous loading, the creep strength equals to

$$\sigma_\gamma = 2\tau_\gamma. \qquad (12.2)$$

The material density equals to

$$\rho = \frac{\rho_o}{[1 + 3\alpha(T - T_o)]}, \tag{12.3}$$

where ρ_o is the density at ambient temperature and α is the coefficient of thermal expansion.

Further on B. Yang et al. [7] define the specific heat according to the Debye model and introduce it into Eq. (12.1). After a series of simplifications, one obtains for the fracture strength the simple equation:

$$\sigma_f = 55 \frac{\Delta T_g}{V_m}, \tag{12.4}$$

where σ_f is the fracture strength, ΔT_g is the temperature difference T_g–T_o, and V_m is the molar volume of the glassy alloy. Presenting the experimental data shown in Fig. 12.1 in the coordinates of Eq. (12.4), one obtains an ordered grouping of data in a linear dependence shown in Fig. 12.3.

It becomes obvious that the unified equation of Yang [7] is really valid for all glassy alloys and allows to predict their mechanical properties by known viscous flow features of the glassy materials.

Figure 12.4 shows a typical veinlike SEM image of $Co_{25}Zr_{75}$ glassy alloy fraction surface after tensile strength tests. This veinlike structure can be explained by the creation of shear bands during the tensile loading of the sample.

On the contrary to the crystalline materials, due to the lack of dislocation systems and grain boundaries, the plastic deformation of amorphous alloys is localized in such shear bands [9–11]. The fraction of the material begins via creation and propagation of cracks along the shear bands. The process is accompanied by intensive heat generation in shear bands. This is an adiabatic process during which rapid increase of the temperature up to the glass transition temperature within the shear bands is observed. After reaching T_g the fraction of the sample happens, due to the rapid decrease of viscosity. This is experimentally observed and monitored with the aid of an infrared camera by B. Yang et al. [12].

Viscous Flow and Mechanical Properties of $Co_{100\text{-}x}$ Zr_x and $Cu_{100\text{-}x}Zr_x$ Glassy Alloys

The viscous flow behavior of nine $Co_{100\text{-}x}$ Zr_x ($7 \leq x \leq 78.5$ at. %) and six $Cu_{100\text{-}x}Zr_x$ ($35.7 \leq x \leq 60$ at. %) ribbonlike glassy alloys, respectively, was described in ▶ Chap. 5 of the present book. All experimentally obtained viscosity temperature dependences were subjected to nonlinear multiparameter regression analysis. The values of the FVM regression parameters η_o, $c_{f,o}$, Q_η , ν_r, Q_r, B, and T_o are presented in ▶ Tables 6.3 and ▶ 6.6, respectively. Figure 12.5 shows as an example the viscosity temperature dependence of the glassy alloy $Co_{91}Zr_9$.

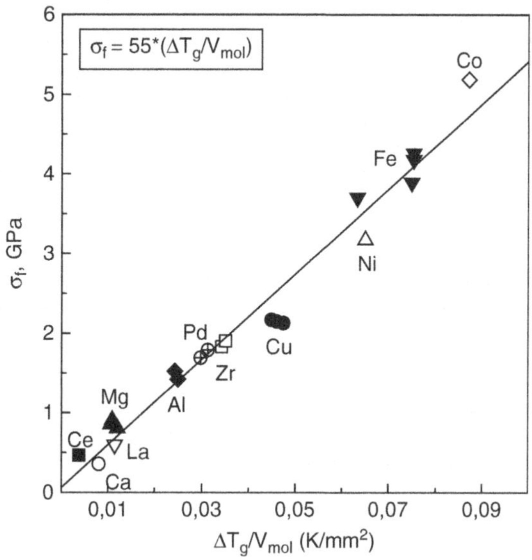

Fig. 12.3 Dependence of the fracture strength σ_f on the glass transition temperature T_g and the molar volume V_{mol} of a series of bulk glassy alloys on the basis of the elements Al, Ca, Ce, Co, Cu, Fe, La, Mg, Ni, Pd, and Zr according to the unified equation of Yang [7]

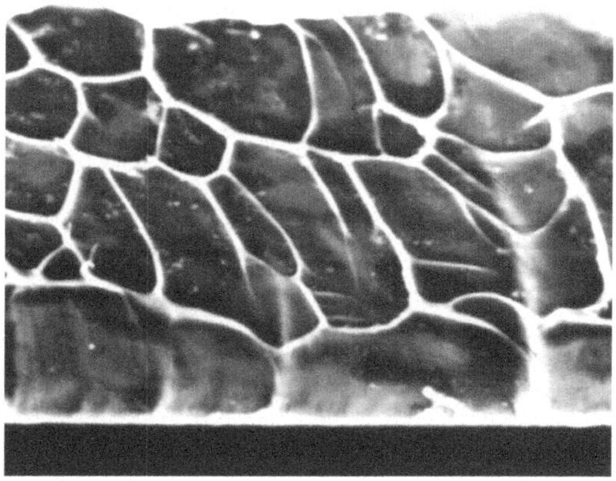

Fig. 12.4 SEM image of typical glassy alloy fraction surface after tensile strength test. $Co_{25}Zr_{75}$ glassy alloy

Figure 12.6 shows the collected experimental data, by W. H. Wang [13], for the fraction strength of 30 different bulk glassy alloys on the basis of Zr, Ti, Cu, Pd, Fe, etc., presented as the function of the glass transition temperature. As a rule an increase of the fraction strength along with the increase of the glass transition temperature T_g is observed. The curve through the experimental points is an empirical and is a guide for the eye. In the same figure (Tables 12.1 and 12.2), the author's data for the fracture strength of the glassy alloys Co_{100-x} Zr_x and $Cu_{100-x}Zr_x$ are plotted. They are calculated according to the unified equation of Yang – Eq. (12.4).

Figure 12.6 shows that a good conformity between the collected data and fraction strength data of W. H. Wang exists. The last ones were calculated according to the unified equation of Yang – Eq. (12.4). The above presentation affords the opportunity to predict the mechanical strength of ribbonlike glassy

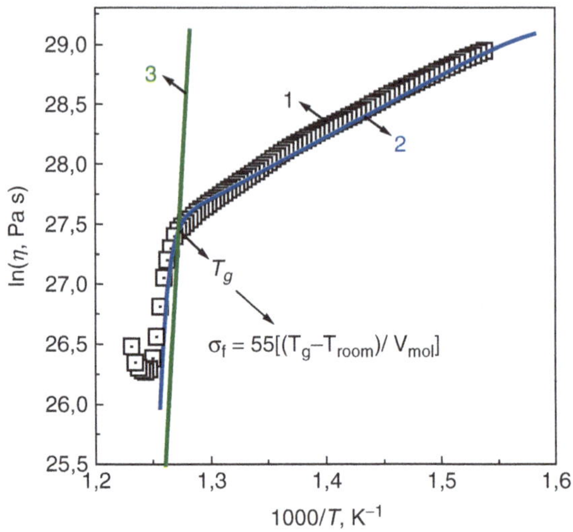

Fig. 12.5 Experimentally obtained at a heating rate of 20 K/min viscosity; temperature dependence of $Co_{91}Zr_9$ (curve 1) (□), fitted based on (▶ Eqs. 5.8) and (▶ 5.17) (▶ Chap. 5), temperature dependence of the nonequilibrium viscosity (curve 2), and temperature dependence of the quasi-equilibrium viscosity η_{eq} calculated based on (▶ Eq. 5.9) (curve 3). The point of intersection of curves 2 and 3 is considered as the glass transition temperature of the glassy alloy studied

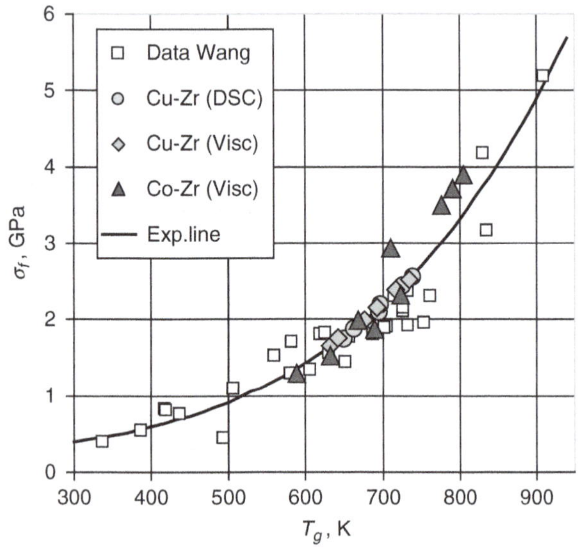

Fig. 12.6 Fracture strength of $Co_{(100\text{-}x)}Zr_x$ and $Cu_{(100\text{-}x)}Zr_x$, compared to a series of fraction strength data from Ref. [13]

alloys only on the basis of their viscosity temperature dependence and glass transition temperature.

Figure 12.7 shows the experimental data for Vickers microhardness as a function of the glass transition temperature T_g for the most glassy alloys cited in [11]. The same figure and Tables 12.1 and 12.2 contain also author's experimental data for the Vickers microhardness for $Co_{100\text{-}x}Zr_x$ and $Cu_{100\text{-}x}Zr_x$ glassy alloys. The microhardness data were obtained by using microhardness meter Micro-Duromat 4000 at 100 g and 10 s loading duration. A relatively good conformity of both types of data is observed again.

Table 12.1 Glass transition temperature T_g, fraction strength σ_f calculated based on Eq. (12.4) and experimentally obtained Vickers microhardness H_{mv} of $Co_{(100-x)}Zr_x$ glassy alloys

Parameters	Dimension	Zr, at.%								
		7.0	9.0	12.0	15.4	47.0	50.0	66.6	75.0	78.5
T_g	K	806	790	775	710	722	668	688	633	588
σ_f	GPa	3.91	3.71	3.49	2.93	2.32	1.98	1.87	1.53	1.30
H_{mv}	GPa	6.71	–	7.91	9.03	6.52	6.10	4.66	4.30	4.19

Table 12.2 Glass transition temperature T_g (defined by viscosity and DSC measurements), fraction strength σ_f, calculated with the aid of Eq. (12.4), and measured Vickers microhardness H_{mv} of $Cu_{(100\text{-}x)}Zr_x$ glassy alloys

		Zr, at.%					
Parameters	Dimension	35.7	38.2	44.0	50.0	54.0	60.0
$T_{g(Visc)}$	K	734	716	692	676	642	632
σ_f	GPa	2.53	2.38	2.16	1.99	1.76	1.65
$T_{g(DSC)}$	K	738	715	695	685	652	640
σ_f	GPa	2.55	2.43	2.18	2.09	1.87	1.74
H_{mv}	GPa	6.37	6.19	5.17	5.23	4.76	4.58

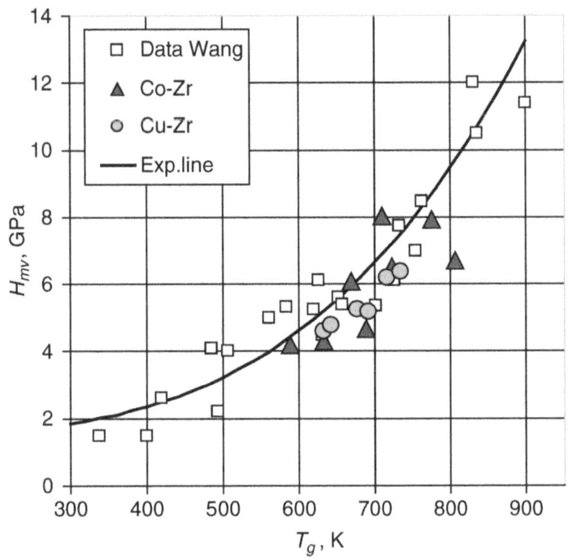

Fig. 12.7 Vickers microhardness of $Co_{(100\text{-}x)}Zr_x$ and $Cu_{(100\text{-}x)}Zr_x$ glassy alloys as function of T_g, together with the experimental data from Ref. [13]

Figure 12.8 presents the relation between the fracture strength calculated with the aid of unified equation of Yang (Eq. 12.4) and the measured Vickers microhardness of $Co_{100\text{-}x}Zr_x$ and $Cu_{100\text{-}x}Zr_x$ glassy alloys, compared with the data for numerous ribbonlike and bulk metallic glasses, experimentally determined by W.H. Wang [13] and H.S. Chen [14]. Irrespective of the considerable data scattering, the totality of data could be represented by a straight line of the type $H_{mv} = 2.93\,\sigma_f$.

It could be concluded that the unified equation of B. Yang et al. can be used as a useful tool for predicting the mechanical strength of glassy metals provided that their viscous flow behavior is studied as proposed by the authors of this book.

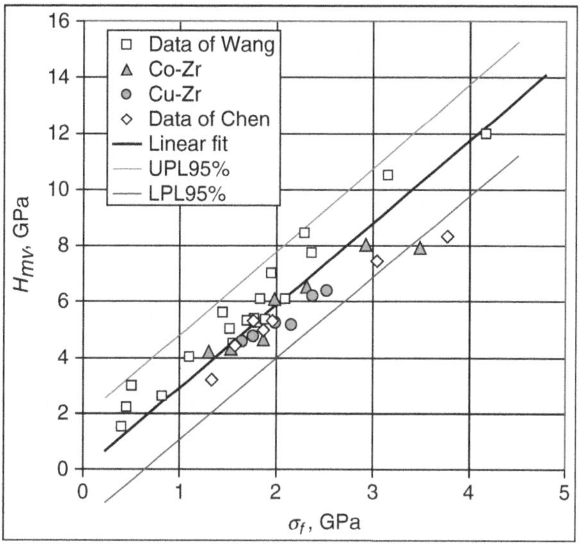

Fig. 12.8 Relation between Vickers microhardness and fracture strength of the $Co_{(100-x)}Zr_x$ and $Cu_{(100-x)}Zr_x$ glassy alloys studied. The values for a large number of ribbonlike and bulk metallic glasses are shown also, experimentally obtained by Wang [13], Chen [14] and Stojanova [15]

References

1. Inoue A (1998) Bulk amorphous alloys, preparation and fundamental characteristics, vol 4, Materials science foundations. Trans Tech Publications, Zürich
2. Peker A, Johnson WL (1993) Appl Phys Lett 63:2342
3. Inoue A, Zhang W, Zhang T, Kurosaka K (2001) Acta Materialia 49(14):2645–2652
4. Shen B, Inoue A (2004) Appl Phys Lett 85:4911
5. Inoue A, Shen BI et al (2004) Acta Mater 53:1631
6. Senkov ON et al (2002) Scr Mater 50:449
7. Yang B, Liu CT, Nieh TG (2006) Unified equation for the strength of bulk metallic glasses. Appl Phys Lett 88:221911
8. Spaepen F (1976) Acta Metall 25:407
9. Davis L et al (1994) Int J Rapid Solidification 8:73–131
10. Xi X et al (2005) Phys Rev Lett 94:125510
11. Xiao X et al (2003) Appl Phys Lett 330:243–247
12. Yang B et al (2005) Appl Phys Lett 86:141904
13. Wang WH (2005) Elastic moduli and behaviors of metallic glasses. J Non-Cryst Solids 351(16):1481–1485
14. Chen HS (1976) Mater Sci Eng 25:59–69
15. Stojanova L (2009) Relation between viscous flow and mechanical properties of Co-Zr and Cu-Zr amorphous alloys In: Proc of Nat Conf Metal Science and Novel Materials'2008, Sofia, p 96

Properties and Applications of Amorphous Metallic Alloys

13

Abstract
An extensive overview upon the application of amorphous metallic alloys in relation to their properties is presented. Their commercial production in the USA, Europe, and Russia with their trademarks and most important application properties in tabular form is presented and discussed.

Keywords
Glassy metals • Amorphous metallic alloys • Properties • Commercial production • Applications

The worldwide interest to the amorphous metallic alloys is provoked by the obvious advantages of their potential applications in various fields of techniques and engineering [1–3]. In order to make clear the prospects of their practical use, however, a relatively long period of time was needed for the performance of wide research programs for the study of the properties and the different aspects of the production of these new materials. The large-scale production and application of glassy metal scale in the USA, Japan, and Western Europe has started in 1978 and in Russia in 1981. In the rest of the European countries, these activities are still in the stage of scientific research projects.

The production and practical introduction of amorphous metallic alloys in the USA is mainly connected with more than ten companies, the most notable of which are Allied Chemical Corp. and General Electric. The companies General Electric, Westinghouse Electric, and Allied Chemical Corp. have carried out a program for the development and production of 1000 transformers with 25 and 50 kVA power with magnetic cores of amorphous metallic alloys, spending for this purpose more than hundreds of million dollars.

In Japan, several companies are dealing with R&D problems of glassy metals united in the corporation Nippon Amorphous Metals. The company Osaka Transformers has produced an experimental series of powerful electro-distribution transformers.

K. Russew, L. Stojanova, *Glassy Metals*,
DOI 10.1007/978-3-662-47882-0_13

In Germany, the leading company is Vacuumschmelze, Hanau. Its main investments are in the field of development and application of metallic glasses for electronic purposes.

For more effective research and development of metallic glasses, the leading companies in the USA and Japan have been united in the consortium "Nippon Amorphous Metals."

For the development of equipments for metallic glass production, the company Allied Chemical Corp. has spent more than 100 million dollars. These equipments function in a continuous regime of work. One of them possesses a productivity output of 1 t/h, while the smaller equipments possess an output of 40 kg/h. High-output equipment for production of soft magnetic amorphous ribbon of 175 mm width has been constructed, which price is more than 10 million dollars. It has been estimated that in 1987 the production of amorphous alloys in the USA has reached 3,000 t and in 2000, 20,000 t.

The consumers of glassy metals in Japan and Western Europe are companies producing articles for electronics, such as TDK, Matsushita Co., Sony, Sanyo, Grundig [4–6], etc.

All over the world, more than 40 kinds of amorphous metallic alloys are produced on a large scale for a variety of applications. In Japan, a national program named Shin Nittetsu Project has been realized. For this project, more than 50 million dollars have been spent. During the same period of time, in the USA 70–80 million dollars have been spent for the same purpose. The leading institutes are Electric Power Research Institute (EPRI) and the laboratories of Empire State Electric Energy Research Corp. (ESEERCO). Amorphous metallic ribbons, produced in the USA, are of 5–150 mm width and of 0.02–0.04 mm thickness, while in Japan these products are of 2–50 mm width by 0.02–0.04 mm thickness.

Amorphous Metallic Alloys Produced in the USA

The amorphous metallic alloys produced in the USA on an industrial scale by the company Allied Chemical Corp. are disseminated under the trademark "METGLAS." The most familiar of them, according to Refs. [7, 8], are:

METGLAS 2605 SC. This amorphous metallic alloy is on the basis of the elements Fe, B, Si, and C. Its soft magnetic properties are better than those of the traditional iron transformer sheet (trafo blech). This alloy finds an application for production of pulse transformers, power transformers, etc. requiring soft magnetic material of high saturation induction and rectangular hysteresis loop. This alloy offers a unique combination of high specific electrical resistance, high saturation induction, and extremely low alternating magnetization losses. This property combination makes this amorphous alloy very suitable for applications in a broad frequency range from low to very high frequencies, respectively. METGLAS 2605 SC is very suitable for pulse applications, which demand high degree of rectangularity of the hysteresis loop. With the aid of this amorphous alloy, high amplitudes of the induced magnetic flux with simultaneously very low losses by

Table 13.1 Magnetic and physical properties of the amorphous alloy METGLAS 2605 SC

	Symbol	As-quenched	After TMT or HT
Magnetic properties			
Saturation induction	B_s	1.57 T	1.61 T
Coercive force	H_c	0.08 Oe	0.04 Oe
Remanent induction	B_r	0.67 T	1.42 T
Induction at H = 80 A/m	B_{80}	0.80 T	1.54 T
Coefficient of magnetostriction	λ_s	30×10^{-6}	
Curie point	T_C	370°C	
Physical properties			
Density	δ	7.32 g/cm^3	
Degree of filling by lamination	S	≥75 %	
Temperature range of exploitation		−50/125 °C	
Specific electrical resistance	ρ	125 μOhm cm	

alternating magnetization can be achieved. Some magnetic and physical properties of METGLAS 2605 SC are given in Table 13.1.

METGLAS 2605 S-2. Due to the fact that this amorphous alloy shows extremely low losses by alternate magnetization at industrial frequencies, it is recommended for elaboration of magnetic core for power transformers. This alloy is based on the elements Fe, B, and Si. METGLAS 2605 S-2 is preferable to the amorphous alloy METGLAS 2605 SC at higher (up to 100 °C) temperatures, typical for the power transformers.

As compared to the METGLAS 2605 SC, this amorphous alloy possesses lower temperature coefficient of the saturation induction. The alternating magnetization losses at induction of 1.4 T by frequency of 60 Hz amount to 0.22 W/kg, that is, only 25 % of the losses typical for the classical Fe-Si transformer sheet iron of M-4 type.

METGLAS 2605 S-2 finds an application mainly in the field of distribution power transformers working at low industrial frequencies. Some of its magnetic and physical properties are given in Table 13.2.

METGLAS 2605 S-3. This amorphous alloy is also on the basis of the elements Fe, Si, and B. The boron content is however higher than the B content in the above discussed alloys. After a suitable thermo-magnetic treatment (TMT), it acquires flat B-H hysteresis loop, which results in very low alternating magnetization losses at high working frequencies (higher than 1 kHz). Due to the substantially higher saturation induction of this alloy as compared to the saturation induction of ferrite- and permalloy-type alloys, its application leads to substantial dimension decrease of the high-frequency devices. The alternating magnetization losses of this alloy are diminished by rectangular form of the alternating induction field. This circumstance makes it perfect for application in the transformers for pulse supply devices. Some magnetic and physical properties of METGLAS 2605 S-3 are presented in Table 13.3.

METGLAS 2605 CO. This amorphous alloy is on the basis of the elements Co, B, and Si. It possesses the highest saturation induction between all industrially

Table 13.2 Magnetic and physical properties of METGLAS 2605 S-2

	Symbol	As-quenched	After TMT or HT
Magnetic properties			
Saturation induction	B_s	1.52 T	1.56 T
Coercive force	H_c	0.17 Oe	0.03 Oe
Remanent induction	B_r	0.4 T	1.3 T
Induction at H = 80 A/m	B_{80}	0.50 T	1.5 T
Coercivity coefficient	λ_s	27×10^{-6}	
Curie point	T_c	415°C	
Physical properties			
Density	δ	7.18 g/cm^3	
Degree of filling by lamination	S	≥75 %	
Temperature range of exploitation		−50/150 °C	
Onset temperature of crystallization	T_x	550 °C	
Specific electrical resistance	ρ	130 μOhm cm	

Table 13.3 Magnetic and physical properties of METGLAS 2605 S-3

	Symbol	As-quenched	After TMT or HT
Magnetic properties			
Saturation induction	B_s	1.54 T	1.58 T
Coercive force	H_c	0.20 Oe	0.1 Oe
Remanent induction	B_r	0.3 T	0.4 T
Coefficient of magnetostriction	λ_s	27×10^{-6}	
Coercivity point	T_C	405°C	–
Physical properties			
Density	δ	7.28 g/cm^3	
Degree of filling by core lamination	S	≥75 %	
Onset temperature of crystallization	T_x	515 °C	
Temperature range of exploitation		−50/150 °C	
Specific electrical resistance	ρ	125 μOhm cm	

produced and commercially available amorphous alloys at the moment. This circumstance combined with the very low alternating magnetization losses makes it very suitable for production of transformers of low weight and big power. This alloy is especially suitable for application in small power supply transformer working by 400 Hz usual in aviation and rocket building. The alternating magnetization losses of METGLAS 2605 CO at 1.6 T induction are more than twice lower as compared to the losses usual for silicon iron sheet. After suitable heat treatment, the hysteresis loop of this amorphous alloy acquires a very high degree of rectangularity. Taking into account also the high specific electrical resistance of METGLAS 2605 CO. Its application in magnetic devices working in pulse mode with high amplitude of induction changes is advisable (Table 13.4).

METGLAS 2705 X. This amorphous alloy is of very high content of Co including also the elements Fe, B, Si, and Mo. It possesses zero coefficient of

Table 13.4 Magnetic and physical properties of METGLAS 2605 CO

	Symbol	As-quenched	After TMT or HT
Magnetic properties			
Saturation induction	B_s	1.75 T	1.8 T
Coercive force	H_c	0.08 Oe	0.05 Oe
Remanent induction	B_r	0.7 T	1.6 T
Induction at H = 80 A/m	B_{80}	0.80 T	1.65 T
Coefficient of magnetostriction	λ_s	35×10^{-6}	
Curie point	T_C	415°C	–
Physical properties			
Density	δ	7.56 g/cm^3	
Degree of filling at core lamination	S	≥75 %	
Onset temperature of crystallization	T_x	430 °C	
Exploitation temperature range		−50/125 °C	
Specific electro-resistance	ρ	130 μOhm/cm	

magnetostriction and high initial and maximal magnetic permeability. The saturation induction of this alloy is moderate together with low coercive force. After optimizing of the alloy properties with the aid of TMT, it becomes suitable for the production of magnetic heads for recording and reproduction of digital information for magnetic shields. Most typical features of METGLAS 2705 X are:

Saturation induction 1.0 T
Coercive force 0.018 Oe
Magnetic permeability at frequency of 1 kHz and 0.01 T, 12,000
Curie point 530 °C
Density 8.06 g/cm^3
Specific electrical resistance 115 μOhm/cm

The above discussed amorphous alloys do not exhaust all kinds of industrially produced amorphous metallic alloys of Allied Chemical Corp.

Amorphous Metallic Alloys Produced in Europe

The most widely known company in Western Europe producing industrial amount of amorphous metallic alloys is Vacuumschmelze GmbH & Co, KG, Hanau, FRG. Its production is oriented to articles of special magnetic applications. The trademark of this company is VITROVAK [9, 10]. The efforts of Vacuumschmelze, Hanau, in the field of glassy metals are first of all focused upon soft magnetic alloys as well as upon alloys with high mechanical properties for application as constructive materials. In order to achieve optimal soft magnetic properties, the amorphous alloys of suitable chemical composition are subjected to different kinds of thermomagnetic treatment (TMT). Depending on the type of TMT, the amorphous alloys

Table 13.5 Soft magnetic features of the VITROVAK series

Alloy	VITROVAK 7505	VITROVAK 4040	VITROVAK 6025
Metallic components	Iron	Iron. Nickel	Cobalt
Saturation induction. T	1.5	0.8	0.55
Curie point. °C	420	260	250
Coefficient of magnetostriction	$+30.10^{-6}$	$+8.10^{-6}$	$<0.3.10^{-6}$
Toroids of rectangular hysteresis loop (Z)			
Coercive force, mA/cm	$\leq$40	$\leq$10	$\leq$4
B_r/B ratio	0.80–0.90	0.80–0.90	0.80–0.90
Initial magnetic permeability μ_4(50 Hz)	$\geq$3,000	$\geq$20,000	$\geq$150,000
Maximal magnetic permeability. μ_{max} (50 Hz)	100,000	250,000	600,000
Toroids of extended (flat) hysteresis loop (F)			
Maximal ΔB . T		0.7	0.4
Maximal impulse magnetic permeability, μ_{pmax}		6,500	50,000
Alternating magnetization losses (20 kHz, 0.2 T), W/kg	10	6	4

acquire different form of the hysteresis loop: rectangular (Z), rounded (R) or extended, or flat (F). Depending on the hysteresis loop type, extremely high values of the maximal magnetic permeability (Z-type hysteresis loops), high initial magnetic permeability (R-type hysteresis loops), or extremely low alternating magnetization losses (F-type hysteresis loops) can be achieved.

Some of the amorphous metallic alloys offered by Vacuumschmelze GmbH & Co, KG, Hanau, are:

VITROVAK 7505. This is a soft magnetic amorphous alloy on a Fe basis, with high saturation induction. It is characterized by low alternating magnetization losses and relatively high coefficient of magnetostriction, respectively.

VITROVAK 4040. This is an amorphous soft magnetic metallic alloy on the basis of the elements Fe and Ni. This alloy possesses moderately high saturation induction and higher magnetic permeability, less coercive force, and substantially lower alternating magnetization losses, respectively, as compared to VITROVAK 7505. Its coefficient of magnetostriction is also lower.

VITROVAK 6025. This is an amorphous soft magnetic metallic alloy on the basis of Co. It is characterized by almost zero coefficient of magnetostriction. Due to this circumstance, its soft magnetic properties are practically insensitive to mechanical impacts. This amorphous metallic alloy possesses very high initial permeability. The alternating magnetization losses are extremely low. Variant of this alloy is the amorphous alloy VITROVAK 6025 X.

VITROVAK 0080. This is an amorphous paramagnetic alloy on the basis of Ni. As compared to its crystalline analogs, it is characterized by very high limit of flow and steadiness to multiple bending. The high mechanical properties of this amorphous alloy make it suitable for application as a constructive material for

Table 13.6 Physical and mechanical properties of amorphous metallic alloys of VITROVAK series

Alloy	VITROVAK 7505	VITROVAK 4040	VITROVAK 0080
Density, g/cm^3	7.1	7.4	8.0
Specific electrical resistance Ohm m	1.3	1.35	0.9
Temperature coefficient of electrical resistance (20–200 °C)	2.10^{-4}	$2.3.10^{-4}$	$1.2.10^{-4}$
Thermal coefficient of expansion, $K^{-1} \times 10^{-7}$	80	120	130
Vickers hardness, $HV_{0.2}$	950	800	850
Young modulus, kN/mm^2	150	150	150
Onset temperature of crystallization, °C	500	450	450
Upper exploitation temperature, °C	120	120	200
Maximal ribbon thickness, mm	0.04	0.04	0.06

Table 13.7 Fields of application of amorphous metallic alloys of VITROVAK series

	VITROVAK				In the form of...
Applications	7505	4040	6025	0080	
Special electrical transducers	Yes	Yes	Yes	No	Ribbons, toroids
Tiny carriers	No	Yes	Yes	No	Cut parts
Power supply transformers working at 400 kHz	Yes	No	No	No	Ring-shaped toroid inductive elements.
Magnetic amplifiers	No	Yes	Yes	No	–
Flexible magnetic screening	No	No	Yes	No	Ribbons for cable screening
Magnetoelastic sensor for displacement, bending, tension, etc.	Yes	Yes	No	No	Ribbon and special cut parts
Magnetic heads	No	No	Yes	No	Ribbons, cut or etched lamellae
Magnetic springs	Yes	Yes	Yes	No	–
Nonmagnetic springs	No	No	No	Yes	–
Markers for protection against theft	Yes	Yes	Yes	No	–

spring elements. Due to the high content of Ni, this alloy is also of high corrosion resistance. Table 13.5 represents the typical soft magnetic features of the VITROVAK series.

Table 13.6 represents some physical and mechanical properties of amorphous metallic alloys of VITROVAK series.

The fields of application of the amorphous metallic alloys of VITROVAK series are given in Table 13.7.

Table 13.8 represents the magnetic features of most widely known amorphous metallic alloys in comparison to their crystalline alternative materials.

Table 13.8 Magnetic features of most widely known amorphous metallic alloys in comparison to their alternative crystalline materials

Composition (trademark) of the alloy	T_C, K	$\mu_{max}.10^3$	H_C, A/m	B_s, T	$\lambda_s.10^{-6}$	ρ, μOhm/cm	T_x, K
METGLAS 2605	647	–	–	1.6	30	140	662
METGLAS 2615	565	–	–	1.5	30	150	600
$Fe_{80}B_{20}$	651	–	–	1.61	29	–	613
$Fe_{80}P_{14}B_6$	617	–	–	1.36	26	–	701
METGLAS 2605 SC[a]	–	–	6.37	1.61	–	125	753
METGLAS 2605 SC[b]	643	8	4.77	1.61	125	753	–
METGLAS 2605 S-2[a]	–	12.5	14	1.58	–	130	823
$Fe_{80}B_{20}$[b]	688	300	3	1.58	–	125	788
METGLAS 2605 S-3[a]	–	–	16	1.58	–	125	788
METGLAS 2605 S-3[b]	678	–	8	1.58	–	125	788
METGLAS 2605 CO[a]	–	–	6.5	1.75	–	130	702
METGLAS 2605 CO[b]	688	200	4	1.75	–	130	703
$Fe_{50}Ni_{30}P_{14}B_6$	607	–	4	1.75	–	–	696
METGLAS 2826	520	–	0.8	0.83	1.04	180	700
METGLAS 2826 MB[a]	–	12	7.9	0.88	0.88	–	683
METGLAS 2826 MB[b]	626	750	1.2	0.88	1	160	683
$Fe_{40}Ni_{40}B_{20}$	669	–	7	1.03	13.5	–	628
$Fe_5Co_{70}Si_{15}B_{10}$	703	–	–	0.67	0.1	134	–
$Fe_3Co_{72}P_{16}B_6Al_3$	513	–	–	0.63	0	–	623
$Fe_{80}B_{16}Zr_4$	531	–	1.5	1.33	27	–	753
$Fe_{75}B_{20}Zr_5$	569	–	2.4	1.32	26	–	782
$Fe_{85}B_{10}Zr_5$	411	–	2.4	1.04	18	–	741
$Fe_{80}B_{10}Zr_{10}$	382	–	1.23	0.75	10	–	850
$Fe_{78}Si_{10}B_{12}$	733	–	2.8	1.44	33	–	751
$Fe_{80}P_{13}C_7$	588	–	6.4	1.42	–	–	683
КСР-А[c] (Russia)	–	–	3.18	–	–	–	–
НПР-А* (Russia)	523	70	4.9	0.80	–	14-	–
НПР-А[d] (Russia)	523	310	16	–	–	–	–
КСР-А[d] (Russia)	703	335	2.2	0.67	–	130	773
НМР-А[d](Russia)	623	750	0.56	0.88	–	160	683
ЖСР-А[e] (Russia)	643	–	4.7	1.6	–	125	753
$Co_{77}Cr_{12}Zr_{11}$	633	34.8	0.6	0.54	–	126	822
$Co_{79}Cr_{11}Zr_{10}$	706	14	0.7	0.67	–	125	813
$Co_{82}Mo_{10}Zr_8$	849	6.8	1.5	0.85	–	123	799
$Co_{84}W_6Zr_{10}$	726	8.8	0.7	0.72	–	137	851
$Co_{80}V_{10}Zr_{10}$	–	12.6	0.6	0.77	–	127	793
Molybdenum permalloy	733	–	1.9	0.78	0	55	–
Supermalloy	673	–	0.4	0.82	0	55	–
Deltamax	753	–	7.9	1.6	40	45	–
Rectangular permalloy	733	–	2.2	0.82	0	55	–

(*continued*)

Table 13.8 (continued)

Composition (trademark) of the alloy	T_C, K	$\mu_{max}.10^3$	H_C, A/m	B_s, T	$\lambda_s.10^{-6}$	ρ, μOhm/cm	T_x, K
Supermendur	1,213	–	14.3	2.3	70	26	–
Silectron	1,003	–	23.8	2.03	4	50	–
Permalloy 80HM[f]	–	250	0.01	–	–	–	58
Permalloy 50H[g]	–	30	0.15	–	–	45	–
Permendur 49KΦ2[h]	–	10	0.375	2.25	–	40	–
Ferrite nickel zink 2000HH	–	–	0.1	0.26	–	–	–

Remarks
[a]As-quenched
[b]Heat treated in magnetic field
[c]Heat treated in vacuum
[d]Heat treated in longitudinal magnetic field
[e]TMT in magnetic field H = 800 A/m at 638 K. 2 h
[f]Heat treated in hydrogen atmosphere at 1,473 K
[g]Heat treated in vacuum at 1,373 K
[h]Heat treated in vacuum at 993 K

Amorphous Metallic Alloys Produced in Russia

Several industrial entertainments in Russia are engaged in the production of amorphous metallic alloys. They are mainly soft magnetic alloys, which are known as 45 HΠP-A, 44 HMP-A, 85 KCP-A, 84 KCP, 81 KCP-A, and 94 KCP-A [11]. These products are in the form of ribbons of width 1–30 mm by thickness of 0.02–0.04 mm. Their basic physical properties in as-quenched state are presented in Table 13.9.

The amorphous metallic alloy 45 HΠP-A is based on the transition metals Fe and Ni. The main feature of this alloy is the high magnetic permeability which after suitable heat treatment reaches values typical for high Ni permalloys. In as-quenched state, this alloy shows an excellent combination of magnetic and mechanical properties. It is recommended for construction of magnetic shields and transformer magnetic cores. In order to achieve optimal exploitation magnetic properties, the alloy 45 HΠP-A is subjected to thermo-magnetic treatment after which it can be used for magnetic cores for patterns of high magnetic permeability, requested to work in the frequency range up to 50 kHz. The alternating magnetization losses in this frequency range are reported [11] to be comparable to the alternating magnetization losses of the 79 HM and ferrite alloys, respectively. The most important magnetic features of the 45 HΠP-A alloy are given in Table 13.10. After suitable thermo-magnetic treatment (TMT), this alloy acquires increased hardness and wear resistance, making it suitable for production of audio recorder magnetic heads.

The amorphous alloy *81 KCP-A*. This amorphous metallic alloy is on a Co basis and is characterized by a high initial and maximal magnetic permeability along with

Table 13.9 Physical properties of some soft magnetic amorphous metallic alloys produced in Russia

Trademark	Saturation induction B_s, T	Density, g/cm^3	Curie point, °C	Onset temperature of crystallization, °C	Maximal exploitation temperature, °C
45 НПР-А	0.78	7.5	250	412	150
85 КСР-А	0.67	7.5	430	500	150
44 НМР-А	0.88	8.0	350	410	125
94 КСР-А	1.60	7.3	370	480	125

The lettering of trademarks in Table 13.9 stems from the Cyrillic alphabet. The meaning of letters is Б, niobium; В, tungsten; Г, manganese; Д, copper; К, Ж, cobalt; Л, beryllium; М, molybdenum; Н, nickel; Р, boron; С, silicon; Т, titanium; Ю, aluminum; Х, chromium; and Ф, vanadium. The letter A at the end of any alloy trademark means precisely defined concentration limits of the chemical composition

Table 13.10 Magnetic features of the Ni-Fe based amorphous metallic alloy 45 НПР-А Ref. [8]

Kind of TMT	Max. magnetic permeability, μ_{max}	Coercive force H_c, A/m	B_r/B_s at $H = 8$ A/m	B_r/B_s at $H = 800$ A/m
As-quenched state	70,000	4.9	0.93	0.57
Heat treatment in longitudinal magnetic field 1,200–1,600 A/m at 300 °C, 1.5 h	310,000	1.6	0.93	0.75

a good combination of magnetic and mechanical properties. It is recommended [2] for production of magnetic heads for recording and reproduction of digital information for magnetic shields, magnetic cores, magnetic amplifiers, and tiny transformers working at increased frequencies, respectively. The optimal magnetic properties of this amorphous alloy are reached after heat treatment in vacuum at 300 °C for 90 min (Table 13.11).

Magnetic features of the amorphous alloy 85 КСР-А are presented in Table 13.12.

The amorphous alloy *94 КСР-А*. This amorphous alloy is on the basis of Fe. It possesses high saturation induction and low alternating magnetization losses (Table 13.13). After suitable heat treatment, it is recommended [2] for production of magnetic cores for power transformers and electro-generators.

The amorphous alloy 44 НМР-А. This amorphous alloy is on a basis of Ni and Fe. It possesses high magnetic permeability. The thermal stability of this alloy exceeds the thermal stability of 45 НПР-А. After heat treatment in longitudinal magnetic field of 800 A/m intensity at 355 °C for 2 h, followed by cooling at a cooling rate of 15 °C/min, this amorphous alloy acquires the soft magnetic properties shown in Table 13.14. These soft magnetic properties are similar to the properties of permalloys of high Ni content. The alloy 44 НМР-А is recommended for production of magnetic cores of patterns working at frequencies up to 50 kHz, at which the alternating magnetization losses are comparable with those of the crystalline alloys of 79 НМ type.

Table 13.11 Magnetic characteristics of the 81 KCP-A amorphous metallic alloy

Kind of heat treatment	Maximum magnetic permeability, μ_{max}	Coercive force H_c, A/m	B_r/B_s at $H = 8$ A/m	B_r/B_s at $H = 800$ A/m
As-quenched	290,000	1.6	0.81	0.47
After heat treatment in vacuum at 300 °C, 90 min	370,000	0.4	0.65	0.50

Table 13.12 Magnetic features of the amorphous alloy *85 KCP-A*

Kind of heat treatment	Maximum magnetic permeability, μ_{max}	Coercive force H_c, A/m	B_r/B_s at $H = 8$ A/m	B_r/B_s at $H = 800$ A/m
As-quenched	125,000	3.2	0.95	0.75
Heat treatment in vacuum at 300 °C. 90 min	190,000	2.6	0.90	0.65
TMT in longitudinal magnetic field 1,200–1,600 A/m at 300 °C, 90 min	335,000	2.2	0.985	0.925

Table 13.13 Magnetic features of *94 KCP-A* amorphous alloy

Kind of heat (HT) or magnetic heat treatment (MHT)	H_c, A/m	B_r, T	B_{80}, T	B_s, T	$P_{1.3/50}$, W/kg
As-quenched	6.4	0.67	0.80	1.6	0.46
MHT in a field of 800 A/m at 365 °C followed by cooling till 100 °C reached by a cooling rate of 15 °C/min.	4.7	1.12	1.36	1.6	0.26

Table 13.14 Magnetic features of 44 HMP-A amorphous alloy

Kind of heat treatment	Maximum magnetic permeability, μ_{max}	H_c. A/m	B_r/B_s at $H = 8$ A/m	B_r/B_s at $H = 800$ A/m
As-quenched	12,000	8.0	–	–
TMT in longitudinal magnetic field 800 A/m. 2 h at 365 °C, followed by cooling at cooling rate of 15C/min till 100 °C	750,000	0.56	0.72	0.68

Invar-Type Amorphous Metallic Alloys Produced in Russia

Series of amorphous metallic alloys on Fe basis possess very low coefficient of thermal expansion α near to 10^{-6} K^{-1}. At room temperature, their properties are similar to the properties of the crystalline Invar 36 X alloy (see Table 13.15).

Table 13.15 Coefficient of thermal expansion α ($\times 10^{-6}$ K^{-1}) depending on the temperature range

Alloy trademark	Coefficient of thermal expansion α×(10^{-6} K^{-1}) depending on the temperature range (°C)						
	20–50°	20–100°	20–150°	20–200°	20–250°	20–300°	20–350°
86 ЖХР-А	4.16	5.75	8.91	10.19	11.00	11.60	12.20
93 ЖХР-А	−0.80	0.48	1.90	2.20	3.00	4.70	6.40
96ЖР-А	−1.06	0.03	0.10	0.22	0.32	0.46	2.02
36 Н	0.41	0.66	1.78	2.82	4.93	6.32	7.85

By some of these amorphous alloys, the very low coefficient of thermal expansion remains preserved till approximately 300 °C, while for the crystalline Invar alloy *36H*, this is true till only 100 °C.

Classification of Amorphous Metallic Alloys with Respect to Their Application as Soft Magnetic Materials

The different kinds of amorphous metallic alloys can be classified with respect to their application as soft magnetic materials [12] as follows:

- Amorphous alloys on the basis of Fe characterized by high saturation induction, relatively low magnetic permeability, and high coefficient of magnetostriction. They are requested to work in the field of electrotechnics for exploitation at industrial frequencies (50–60 Hz). This way of exploitation demands the usage of relatively big amounts of material, mainly for production of magnetic cores for power transformers. This circumstance automatically raises the question for acceptable price of these new materials.
- Amorphous metallic alloys on the basis of Fe, Ni, and to more limited extent of Co, which are characterized by moderately high saturation induction and moderately high magnetic permeability, preserving their soft magnetic properties to acceptably high level up to frequencies of 100–200 kHz, respectively. They find an application for production of pulse power supply toroids for pulse linear accelerators, etc.
- Amorphous metallic alloys on a Co basis characterized by relatively low saturation induction but simultaneously with excellent high-frequency soft magnetic properties and almost zero coefficient of magnetostriction. Due to their high price, these alloys are suitable only for special applications, demanding low mass – such as magnetic heads for recording and reproduction digital information, magnetic sensors, etc.

In the bibliography, one can find a lot of sources concerning the composition, the soft magnetic properties, and applications of amorphous metallic alloys in their interrelation. Most of the discussed alloys could be classified in the abovementioned groups. Very often, however, substantial improvements of soft magnetic properties via modification of already known compositions can be achieved, leading to further expansion of their fields of practical application. For example, Yamasaki, J, et al. [13] reported for achievement of very high coefficient of magnetostriction amounted to 45×10^{-6}, via inserting of small additions of Pt into Fe-B amorphous alloy. Fish, GE, [14] reported for substantial improvement of high-frequency soft magnetic properties of amorphous metallic alloys on the basis of the elements Fe, B, and Si (similar to the amorphous alloy METGLAS 2605 S-2) via replacement of 2 at.% Cr through Mo. Hilzinger, HR, et al. [15] report for a new type amorphous alloys on the basis of the transition metals Co and Mn, characterized by extremely low alternating magnetization losses at high frequencies.

Amorphous Metallic Alloys on Fe Basis for Application in Electrotechnics at Frequencies of 50–60 Hz

The distribution transformers are used in the final stage of decrease of voltage in the electric net and energy supply to the final user. The main feature of distribution transformers as compared to the other kinds of electrical transformers is the circumstance that the period of time of their uninterrupted work is about 25–40 years. During this period of time, the primary coil of the transformers is continuously fed with electrical energy. Under such a regime of exploitation, the magnetic core connecting the primary and the secondary coils is uninterruptedly kept on remagnetization with the frequency of the electric net (50–60 Hz). This causes loss of a definite part of the energy due to the hysteresis losses and eddy current losses. Estimations of these losses have been made in the USA for a standard distribution transformer with magnetic core made of crystalline iron sheet laminated transformer core (electrotechnics Fe-Si steel) with power stage of 25 kVA. By average core losses of 35 W for 35 years of uninterrupted exploitation, the net losses amounted to 26 MWh that corresponds to losses of 1,500US$. The usage of the bulk amorphous alloy $(Fe,Co,Ni)_{70}Zr_{10}B_{20}$ in the form of lamellae of 0.1 mm thickness with high degree of magnetic core filling [16] provides the possibility to decrease the alternating magnetization losses up to 75 %. It is nowadays believed [17] that the creation of electric distribution transformers with efficiency higher than 99 % is only possible on the basis of iron-containing amorphous alloys. Irrespective of the lower saturation induction of amorphous alloys as compared to the classical Fe-3.3 % Si transformer steel (1.6–2 T, respectively), it is completely enough for their usage by power transformers. In the modern constructions of power distribution transformers, the magnetic induction does not exceed 1.2 T [18]. Table 13.16 presents comparative data for 25 kVA distribution transformer and an experimental sample of toroidal core produced by using METGLAS 2605 S-2.

As is seen, the application of amorphous metallic alloys in the distribution transformers allows the achievement of substantially lower core losses. It must be taken into account, however, that the standard nowadays constructions and schemes

Table 13.16 Comparative data for 25 kVA distribution transformer and an experimental sample of toroidal core produced by using METGLAS 2605 S-2

Data for comparison	Industrial version	Experimental sample with amorphous core
Core material	Textured iron sheet M-4 (ASA)	METGLAS 2605 S-2
Losses, watt		
In the core	85	16
In the coil	240	235
Core weight, kg	65	77
Overall weight, kg	182	164

of production are unsuitable. In order to overcome these difficulties, one has to seek for new constructive and technological solutions.

Power Transformers Working at Frequencies of 400 Hz

This type of transformers is widely used in aviation, shipbuilding, and military technics. Due to the low alternating magnetization losses in the cores made by amorphous metallic alloys, they can be used to reduce the core weight and volume. The multiple lower losses of alternating magnetization as compared to the losses in the classical Fe-Si sheet of 0.1 mm thickness at 400 Hz and induction of 1 T allow to increase substantially the losses in the copper coils without going beyond the overheating temperature of the transformer. As a result, the copper coil volume can be reduced by using thinner wire. This circumstance is even more important than the effect of reduction of the magnetic core size.

According to Ref. [19], the application of amorphous alloys for production of transformer magnetic cores for 400 Hz has also additional advantages. The maximal converted power at a temperature of 120 °C in the toroidal cores of equal size, made by the amorphous alloy METGLAS 2605 SC, and textured Fe-Si sheet, respectively, is up to 60 % bigger. The relatively small dimensions of these transformers decrease the difficulties by production of toroidal cores. No doubt, the usage of amorphous metallic alloys for production of 400 Hz power transformers will find a wide application.

Applications in Electromotors

Except the transformers, the electromotors in number and universality of application are the greatest consumers of soft magnetic materials in electrotechnics. As far as the specific power losses by alternating magnetization in electromotors are usually much higher than the losses in distribution transformers, their efficiency is lower (about 85 %). Consequently, at least one half of the energy losses in electrotechnics are due to the different kinds of electromotors. While by the power transformers the direction of magnetic flux can be aligned to coincide with the magnetic anisotropy of the core material, by the electromotors this is practically impossible. This is why the usage of textured Fe-Si sheet by electromotors is meaningless. Amorphous metallic alloys provide the unique possibility to achieve maximal high magnetic efficiency of electromotors, as far as the in this case necessary anisotropy of the core material can be determined in the last stage of core heat treatment. In this way, it becomes possible to ensure low magnetic resistance exactly in the core sections in accordance with the configuration of the magnetic flux [20]. This is why the application of amorphous metallic alloys in electromotors allows to reduce the power losses with more than 90 %. These losses are especially great in industrial alternate current electromotors with uninterrupted regime of exploitation. The application of amorphous alloy is however limited due to the sophisticated constructive and technological problems by the production of lamellae with complicated configuration, from a material as hard and thin as the amorphous alloys.

Application of Amorphous Metallic Alloys on Fe-Ni Basis with Improved High-Frequency Properties: Transformers for Pulse Power Supply Devices (Secondary Sources for Voltage Power Supply)

The secondary power supply sources en masse more and more replace the less efficient power supply sources directly from the electric net. For this purpose, more desirable are electron devices working in the frequency range of 10–200 kHz. To these devices belong high-frequency transformers, inductive elements, and magnetic switches for use as regulators for initial voltage. For the purposes of this application, the soft magnetic materials are requested to possess high saturation induction combined with low losses by alternating magnetization and high Curie point, allowing the transformers to work at maximally high induction. The soft magnetic materials for production of inductive filters have to possess low effective magnetic permeability, while the switching materials should possess high remanent induction, combined with low losses of alternating magnetization. A series of amorphous alloys on Fe-Ni and Co basis possess soft magnetic properties, which make them prospective for their usage as components for secondary voltage power supply sources. High-frequency losses are caused first of all by the eddy currents of Foucault [21]. The specific electric resistance of amorphous metallic alloys is 2–3 times higher than the typical values of the specific resistance of crystalline ferromagnetic materials used in the high frequency range (130 μOhm/cm for amorphous materials against 45–60 μOhm/cm for permalloys, see Table 13.15). The thickness of as-quenched amorphous ribbons is about 0.02–0.04 mm. The same thickness of permalloy materials can be achieved only after complicated and high-costing technological procedures. These two circumstances only determine the amorphous metallic alloys as highly prospective for the above discussed applications [22]. Comparing the glassy metals on Fe-Ni and Co basis with different kinds of permalloys and ferrites, it can be stated that they are comparable with permalloys and are superior to the ferrites, when used as high-frequency materials at frequencies up to 100 and more kHz [23, 24]. Secondary power supply sources with power of about 1 kVA are already produced on a large scale.

Application of Amorphous Metallic Alloys on Co Basis of Zero Magnetostriction and Good High-Frequency Properties

Flexible Magnetic Shields The combination of high magnetic permeability with zero coefficient of magnetostriction together with high toughness makes possible the application of amorphous metallic alloys for production of flexible magnetic shielding. One of the first commercial products of Allied Chemical Corp [25] was a knitted 1-m-wide shield made of 3-mm-wide ribbon of $Fe_{40}Ni_{40}P_{14}B_6$ (METGLAS 2836) amorphous alloy. The magnetic shielding is used in order to limit the hazardous influence of static and high-frequency external magnetic fields. In this case, the usage of Al or Cu shields is ineffective. They are suitable for shielding of high-frequency perturbing fields. The higher the magnetic permeability of the shielding material, the higher is its shielding efficiency. Amorphous metallic alloys on Co basis are especially suitable for this purpose. They combine high magnetic permeability with practically zero coefficient of magnetostriction. This circumstance allows the crude mechanical deformation, which is inevitable by shielding of cables of low cross section, without deterioration of the shield characteristics. The low sensitivity of glassy metals on Co basis against bending makes them unique materials for shielding of cables of very accountable function. Hilzinger, HR, et al. [15] reported for shielding factor of 100 of twofold shield made from the amorphous alloy VITROVAK 6025 X by a distance of 1 mm between the separate windings of the shield. Such a shielding factor is about eight times higher than in the shields made of conventional permalloy.

Magnetic Heads The magnetic heads also belong to the first series of commercially available applications of amorphous magnetic alloys. The materials for magnetic heads used nowadays for audio recording and reproduction, as well as for recording and reproduction of digital information, should possess combination of excellent magnetic and mechanical characteristics. The most important magnetic characteristics are:

- High saturation induction, because the high coercive materials for recording materials (ribbons, discs) require high density of the magnetic flux in order to achieve high fidelity of the recordings.
- High initial magnetic permeability. This enables high efficiency of recording and reproduction of information due of the reduced magnetic flux dissipation.
- Low or even zero magnetostriction coefficient in order to avoid the deterioration of the material magnetic properties of the core as a result of the magnetoelastic interaction during the production of core lamellae or as a result mechanical vibrations and impacts of moving recording tape or disc.

- Low losses, i.e., high specific electrical resistance and low coercive force.
- Thermal stability of the soft magnetic properties. Due to the friction between the magnetic head and the tape/disc, the temperature of the head can raise significantly.

The most important mechanical requirements are high wear resistance and hardness, which determine the long-term exploitation period of the head. Amorphous metallic alloys on Co basis with almost zero coefficient of magnetostriction satisfied practically all abovementioned requirements. The only shortcoming of these alloys is the necessity to avoid the heating of the magnetic material above the onset temperature of crystallization (400–500 °C) by the preparation of the head. This limitation does not exist by the conventional crystalline materials sendust and permalloy. In order to avoid the possible crystallization of the soft magnetic amorphous core material, the firm SONY introduced laser point welding of the lamellae. Such kind of magnetic heads is sold under the name "laser amorphous heads."

Application of Amorphous Metallic Alloys for the Production of Sensors

The development of high precious sensors for acquisition and measurement of a definite type signals to a great extent determines the accuracy and action fastness of the regulating and command systems in the industry. Due to this circumstance, this field of R&D is especially actual and important.

Amorphous metallic alloys can be used for production of inductive and magnetic sensors. They possess the unique advantage to be obtained, via suitable choice of the chemical composition, with a selected combination of properties. It is easy to produce soft magnetic amorphous alloys with high magnetostriction coefficient, as well as with a vanishing low coefficient of magnetostriction. The first group of glassy metals is extremely suitable materials for magnetoelastic sensors, while the second group is suitable for magnetic sensors.

Other types of amorphous alloy properties, for example, the high sensitivity of the elasticity module or of their electrical resistance to the intensity of the magnetic field, can be used as sensor for elongation measurements, etc.

Table 13.17 represents different types of amorphous alloys with indications for their possible application in comparison with some convenient crystalline alloys for analogous applications.

Sensors for Magnetic Fields

One of the well-known methods for detection and measurement of stationary magnetic fields is based on magnetization of alternating magnetic probes and registration of the caused hysteresis curve displacements of the ferromagnetic

Table 13.17 Amorphous and crystalline metallic alloys, suitable for the production of magnetic and magnetoelastic sensors [26–29]

Kind of material	Alloy composition	Properties					Sensor type	
		$\lambda_s.10^{-6}$	H_c .A/m	B_s.T	HV.N/mm^2	R_p.N/mm^2	Magnetic	Magnetoelastic
Amorphous	$Fe_{80}B_{14}Si_6$	+30	0.04	1.5	950	–	No	Yes
	$Fe_{40}Ni_{38}(Mo,Si,B)_{22}$	+8	0.03	0.8	800	1,500	No	Yes
	$(Co,Fe)_{70}(Mo,Si,B)_{30}$	0	0.005	0.55	1,000	2,000	Yes	No
	$Co_{75}Si_{15}B_{10}$	−3.5	0.025	0.7	1,000	–	Yes	Yes
Crystalline	97Fe.3Si	+9	0.1	2.0	180	350	No	Yes
	77Ni,15Fe,Mo,Cu	±1	0.01	0.8	100	150	Yes	No
	77Ni,15Fe,							
	Mo,CuTi,Nb spring steel	0.05	0.025	0.45	220	500	No	Yes
	spring steel	−1	15	2.1	550	1,500	No	Yes

Remark: λ_s, coefficient of magnetostriction; H_c, coercive force; B_s, saturation induction; HV, Vickers hardness; R_p, yielding toughness

materials. The caused response is as stronger as the more soft magnetic is the probe material, i.e., the easier is the magnetization of the probe material. Due to this circumstance along with the used till now Ni-Fe permalloy alloys, Co-based amorphous alloys find an application for this purpose. They are characterized by vanishing low coefficient of magnetostriction, very high magnetic permeability, and very low coercive force. These amorphous alloys are insensitive to mechanical deformations and, because of this, are especially suitable for the preparation of highly sensitive sensors for magnetic fields [30, 31]. The construction of this type of sensors is based on two different principles:

- Exerting of influence from the external magnetic field on the reversible magnetic permeability. The different constructions based on this principle are described in Refs. [32, 33].
- Displacement of the hysteresis loop in a regime of a full magnetic saturation as a result of the influence of the external magnetic field – the object of measurement. The different constructions based on this principle of measurement are described in Refs. [34–36]. The sensitivity of both types of magnetic sensors is about 10^{-2} mA/cm.

A special type of magnetic field sensors is the safety markers against theft in universal stores [37, 38]. They represent small strips of soft magnetic material of high magnetic permeability (amorphous alloy on Co basis), positioned in imperceptible place into the guarded object. The magnetic system for registration of safety markers usually consists of combination of one inductive and one detective coil. The alternating electromagnetic field of the inductive coil causes magnetization of soft magnetic safety strip up to saturation. The detective coil causes extinction of the basis frequency of the inductive field and registers the overtone frequencies of the safety marker, which are as high as up to 100 kHz. This is especially important because the high-frequency overtones are generated by inductive magnetic fields of small intensity. In this way, it is possible to avoid the danger of possible upset of such sensitive electron devices as pacemakers for regulation of the heart rhythm of some visitors with heart health problems.

Magnetoelastic Sensors

The working principle of these sensors is based on the magnetostriction effect. Such kind of sensors is used for measurement of the rotating moment of shafts or forces. In this case, the mechanical quantities are converted into proportional electrical signals by using the magnetostriction interaction. The dimension (length) change of the sensor causes change in its magnetic permeability or deforms its magnetization hysteresis curve [39]. In order to be capable for use for magnetoelastic gauges, a definite material has to combine suitable mechanical properties (high Young modulus and high yielding toughness) with the necessary magnetic properties. A limited number of crystalline magnetostriction materials satisfy this condition. On the contrary, a lot of magnetostriction amorphous metallic alloys are characterized by excellent combination of suitable for this purpose magnetic and mechanical properties (see Table 13.17). Mohri, K, et al. [40] and Meydan, T, et al. [41] describe different constructive solutions for preparation of such kinds of sensors.

Amorphous Metallic Alloys Used for Sensor Based on Other Measurement Principles

Sensor Based on the ΔE Effect The ΔE effect means a strong dependence of Young modulus of a magnetostriction material on the intensity of the magnetization field. The Young modulus itself is measured via determination of propagation velocity of ultrasound in the sensor material. The velocity change gives information about the intensity of the magnetic field causing the effect. Sensors of this type made by amorphous metallic alloys are described in Refs. [42, 43]. Suitable amorphous alloy for similar applications is $Fe_{40}Ni_{38}Mo_4B_{18}$, for example.

Sensors Based on Magnetoresistive Effect This effect is based on the significant change of electrical resistance of some soft magnetic alloys as a result of applied transverse of longitudinal magnetic field. This effect is observed both in soft magnetic crystalline and amorphous metallic alloys. Depending on the type of material, the relative change of electrical resistance could be as high as several thousandths or up to several percent of the initial resistance. Suitable amorphous metallic alloy for this purpose is $Fe_{80}B_{14}Si_6$ [44].

Sensors Based on the Change of Electrical Resistance as a Result of Elastic Deformation of the Material

Strips of elastic metallic material are used, which are capable to convert their mechanical elongation to proportional respective change of their electrical resistance. By some amorphous metallic alloys with very high yielding, toughness (what means a broad range for measurement) is established K-factor equal to 4. The K-factor in this case means a material constant representing the dependence of the relative change of electrical resistance ΔR/R on the change of sensor length ΔL/L. Amorphous alloys with chemical composition $Fe_{57.5}Cr_{30}Si_{12.5}$ possess K-factor, which equals 4. For comparison, the suitable for similar application crystalline alloy $Ni_{80}Cr_{20}$ possesses K-factor equal to 2 [45]. Additional advantage of the abovementioned amorphous alloy is the circumstance that after suitable heat treatment, it obtains zero temperature coefficient of electrical resistance in a much broader temperature range than the crystalline alloy.

Brazing Amorphous Metallic Alloys

The soldering process, in contrast to the welding, consists in joining of metallic details with the aid of an alloy which melting temperature is lower than that of the details that should be joined. The brazing alloy must be/is capable to wet the details very well and to spill over them into a thin layer. After the crystallization, the soldering alloy ensures very strong joint between the details via adhesive forces.

It should be pointed out that during all soldering stages, the temperature(s) of the main metal(s) for joining remain(s) lower than its (their) temperature(s) of melting [46]. The ribbon-like soldering alloys produced via rapid solidification from the melt are extremely suitable for different kinds and techniques for solder joining. The ribbon could be cut off keeping conformity with the thin contact area for brazing between the machine details. In this way, one can presume on the capillary forces only for the good wetting of the contact surfaces. The usage of amorphous ribbons for soldering of details with responsible function provides a series of advantages as compared to the traditional ways of brazing. The first advantage is the simple technology or production of a thin ribbon from an alloy with complicated chemical composition directly from the melt. This amorphous ribbon is fully homogeneous, can be easily cut off, bended, and configured into complicated forms via suitable artificially caused bending stresses, followed by heat treatment under isothermal or continuous heating conditions [47]. Due to the enormous rate of melt cooling, amorphous and/or microcrystalline ribbon-like homogenous materials can be produced, which existence in conventional crystalline state is impossible. It has to be mentioned also the great corrosion resistance of this kind of soldering alloys. Amorphous and microcrystalline brazing alloys are subdivided to high-temperature solders on Ni and Ni-Pd basis and low-temperature solders on Cu and Al basis. The high-temperature soldering alloys are requested for use on a first place in aviation and space technics. The *Pratt & Whitney* Company, USA, makes use of nickel-based solders for high-temperature solder joints [48]. These nickel-based solder compositions have improved wetting behavior, contain aluminum and chromium, and one or more of the metals zirconium, niobium, or titanium added to the solder in amounts of from 0.1 to 5 at.%. Particularly effective soldering alloys contain 0–4 at.% iron, 0–21 at.% chromium, 0–19 at.% boron, 0–12 at.% silicon, 0–22 at.% phosphorus, 0–3 at.% carbon, and 0.1–5 at.% niobium, zirconium, or titanium; the remainder being nickel. The boron, carbon, silicon, and phosphorus concentrations varied from 16 to 24 at.%. As an example of the invention, 0.05-mm-thick foils of a composition of Fe-23 % Cr-5 % Al foils have been soldered with solder foils having a thickness of 0.025 mm. The base used was a standard solder known as L-Nila having a composition of $Ni_{60}Cr_{14}Fe_4Si_8B_{14}$. To this solder have been added 0.5 and 1 at.% by weight titanium. It was found that the standard solder by itself exhibited an imperfect wetting of the Fe-Cr-Al foils, whereas the two solder foils having the Ti additive produced a significantly improved wetting. These solders can be used for jet turbine blade joints of very high quality.

Bulk amorphous alloys based on quaternary Ni-Zr-Ti-Al alloy system, the extension of this quaternary system to higher order alloys by the addition of one or more alloying elements, methods of casting such alloys, and articles made of such alloys are provided by Allied Chemical Corp. [49].

Table 13.18 shows the composition and properties of some produced by Allied Chemical Corp. high-temperature Ni-based soldering alloys together with their characteristics, important for their application.

Table 13.18 Amorphous metallic alloys produced by Allied Chemical Corp. for soldering purposes

Mark	Composition in wt %	Temperature, K		
		$T_{solidus}$	$T_{liquidus}$	$T_{soldering}$
MB-10	14Cr,4.5Fe,3.2B,0.06C,4.5Si, the remainder, Ni	1,243	1,348	1,448
MBF-15 MBF-15A	14Cr,4.12Fe,2.8B, 0.03C,4.5Si, the remainder, Ni	1,233	1,398	1,448
MBF-20 MBF-20A	7Cr, 3Fe,3.2B,0.06C,4.5Si, the remainder, Ni	1,243	1,273	1,313
MBF-30 MBF-30A	3.2B,0.06C,4.5Si, the remainder, Ni	1,253	1,313	1,313
MBF-60	11P,0.1C, the remainder, Ni	1,153	1,153	1,253
MBF75 MBF75A	10Cr,23Co,5.5Fe,3.7Mo,3.5B, the remainder, Ni	1,330	1,438	1,473
MBF-80 MBF-80A	15.2Cr,4B, the remainder, Ni	1,293	1,338	1,448

Low-Temperature Brazing Amorphous Alloys on the Basis of Cu and P They are suitable substitutes of Ag-Cd conventional soldering alloys. Especially prospective is the Cu-Ni-Cd-P alloy with eutectic chemical composition. However, it is not suitable for brazing of alloys of Fe. For such kind of alloys, soldering alloys on the basis of the Cu-Mn-Si are developed. The system 89 wt.% Cu + 20 wt.% Cd is also suitable in the form of rapidly solidified brazing alloy.

Low-Temperature Solders on the Basis of Al Alloys The traditional brazing alloys for Al and its alloys possess compositions very near to the eutectic composition of Al-Si system (12. 5 wt% Si), with additions of Mg, Ba, and Sr. The additions improved the solders' wetting capability. The additives also remove the thin oxide layer typical for Al surfaces, which hinders the quality bonding. The main shortcoming of the conventional solders is the high reactivity (light oxidation) of the ennobling additives. The rapid solidification from the melt makes possible the production of thin ribbon-like solders on Al-Si-Mg basis. They possess improved solder exploitation characteristics and as consequence – high quality of the bonding layer.

Brazing Alloys for Application in Electronics Traditional brazing alloys for soldering of Si chips in microelectronics are nowadays the brazing alloys on Au basis. They ensure good electrical contact. Their alternatives are solders on the basis of Sn, In, and Ag and have substantial shortcomings [50]. Extended survey on the topics is given in Ref. [51].

Brazing Alloys for Ceramic Materials The employment of ceramic materials and composites in the industry attracts an increased attention during the last years. Because of the high fragility of ceramics almost always, it is necessary to work with combinations ceramics with metallic details. The realization of strong and reliable joints between ceramics and metal provides substantial problems. Several methods for production of suitable brazing alloys for this purpose exist [52, 53]. Duhaj, P, et al. [54] have shown that very suitable in this case are rapidly solidified from the melt alloys on the basis of Ag-Cd-Ti. Such kinds of solders are, for example, $Ag_{80}Cu_{37}Ti_3$, $Ag_{40}Cu_{57}Ti_3$, and $Ag_{20}Cu_{75}Ti_5$. Their melting temperatures are 1,051, 1,133, and 1,202 K, respectively. Amorphous metallic alloys in the beryllium-titanium-zirconium system are also proposed as metal-ceramics brazing alloys by Allied Chemical Corp. [55].

Amorphous Metallic Alloys for Construction Use The high mechanical properties of amorphous metallic alloys make them prospective for use as constructive elements in different kinds of machines and devices. In the bibliography [56] are mentioned elinvar applications of the nonmagnetic Ni-Si-B amorphous alloy. Amorphous elinvar alloys are used for fabrication of seismic gauges, membranes for manometers, timepiece springs, balances, and other kinds of spring elements. The firm Vacuumschmelze, FRG, produces the amorphous alloy VITROVAK 0080, which is recommended for production of springs, membranes, and contact elastic elements for computer keyboards. The mechanical properties of these amorphous alloys are given in Table 13.13 of this survey. In the bibliography [57], one can find also information about the use of amorphous metallic alloys as armor elements for high-pressure hoses or automotive tires.

Prospects

The main prospective directions for development of science and technology in the field of amorphous metallic alloys include the development of satisfactory theory for prediction of glass-forming ability of metallic alloys with complicated multicomponent composition for production of bulk amorphous alloys by low melt cooling rates of about several degrees per minute. In the last two decades, a series of bulk amorphous alloys were discovered [58, 59] that can be obtained in amorphous state at moderate melt cooling rates of 1–100 K/s. This is of great practical and fundamental scientific interest, as such kind of amorphous alloys can be mold in the form of bulk details sizing from several millimeter to several centimeter. For example, such metallic alloys are RE-Al-TM (RE = rare earth element, TM = transition metal) [60–63], Mg-RE-TM [64], Zr-Al-TM [65, 66], Zr-Al-Cu-Ni [67], and Zr-Ti-Cu-Ni [68].

Of late years, A. Inoue and T. Masumoto have carried out deliberate fundamental research [69–72] and developed a series of amorphous metallic alloys with very broad temperature range between the glass transition temperature T_g and the onset temperature of crystallization T_x. This temperature range $\Delta T_{x,g} = T_x - T_g$ amounts up to ≈90 K. This circumstance enables the possibility to bring these materials into a state of deeply undercooled melt. They remain in this state long enough time in order to be mechanically compacted or even to be mold into machine details by relatively low cooling rates of about 10^3 K/s. The possibility to produce bulk amorphous metallic alloys with very high glass-forming ability (GFA) makes possible the extended study of glass transition process and the viscous flow behavior, and the structural relaxation and quite probably will provide useful information about the reasons for the appearance of broad temperature range $\Delta T_{x,g}$ of deeply undercooled metallic melts.

References

1. Gilman JJ (1975) Phys Today 28(5):46
2. Luborsky FE (1978) IEEE Trans Magn 14:1008
3. Wfarlimont H (1980) Phys In Technol 11:28
4. Hilzinger HR (1985) IEEE Trans Magn 21:2020
5. Makino Y (1985) In: Steeb S, Warlimont H (eds) Proceedings of the 5th international conference on rapidly quenched metals (RQ5). Würzburg, Germany, p 1699
6. Kotter KH (1985) Funkschaw 57:42
7. Roy R, Majumdar AK (1981) J Magn Magn Mater 25(1):83
8. Liebermann HH (1993) Rapidly Solidified Alloys: Processes, Structures, Properties, Applications. Marcel Dekker, Inc, NY
9. Amorphous Metals VITROVAC (1989) Catalogue, Edition 7, Vacuumschmelze GmbH, Hanau, Germany
10. Advanced Materials – The Key to Progress (2002), Amorphous and Nano-Crystalline Alloys. Vacuumschmelze GmbH & Co. KG, Hanau
11. Molotilov B (1983) Special alloys. In: Metallurgy. Reference book, Moscow, p 344 (in Russian)
12. Warlimont H (1985) Proceedings of the 5th international conference on rapidly quenched metals (RQ5), vol 2. Germany, p 1599, Sept 1984
13. Yamasaki J et al (1985) IEEE Trans Magn 21(2035):65
14. Fish GE (1985) J Phys 46(C6):207
15. Hilzinger HR et al (1984) IEEE Trans Magn 20:1323
16. Inoue A, Zhang, Itoi T, Takeuchi A (1977) Mater Trans JIM 38:359
17. Werner FE (1981) Energy Efficient Electrical Steels. TMS-AIME, Warrendale, p 1
18. Douglas J et al (1982) Electr Power Res Inst J 7(1):14
19. Boll R et al (1981) IEEE Trans Magn 17:3053
20. Mischler WR, Rosenberry GM, Frischmann PG, Tompkins RE (1981) IEEE PES Winter Meeting, Atlanta, GA, paper 81 WM I89-0
21. Pry RH et al (1958) J Appl Phys 29:532
22. Egami T et al (1979) Proceedings of the 9th annual conference on magnetism and magnetic materials, San Francisco, CA
23. Warlimont H, Boll R (1982) J Magn Magn Mater 26:97
24. Schwarzschild BM (1982) Phys Today 35(2):20
25. Hermansfeldt W (1979) Heavy ion fusion program, University of California, Lawrence Berkeley Laboratory, p 5031
26. Russew K et al (1987) Proceedings of the 1st international conference of COMECON countries on rapidly solidified metallic alloys, Varna, p 173
27. Boll R (1977) Weichmagnetische Werksioffe, Vacuumschmelze GmbH, Hanau, Verlag Siemens AG, Berlin, Muenchen
28. Holmes L (1982) Metallic glasses in sensors and transducers. Electronics & Power 28(2):180 IET Digital Library DOI: 10.1049/ep.1982.0074
29. Boll R et al (1981) Magnetic sensors of New Materials, Siemens Forschungs- und Entwicklungsberichte. Würzbrug 10(2):83
30. Hilzinger HR et al (1979) Magnetostriction Measurements on Co-based Amorphous Alloys. Phys Status Solidi A 55:73
31. Mohri K (1984) Review on Recent Advances in the Field of Amorphous Metal Sensors and Transducers. IEEE Trans Magn 20:942

32. Scarzello JF et al (1977) A Magnetic Vehicle Detection System Using a Low Power Magnetometer. IEEE Trans Magn 13(et al):574
33. Scarzello LF et al (1977) A Low Power Magnetometer for Vehicle Detection. IEEE Trans Magn 13:1101
34. Foerster F (1955) Ein Verfahren zur Messung von Magnetischen Gleichfeldern und Gleichfelddifferenzen und seine Anwendung in der Metallforschung und Technik. Zs Metallkde 46:358
35. Luz H, Luz H (1968) Magnetfeld-messung mit Foerster-sonden und Hallgeneratoren. Elektronik 8:247
36. Takeuchi S et al (1984) IEEE Trans Magn 20:1723
37. Fearon E (1974) Open-strip ferromagnetic marker and method and system for using same Patent US 3,790,945 A
38. Fearon E (1983) Antitheft system, Patent WO 1983002027 A1
39. Vazkues N et al (1983) Phys Status Solidi A 80:195
40. Mohri K et al (1978) IEEE Trans Magn 14:1071
41. Meydan T et al (1981) IEEE Trans Magn 17:3376
42. Webb DC et al (1979) IEEE Trans Magn 15:1410
43. Toerok E et al (1979) J Magn Magn Mater 10:303
44. Yao YD et al (1982) J Appl Phys 53:2258
45. Tachara T et al (1984) RQ 5, vol 2. Wuerzburg, Germany, p 1312
46. DeCristofaro NJ et al (1984) RQ 5, vol 2. Wuerzburg, Germany, p 1715
47. Russew K, de Hey P, Sietsma J, van den Beukel A (1997) Viscous Flow of Amorphous $Fe_{40}Ni_{40}Si_6B_{14}$ Studied by Direct Creep measurements and Relaxation of Bend Stresses under Nonisothermal Conditions. Acta Mater 45(5):2129–2137
48. Emmerich K (1990) Nickel-base solder for high-temperature solder joints US 4900638 A
49. Xu D, Johnson WL (2009) Bulk amorphous refractory glasses based on the Ni(−Cu-)-Ti (−Zr)-Al alloy system US 7591910 B2, 22 Spt
50. Kank S et al (1977) IEEE Trans PHP 13:318
51. Mulugeta A, Guna S (2000) Lead-free Solders in Microelectronics. Mater Sci Eng Rep 27:95–141
52. Akselsen OM (1992) J Mater Sci 27:1989
53. Nicholas MG (1990) Joining of Ceramics. Chapman and Hall, London
54. Janickovic DP, Sebo P, Duhaj P (2001) Svec Mat Sci Eng A304–306:569–573
55. Tanner LE et al (1977) Amorphous metal alloys in the beryllium-titanium-zirconium system US 4,050,931 A
56. Kikuchi A et al (1983) Sci Rep RITU A 31(1):99
57. Schneider I et al (1979) Newe Hütte 24(10):379
58. Greer AL (1993) Nature 366:303
59. Greer AL (1995) Science 267:1947
60. Inoue A et al (1990) Mater Trans Jpn Inst Met 31:425
61. Inoue A et al (1993) Mater Trans Jpn Inst Met 34:351
62. He Y et al (1994) Philos Mag Lett 70:371
63. Inoue A et al (1991) Mater Trans Jpn Inst Met 32:609
64. Inoue A et al (1992) Mater Trans Jpn Inst Met 33:937
65. Zhang T et al (1991) Mater Trans Jpn Inst Met 32:1005
66. Inoue A et al (1993) Mater Trans Jpn Inst Met 34:1234
67. Kim YJ et al (1994) Appl Phys Lett 65:2136
68. Yi He et al (1996) J Non-Cryst Solids 205–207:602
69. Inoue A et al (1992) J Non-Cryst Solids 150:396
70. Zhang T et al (1993) J Mater Sci Lett 12:700
71. Inoue A et al (1993) J Mater Sci Lett 12:946
72. Inoue A et al (1993) J Non Cryst Solids 156–158:598

Quick Reference to the Viscous Flow Behaviour of Glassy Alloys at a Heating Rate of 20 K/min

14

Abstract

This chapter provides selected information about the viscosity temperature dependence at a heating rate of 20 K/min of about 30 binary and multicomponent amorphous metallic alloys, representative for the variety of glassy metals studied and considered in this book. Numerical data for the values of FVM parameters are also given, which makes it possible to calculate the viscosity temperature dependences of these glassy alloys at other heating rates chosen by the user of this book.

Keywords

Free volume model • Glassy metals • Amorphous metallic alloys • Relaxation • Rheology • Viscosity • Glass transition temperature • Nonisothermal

K. Russew, L. Stojanova, *Glassy Metals*,
DOI 10.1007/978-3-662-47882-0_14

Viscous Flow Behavior of Binary Amorphous Alloys at Heating Rate of 20 K/min

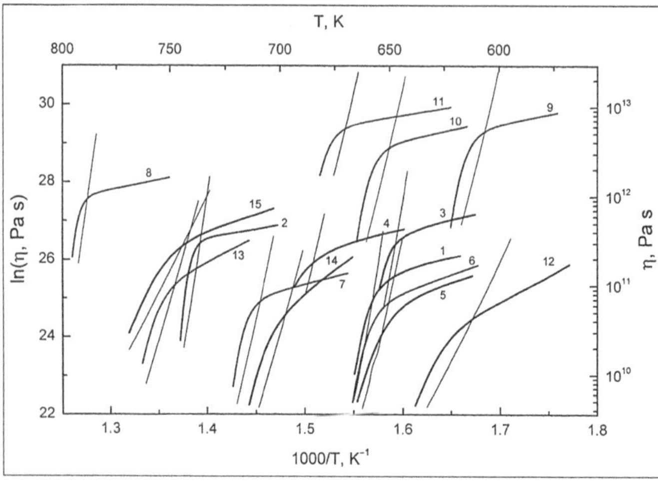

Curve numb.	Binary alloys	T_g, K
1	$Pd_{82}Si_{18}$	638
2	$Fe_{82}B_{18}$	720
3	$Fe_{86.3}P_{13.7}$	628
4	$Fe_{83.8}P_{16.2}$	660
5	$Fe_{25}Zr_{75}$	628
6	$Co_{25}Zr_{75}$	633
7	$Co_{33.4}Zr_{66.6}$	688
8	$Co_{91}Zr_9$	790
9	$Ni_{83.2}P_{16.8}$	588
10	$Ni_{81.3}P_{18.7}$	623
11	$Ni_{79.2}P_{20.8}$	648
12	$Ni_{25}Zr_{75}$	594
13	$Cu_{64.3}Zr_{35.7}$	734
14	$Cu_{50}Zr_{50}$	676
15[1]	$Cu_{40}Hf_{60}$	728

FVM parameters of binary alloys studied

FVM parameters, dimension							
Alloy	ν_r, 1/s	Q_r, kJ/mol	Q_η, kJ/mol	$c_{f,o}$,	T_o, K	B, K	η_o, Pa s/K
$Pd_{82}Si_{18}$	6.8×10^{19}	120	191	2.5×10^{-11}	557	2130	5×10^{-20}
$Fe_{82}B_{18}$	4×10^{23}	181.2	230	4.9×10^{-11}	615	3017	2.7×10^{-11}
$Fe_{86.3}P_{13.7}$	3.4×10^{24}	144.4	200.5	2.1×10^{-11}	455	6052	5.8×10^{-24}
$Fe_{83.8}P_{16.2}$	3.4×10^{24}	175.6	240.5	2.1×10^{-11}	492	5146	1.3×10^{-24}
$Fe_{81}P_{19}$	3.4×10^{24}	145.6	205.5	2.1×10^{-11}	476	6029	4.8×10^{-24}
$Fe_{25}Zr_{75}$	7×10^{19}	89	154	3.2×10^{-14}	460	5749	5.9×10^{-20}
$Co_{91}Zr_9$	1.4×10^{30}	209	270	2.5×10^{-4}	598	7850	1.1×10^{-27}
$Co_{88}Zr_{12}$	1.8×10^{24}	193	281	3.7×10^{-4}	550	7202	3.9×10^{-24}
$Co_{84.6}Zr_{15.4}$	6.6×10^{27}	227	280	3.7×10^{-4}	505	6300	1.3×10^{-24}
$Co_{53}Zr_{47}$	9.9×10^{21}	184	270	6.9×10^{-6}	475	6700	4.7×10^{-22}
$Co_{50}Zr_{50}$	8.5×10^{23}	151	204	6.1×10^{-6}	452	6730	1.4×10^{-21}
$Co_{33.4}Zr_{66.6}$	5.7×10^{24}	150	220	1.7×10^{-7}	500	6640	1.3×10^{-24}
$Co_{25}Zr_{75}$	9.4×10^{19}	101	191	6.1×10^{-6}	455	5500	5.4×10^{-22}
$Co_{21.5}Zr_{78.5}$	4.5×10^{23}	130	190	7.6×10^{-7}	430	4968	5.2×10^{-22}
$Ni_{83.2}P_{16.8}$	7.5×10^{16}	88.2	145	8.1×10^{-10}	490	2640	1.3×10^{-14}
$Ni_{81.3}P_{18.7}$	1.9×10^{17}	100.3	162	2.5×10^{-9}	515	2898	2.5×10^{-15}
$Ni_{80.9}P_{19.1}$	1.8×10^{16}	88.6	158	5.6×10^{-9}	521	3211	5.7×10^{-15}
$Ni_{79.2}P_{20.8}$	1.5×10^{17}	97.9	150	3.1×10^{-8}	523	3287	3.8×10^{-14}
$Ni_{25}Zr_{75}$	9.5×10^{13}	108	162	1.4×10^{-6}	428	2463	1.4×10^{-11}
$Cu_{64.3}Zr_{35.7}$	2.8×10^{15}	116	238	2.5×10^{-7}	529	4290	1.1×10^{-18}

(continued)

FVM parameters, dimension							
Alloy	ν_r, 1/s	Q_r, kJ/mol	Q_η, kJ/mol	$c_{f,o}$,	T_o, K	B, K	η_o, Pa s/K
$Cu_{61.8}Zr_{38.2}$	1.1×10^{17}	135	248	2.9×10^{-7}	510	4297	1.5×10^{-19}
$Cu_{56}Zr_{44}$	4×10^{16}	95	334	8×10^{-8}	481	5620	4×10^{-25}
$Cu_{50}Zr_{50}$	8×10^{16}	102	275	7.8×10^{-7}	460	5484	4.2×10^{-25}
$Cu_{46}Zr_{54}$	2×10^{14}	96	166	7×10^{-5}	409	4477	3×10^{-14}
$Cu_{40}Zr_{60}$	4.9×10^{15}	98	171	1.7×10^{-5}	400	5038	3.4×10^{-16}
$Cu_{40}Hf_{60}$ [1]	1.2×10^{12}	116	186	3.4×10^{-4}	470	3450	2.1×10^{-11}

Viscous Flow Behavior of Ternary Amorphous Alloys at Heating Rate of 20 K/min

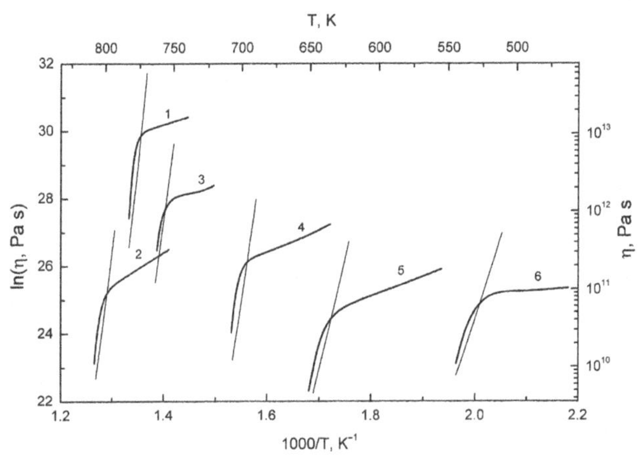

Curve numb.	Ternary alloys	Tg, K
1	$Fe_{73.8}Cr_{8.2}B_{18}$	744
2	$Fe_{73}V_9B_{18}$	758
3	$Fe_{40}Ni_{40}B_{20}$	710
4	$Pd_{77.5}Cu_6Si_{16.5}$	645
5	$Pd_{40}Ni_{40}P_{20}$	578
6	$Al_{25}La_{50}Ni_{25}$	495

[1] Stojanova L, Fazakas E, Varga LK, Yankova S, Russew K Thermal stability, viscosity and glass forming ability of $Hf_{(100-x)}Cu_x$ (x = 30 and 40 at.%) amorphous metallic alloys. XXV Nat Conf on Non-Destructive Testing NDT-2010, Sozopol. Sci Comm of STU Machine Building vol 18 XVII No 5(115) p. 186

FVM parameters of ternary alloys studied

FVM parameters, dimension							
Alloy	ν_r, 1/s	Q_r, kJ/mol	Q_η, kJ/mol	$c_{f,o}$,	T_o, K	B, K	η_o, Pa s/K
$Fe_{73.8}Cr_{8.2}B_{18}$	2.6×10^{16}	95	157	2×10^{-8}	603	3807	9.6×10^{-14}
$Fe_{73}V_9B_{18}$	1.8×10^{18}	151	252	9.7×10^{-8}	630	3400	1.2×10^{-19}
$Fe_{40}Ni_{40}B_{20}$	7.3×10^{22}	150.6	210.4	5.9×10^{-10}	543	5355	1.4×10^{-20}
$Pd_{77.5}Cu_6Si_{16.5}$	5.3×10^{21}	100	155	7.7×10^{-15}	480	5712	2.6×10^{-20}
$Pd_{40}Ni_{40}P_{20}$	1.8×10^{23}	126	196	5.1×10^{-12}	355	6900	3×10^{-23}
$Al_{25}La_{50}Ni_{25}$	8.6×10^{21}	134.4	149.6	1.8×10^{-7}	320	3770	1.2×10^{-17}

Viscous Flow Behavior of Quaternary Amorphous Alloys at Heating Rate of 20 K/min

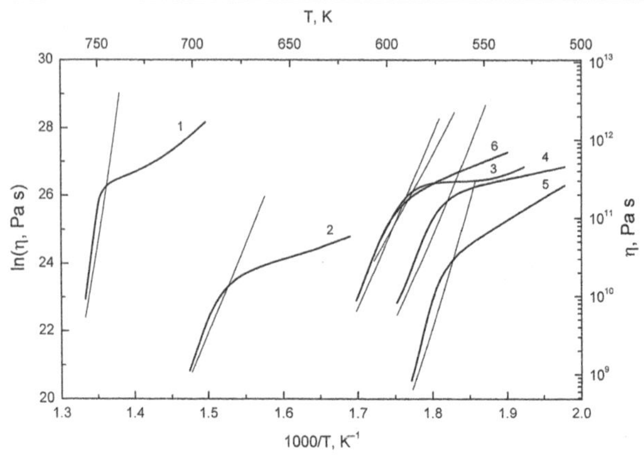

Curve numb.	Quaternary alloys	Tg, K
1	$Fe_{40}Ni_{40}Si_6B_{14}$	729
2	$Al_{7.5}Cu_{17.5}Ni_{10}Zr_{65}$	665
3	$Al_{85}Ni_5Co_2U_8$	564
4	$Al_{85}Ni_5Co_2Ce_8$	545
5	$Al_{85}Ni_5Co_2Gd_8$	548
6	$Pd_{40}Cu_{30}Ni_{10}P_{20}$	562

FVM parameters of quaternary alloys studied

FVM parameters, dimension							
Alloy	ν_r, 1/s	Q_r, kJ/mol	Q_η, kJ/mol	$c_{f,o}$,	T_o, K	B, K	η_o, Pa s/K
$Fe_{40}Ni_{40}Si_6B_{14}$	1.2×10^{23}	177	224	3.6×10^{-12}	603	3612	1.7×10^{-20}
$Al_{7.5}Cu_{17.5}Ni_{10}Zr_{65}$	3.1×10^{18}	137.2	195.9	3.1×10^{-8}	350	6252	2.6×10^{-18}
$Al_{85}Ni_5Co_2Ce_8$	2.7×10^{17}	135	182	9.7×10^{-7}	411	1990	1.5×10^{-11}
$Al_{85}Ni_5Co_2Gd_8$	5.6×10^{16}	105	230	1.7×10^{-8}	424	2390	2.6×10^{-10}
$Al_{85}Ni_5Co_2U_8$	8.6×10^{21}	204	175	5.9×10^{-5}	397	1710	2.1×10^{-11}
$Pd_{40}Cu_{30}Ni_{10}P_{20}$	1.3×10^{21}	122	208	1.5×10^{-9}	311	5485	2.7×10^{-22}

Appendix 1: Quick Reference to Most Frequently Used Important FVM Related Equations

Chapter 3

Equation of Uhlmann for the time needed for a small volume fraction ξ to crystallize:

$$t = \left[\frac{9.3\eta(T)}{kT}\right]\left\{\left(\frac{a_o\xi}{f^3N_v}\right)\frac{\exp\left(\frac{1.024}{T_r^3\Delta T_r^3}\right)}{\left[1-\exp\left(\frac{-\Delta H_m \Delta T_r}{RT}\right)\right]^3}\right\}^{0.25}, \tag{3.1}$$

where $\Delta T_r = (T_m - T)$, $T_r = T/T_m$, a_o is the average atomic diameter, N_v is the number of atoms per unit volume, f is the fraction of sites at the melt/crystalline interface where atoms are preferentially added or removed, and ΔH_m is the molar enthalpy of fusion.

Critical cooling rate for vitrification is

$$\dot{T}_{cr} = (T_m - T_n)/t_n, \tag{3.2}$$

where T_n and t_n are the temperature and the time at the nose of the time-temperature-transformation diagram.

Arrhenian type of viscosity temperature dependence:

$$\eta = \eta_o \exp\left(\frac{Q_a}{RT}\right). \tag{3.3}$$

Vogel-Fulcher-Tammann type of viscosity temperature dependence:

$$\eta = \eta_o^{VFT} \exp\left(\frac{B}{T - T_o}\right). \tag{3.4}$$

where η_o^{VFT} and B are empirical constants and T_o is known as ideal glass transition temperature.

Newtonian relation of homogeneous viscous flow:

K. Russew, L. Stojanova, *Glassy Metals*,
DOI 10.1007/978-3-662-47882-0

$$\eta = \frac{\tau}{\dot{\varepsilon}}, \tag{3.5}$$

where $\dot{\varepsilon}$ is the strain rate of the specimen under applied shear stress τ.

Chapter 4

Einstein equation for flow of mixtures:

$$\eta_{eff} = \eta(1 + 2.5\varsigma). \tag{4.1}$$

where ζ is the volume fraction of the suspended particles, η_o is the viscosity of the viscous medium, and η_{eff} is the viscosity of a mixture consisting of a small volume fraction of spherical particles suspended in the viscous medium.

Chapter 5

Probability for appearance of a void of volume υ' is

$$P(\upsilon') = \frac{\gamma}{\upsilon_f}\exp\left(\frac{-\gamma\upsilon'}{\upsilon_f}\right), \tag{5.1}$$

where γ is geometric overlap factor, which value is between 0.5 and 1. The full probability $P(\upsilon^*)$ for appearance of a void of a volume greater than υ^* is $P(\upsilon^*) = \int_{\upsilon^*}^{\infty} P(\upsilon^*)d\upsilon = \exp(-\gamma\upsilon^*/\upsilon_f)$. This probability is called concentration of structural defects c_f:

$$c_f = \exp\left(\frac{-\gamma\upsilon^*}{\upsilon_f}\right) = \exp\left(\frac{-1}{x}\right), \tag{5.2}$$

where $x = \frac{\upsilon_f}{\gamma\upsilon^*}$ is the so-called reduced free volume.

Quasi-equilibrium atomic free volume:

$$\upsilon_{f,eq}(T) = x_{eq}(T)\gamma\upsilon^* = \frac{T - T_o}{B}\gamma\upsilon^*, \tag{5.3}$$

where B and T_o are two model parameters.

The rate of tangential deformation (viscous flow):

$$\dot{\varepsilon} = \frac{c_f}{\Omega} \upsilon_o \varepsilon_o k_r, \tag{5.4}$$

where c_f is the concentration of structural flow, υ_o is the volume of the flow defect, and k_r is a frequency factor of atomic rearrangements in the vicinity of flow defects.

Temperature dependence of k_r:

$$k_r = \nu_r \exp\left(-\frac{Q_f}{kT}\right) \sinh\left(\frac{\tau \varepsilon_o \upsilon_o}{kT}\right), \tag{5.5a}$$

where ν_r is vibrational (attempt) frequency and Q_f is the activation energy for atomic jump over the potential barrier. At $\left(\frac{\tau \varepsilon_o \upsilon_o}{kT}\right) << 1$, Eq. 5.5a reduces to

$$k_r = \nu_r \left(\frac{\tau \varepsilon_o \upsilon_o}{kT}\right) \exp\left(-\frac{Q_f}{Kt}\right). \tag{5.5b}$$

The viscosity η:

$$\eta = exp\left(\frac{Q_f}{kT}\right) \frac{kT\Omega}{(\varepsilon_o \upsilon_o)^2 \nu_r} \frac{1}{c_f}. \tag{5.6}$$

The product $(\varepsilon_o \upsilon_o)$:

$$\varepsilon_o \upsilon_o = A\exp\left(\frac{-Q_s}{kT}\right). \tag{5.7}$$

General form of the viscosity temperature dependence:

$$\eta = \eta_o T\exp\left(\frac{Q_n}{RT}\right)\left(\frac{1}{c_f}\right). \tag{5.8}$$

The hybrid equation for quasi-equilibrium viscosity is

$$\eta_{eq} = \eta_o T\exp\left(\frac{Q_n}{RT}\right)\exp\left(\frac{B}{T - T_o}\right). \tag{5.9}$$

Time dependence of the concentration c_f of flow defects:

$$\frac{dc_f}{dt} = -k_r c_f^2, \tag{5.10}$$

where $k_r = \nu_r \exp\left(-\frac{Q_r}{RT}\right)$ and Q_r is the activation energy of annihilation. Equation 5.10 supposes that the process of defect annihilation is a bimolecular reaction:

$$\frac{dc_f}{dt} = -k_r c_f^2 + k_p P, \tag{5.11}$$

where P is a constant and k_p is the rate constant of defect production process. Objectively, it should be equal to k_r. When the system comes to equilibrium:

$$\frac{dc_f}{dt} = -k_r \left(c_f^2 - c_{f,eq}^2 \right). \tag{5.12}$$

Supposing that the production of defects occurs at already existing defects:

$$\frac{dc_f}{dt} = -k_r c_f \left(c_f - c_{f,eq} \right). \tag{5.13}$$

Supposing that the sites for annihilation and production of defects depend upon the number of defects in excess:

$$\frac{dc_f}{dt} = -k_r \left| c_f - c_{f,eq} \right| \left(c_f - c_{f,eq} \right). \tag{5.14}$$

$$\varepsilon(t) = \frac{\tau \exp\left(\frac{-Qn}{RT}\right)}{\eta_o T} \left(c_{f,eq} t + \frac{1}{k_r} \ln\left(1 - \frac{c_{f,o} - c_{f,eq}}{c_{f,o}} \right) \exp\left(-k_r c_{f,eq} t \right) - \frac{1}{k_r} \ln \frac{c_{f,eq}}{c_{f,o}} \right). \tag{5.15}$$

Under continuous heating conditions with a constant heating rate q, the differential describing the concentration change of structural flow defects as a function of the isothermal annealing time:

$$\frac{dc_f}{dT} + P(T) c_f = c_f^2 Q(T), \tag{5.16}$$

where $P(T,q) = -\frac{\nu_r}{q} \exp\left(-\frac{Q_r}{RT} - \frac{B}{T - T_0} \right)$ and $Q(T,q) = -\frac{\nu_r}{q} \exp\left(-\frac{Q_r}{RT} \right)$.

Its solution is

$$c_{f,high}^{-1}(T,q) = \left(c_{f,0}^{-1} - \int_{T_0}^{T} Q(\theta,q) \exp\left(-\int_{T_0}^{\theta} P(\theta'',q) d\theta'' \right) d\theta'' \right) \exp\left(\int_{T_0}^{T} P(\theta',q) d\theta' \right). \tag{5.17}$$

The parameter T_o is considered as the starting temperature of heating. The combination of Eq. 5.8 with Eq. 5.17 represents the FVM description of glassy alloys' viscosity temperature dependence under continuous heating conditions with a constant heating rate q.

At temperatures, considerably lower than the glass transition temperature T_g,

$$\frac{dc_f}{dt} = -k_r c_f\left(c_f - c_{f,eq}\right) \cong -k_r c_f^2. \tag{5.18}$$

Its solution for nonisothermal annealing conditions with a constant heating rate q is

$$c_{f,low}^{-1}(T,q) = c_{f,0}^{-1} + \frac{\nu_r}{q}\int_{T_B}^{T}\exp\left(-\frac{Q_r}{RT'}\right)dT', \tag{5.19}$$

where T_B is the starting temperature of heating.

At temperatures, considerably lower than the glass transition temperature T_g,

$$\eta(T,q) = \eta_0 T\exp\left(\frac{Q_\eta}{RT}\right)\left[c_{f,0}^{-1} + \frac{\nu_r}{q}\int_{T_B}^{T}\exp\left(-\frac{Q_r}{RT'}\right)dT'\right]. \tag{5.20}$$

Chapter 6

Classical definition of Angell's melt fragility number:

$$m_A = \left[\frac{d(\log\eta)}{d\left(\frac{T_g}{T}\right)}\right]_{T=T_g}. \tag{6.1}$$

FVM definition of Angell's melt fragility number:

$$m_A = 0.434\left[\frac{BT_g}{\left(T_g - T_o\right)^2} + \frac{Q_\eta}{RT_g} - 1\right]. \tag{6.2}$$

Moynihan's interpretation of the melt fragility number:

$$\frac{\Delta T_g}{T_g} \cong \frac{2}{m_M}, \tag{6.3}$$

where $\Delta T_g = T_g - T_{on}$, T_g is the glass transition temperature, and T_{on} is the onset temperature of crystallization of the glass forming material.

FVM interpretation of Moynihan's melt fragility number:

$$m'_M = 0.434\left[\frac{BT_{on}}{(T_g - T_o)(T_{on} - T_o)} + \frac{Q_n}{R}\left(\frac{2}{T_g} - \frac{1}{T_{on}}\right) - \mu - 1\right]. \tag{6.4}$$

Chapter 7

The maximal value of bending stress at the surface of the ribbon:

$$\sigma_o = E\frac{d}{2r_o}, \tag{7.1}$$

where E denotes the Young modulus and d is the ribbon thickness.

The initial overall (elastic only) deformation:

$$\varepsilon_o - \frac{\sigma_o}{E} = \frac{d}{2r_o} \tag{7.2}$$

Under isothermal conditions:

$$\varepsilon_0 = \frac{\sigma(t)}{E} + \varepsilon_f(t) + \varepsilon_a(t) \tag{7.3a}$$

where $\sigma(t)/E$ is the elastic deformation, $\varepsilon_f(t)$ is the plastic deformation, and $\varepsilon_a(t)$ is the anelastic deformation.

For nonisothermal experimental conditions:

$$\varepsilon_0 = \frac{\sigma(T_F, q)}{E} + \varepsilon_f(T_F, q) + \varepsilon_a(T_F, q) \tag{7.3b}$$

where $\varepsilon(T_F, q)/E$ is the elastic deformation, $\varepsilon_f(T_F,q)$ is the plastic deformation, $\varepsilon_a(T_F,q)$ is the anelastic deformation, T_F is the final temperature of heating, and q is the heating rate.

The experimentally determined ratios under isothermal and nonisothermal conditions are

$$\frac{\varepsilon_f(t_1) + \varepsilon_a(t_1)}{\varepsilon_o} = \frac{r_o}{r_1} \tag{7.4a}$$

and

$$\frac{\varepsilon_f(T_{F1}, q_1) + \varepsilon_a(T_{F1}, q_1)}{\varepsilon_o} = \frac{r_o}{r_1}, \text{respectively} \tag{7.4b}$$

The dependence of bend stress relaxation σ/σ_o is

$$\frac{\sigma(t_1)}{\sigma_o} = 1 - \frac{r_o}{r_1}, \text{ and} \tag{7.5a}$$

$$\frac{\sigma(T_{F1}, q_1)}{\sigma_o} = 1 - \frac{r_o}{r_1}. \tag{7.5b}$$

The irreversible bend strain caused by the viscous flow:

$$\frac{\varepsilon_f(t)}{\varepsilon_o} = \frac{E}{3\eta_o \exp\left(\frac{Q_\eta}{RT}\right)} \int_o^t \left(1 - \frac{r_o}{r_1(t')}\right) \frac{dt'}{t'}. \tag{7.6a}$$

For nonisothermal constant heating rate conditions:

$$\begin{aligned} \varepsilon_f(T_F, q) &= \frac{1}{q} \int_{T_B}^{T_F} \frac{\sigma(T', q)}{3\eta(T', q)} dT', \text{ and} \\ \frac{\varepsilon_f(T_F, q)}{\varepsilon_o} &= \frac{E}{q} \int_{T_B}^{T_F} \left(1 - \frac{r_o}{r_1}\right) \frac{1}{3\eta(T, q)} dT, \end{aligned} \tag{7.6b}$$

where T_F denotes the maximal temperature of annealing and T_B denotes the starting temperature of the nonisothermal heat treatment.

By isothermal or nonisothermal annealing, respectively, the deformations due to the viscous flow only are

$$\frac{\varepsilon_f(t_1)}{\varepsilon_o} = \frac{r_o}{r_2} \quad \text{and} \tag{7.7a}$$

$$\frac{\varepsilon_f(T_{F1}, q_1)}{\varepsilon_o} = \frac{r_o}{r_2}. \tag{7.7b}$$

The anelastic part of bend deformation is

$$\frac{\varepsilon_a(t_1)}{\varepsilon_o} = r_o \left(\frac{1}{r_1} - \frac{1}{r_2}\right) \quad \text{and} \tag{7.8a}$$

$$\frac{\varepsilon_a(T_{F1}, q_1)}{\varepsilon_o} = r_o \left(\frac{1}{r_1} - \frac{1}{r_2}\right). \tag{7.8b}$$

The time derivative of Eqs. 7.3a and 7.3b is

$$\frac{d\sigma(t)}{dt} = -E\dot{\varepsilon}_f = -E\frac{\sigma(t)}{3\eta(t)} \quad \text{or} \tag{7.9a}$$

$$\frac{d\sigma(T_F,q)}{dt} = -E\dot{\varepsilon}_f = -E\frac{\sigma(T_F,q)}{3\eta(T_F,q)}. \tag{7.9b}$$

The integration of Eq. 7.9a for isothermal experimental conditions results in

$$\ln\left(\frac{\sigma(t)}{\sigma_o}\right) = -\frac{E}{3k_r\eta_o T\exp\left(\frac{Q_\eta}{RT}\right)}\ln\left(1 + k_r c_{f,o} t\right), \tag{7.10a}$$

where $k_r = \nu_r\ exp(-Q_r/RT)$ denotes the rate constant of relaxation.

For nonisothermal conditions of the bend stress relaxation experiment under constant heating rate q, the solution of Eq. 7.9b is

$$\ln\left(\frac{\sigma(T_F,q)}{\sigma_o}\right) = -\frac{E}{3q\eta_o}\int\limits_{T_B}^{T_F}\left[c_{f,o}^{-1} + \frac{\nu_r}{q}\int\limits_{T_B}^{T}\exp\left(-\frac{Q_r}{RT'}\right)dT'\right]^{-1}\left[T\exp\left(\frac{Q_\eta}{RT}\right)\right]^{-1}dT \tag{7.10b}$$

where T_B and T_F are the starting temperature of heating and the maximal temperature.

The ratio is

$$\frac{\sigma_2(t)}{\sigma_o} = 1 - \frac{r_o}{r_2} \quad \text{and} \tag{7.11a}$$

$$\frac{\sigma_2(T_F,q)}{\sigma_o} = 1 - \frac{r_o}{r_2}, \tag{7.11b}$$

where r_2 denotes the sample radius of curvature after the second stress-free annealing of the amorphous sample.

The empirical equation of Kohlrausch-Williams-Watts (KWW) describing the anelastic deformation is

$$\frac{\varepsilon_a(t)}{\varepsilon_o} = \exp\left[-\left(\frac{t}{\tau_r}\right)^b\right], \tag{7.12}$$

where

$$\tau_r = \tau_o\exp\left(\frac{Q_r}{RT}\right). \tag{7.13}$$

Under nonisothermal experimental conditions, equation of Kohlrausch-Williams-Watts is

$$\frac{\varepsilon_a(T,q)}{\varepsilon_o} = \exp\left[-\left(\frac{(T-T_B)/q}{\tau_o \exp(Q_r/RT)}\right)^b\right], \tag{7.14}$$

where T_B is the starting temperature of annealing and q is the heating rate.

As a function of the time t and the heating rate q used

$$\frac{\varepsilon_a(t,q)}{\varepsilon_o} = \exp\left[-\left(\frac{t}{\tau_o \exp\left[\frac{Q_r}{R(qt+T_B)}\right]}\right)^b\right]. \tag{7.15}$$

with fraction exponent $0.1 < \mathrm{b} < 1$.

Chapter 9

Low temperature linear temperature dependence of sample length:

$$L_0(T) = L_0(T_B)\left[1 + \alpha_l^0(T - T_B)\right], \tag{9.1}$$

where T_B is the starting temperature of heating, $L_o(T_B)$ is the initial sample length at T_B, α_l^o is the coefficient of linear thermal expansion in the low-temperature range $T_B \div T_o$, and T_o is the ideal glass temperature.

Deviation from the real temperature dependence of the sample length:

$$\Delta L_f(T,q) = L_f(T,q) - L_0(T). \tag{9.2}$$

Taking into account that the x, Refs. 2–5, correlates to the real mean free volume υ_f according to the expression $x = \upsilon_f/\gamma\upsilon^*$, change of the reduced free volume along with increasing temperature:

$$x(T,q) - x_0 = \frac{L_f^3(T,q) - L_0^3(T)}{\gamma\upsilon^* N}, \tag{9.3}$$

where N is the number of atoms in a cube of the glassy alloy studied with edge length $L_0(T_B)$ at temperature $T = T_B$.

$$N = \frac{L_0^3(T_B)}{V_{mol}} N_A, \tag{9.4}$$

When expanding Eq. 9.3 into a power series, one obtains

$$x(T,q) - x_0 \cong 3L_0^2(T)\Delta L_f(T,q)\frac{V_{mol}}{\gamma v^* N_A L_0^3(T_B)}. \tag{9.5}$$

Or

$$3L_0^2(T)\Delta L_f(T,q)\frac{V_{mol}}{\gamma v^* N_A L_0^3(T_B)} = \frac{1}{\ln c_{f,0}} - \frac{1}{\ln c_f(T,q)}, \tag{9.6}$$

where $c_f(T,q)$ is defined by ▶ Eq. 5.17.

The temperature dependence for the experimentally observed sample length by heating the sample with constant heating rate q is

(a) $T < T_o$:

$$L_f(T,q) = L_0(T) = L_0 T_B\left[1 + \alpha_l^0(T - T_B)\right]; \tag{9.7}$$

(b) $T > T_o$:

$$L_f(T,q) = L_0(T_B)\left[1 + \alpha_l^0(T - T_B)\right] + \frac{\gamma v^* N_A L_0(T_B)}{3V_{mol}\left[1 + \alpha_l^0(T - T_B)\right]^2} \times \left[\frac{1}{\ln c_{f,0}} + \frac{1}{\ln\left(\frac{1}{c_{f,0}} - \int_{T_0}^{T} Q(\theta,q)e^{-\int_{T_0}^{\theta} P(\theta',q)d\theta'}d\theta\right) + \int_{T_0}^{T} P(\theta',q)d\theta'}\right], \tag{9.8}$$

where the functions $P(T,q)$ and $Q(T,q)$ were defined in ▶ Chap. 5.

The "anomaly" of thermal expansion

$$\Delta L_f(T,q) = \frac{\gamma v^* N_A L_0(T_B)}{3V_{mol}\left[1 + \alpha_l^0(T - T_B)\right]^2} \times \left[\frac{1}{\ln c_{f,0}} + \frac{1}{\ln\left(\frac{1}{c_{f,0}} - \int_{T_0}^{T} Q(\theta,q)e^{-\int_{T_0}^{\theta} P(\theta',q)d\theta'}d\theta\right) + \int_{T_0}^{\theta} P(\theta',q)d\theta'}\right]. \tag{9.9}$$

The "anomalous" contribution ΔC_p to the specific heat C_p is presented as

$$\Delta C_p(T,q) = \beta \frac{dx}{dT}, \tag{9.10}$$

where β is a material-specific coefficient of proportionality.

In the temperature range considerably lower than the glass transition temperature T_g,

$$\Delta C_{p,\,low}(T,q) = \beta \frac{c_{f,\,low}(T,q)Q(T,q)}{\left(\ln c_{f,\,low}(T,q)\right)^2}. \tag{9.11}$$

In the temperature range in the vicinity of the glass transition temperature T_g,

$$\Delta C_{p,\,high}(T,q) = \beta \frac{c_{f,\,high}(T,q)Q(T,q) - P(T,q)}{\left(\ln c_{f,\,high}(T,q)\right)^2}. \tag{9.12}$$

The functions $P(T,q)$ and $Q(T,q)$ were defined in ▶ Chap. 5.

Chapter 10

The change of structural defects concentration along with increasing the time of annealing at constant temperature T is

$$\frac{dc_f(T,t)}{dt} = -k_r c_f\left(c_f - c_{f,e}\right), \tag{10.5}$$

where $k_r = \nu_r \exp\left(-\frac{Q_r}{RT}\right)$ is the rate constant of relaxation, $c_{f,e}(T)$ is the equilibrium defect concentration, and R is the universal gas constant.

Its solution is

$$c_f(T,t)^{-1} = c_{f,e}^{-1} + \left(c_{f,o}^{-1} - c_{f,e}^{-1}\right)\exp\left(-k_r c_{f,e} t\right); \quad c_f(t = 0s) = c_{f,0}. \tag{10.6}$$

$c_{f,o}$ is here the defect concentration at the start of the isothermal heat treatment.

The temperature dependence of $c_{f,e}$(T) is

$$c_{f,e}(T) = exp\left(-\frac{B}{T - T_o}\right), \tag{10.7}$$

where B and T_o are constants.

The density change of a cube of initial edge length L_o, caused by nonisothermal or isothermal structural relaxation, is presented as

$$\Delta\rho(T,q) = \rho(T,q) - \rho_o = \frac{W}{\left(L_o + \Delta L_f(T,q)\right)^3} - \frac{W}{L_o^3}$$
$$= -\frac{3W\left(\Delta L_f(T,q)/L_o\right)}{L_o^3\left(1 + 3\Delta L_f(T,q)/L_o\right)}, \tag{10.8}$$

$$\frac{\Delta\rho(T,q)}{\rho_o} = -\frac{3\left(\Delta L_f(T,q)/L_o\right)}{1 + 3\left(\Delta L_f(T,q)/L_o\right)}, \tag{10.9}$$

under nonisothermal relaxation conditions, and

$$\frac{\Delta\rho(t)}{\rho_o} = -\frac{3\left(\Delta L_f(T,t)/L_o\right)}{1 + 3\left(\Delta L_f(T,t)/L_o\right)}, \tag{10.10}$$

under isothermal relaxation conditions.

Chapter 11

Crystal growth rates G:

$$G = \frac{fkT}{3\pi a_o \eta}[1 - exp(-\Delta G_m/RT)], \tag{11.1}$$

where η denotes the coefficient of viscous flow responsible for atomic motion required for growth, a_o is the mean atomic diameter, ΔG_m is the molar free energy change, and f is the fraction of sites at the interface where crystal growth occurs.

General form of the equation of Kolmogorov, Avrami, Johnson, and Mehl:

$$\zeta = 1 - \exp(-V_e), \tag{11.8}$$

where $\xi = V_{cr}/V_o$ is the volume fraction of crystallized regions, V_{cr} is the overall volume of crystallized regions, V_o is the total volume of the sample, and V_e is the so-called extended volume of growing crystalline particles per unit volume of the matrix.

The extended volume of growing crystalline particles is

$$V_e = \frac{4}{3}\pi N(t)G^3(T)t^3, \tag{11.9}$$

where t is the time of crystallization and $G(T)$ is the crystal growth rate at a temperature T.

Arrhenian type equation of the crystal growth rate:

$$G(T) = G_o \exp\left(-\frac{Q_{eff}}{RT}\right), \tag{11.10}$$

where G_o is a pre-exponential factor and Q_{eff} is the effective activation energy of crystallization.

By nucleation site saturation:

$$\zeta = 1 - \exp\left(-K(T)t^3\right), \tag{11.11}$$

where $K(T) = N\, G^3(T)$.

Its temperature dependence can be presented as

$$\ln(K(T)) = \ln\left(NG_o^3\right) - 3\frac{Q_{eff}}{RT}. \tag{11.12}$$

Chapter 12

The energy density per unit volume:

$$\tau_\gamma \gamma_o = \int_{T_o}^{T_g} \rho C_p dT, \tag{12.1}$$

where τ_γ is the maximal stress of share deformation yielding deformation $\gamma_o \approx 1$ of the basic share unit, ρ is the density of the material, C_p is the molar specific heat, T_g is the glass transition temperature, and T_o is the ambient temperature.

The creep strength equals

$$\sigma_\gamma = 2\tau_\gamma. \tag{12.2}$$

The material density:

$$\rho = \frac{\rho_o}{[1 - 3\alpha(T - T_o)]}, \tag{12.3}$$

where ρ_o is the density at ambient temperature and α is the coefficient of thermal expansion.

Equation of B. Yang for the fracture strength:

$$\sigma_f = 55\frac{\Delta T_g}{V_m}, \tag{12.4}$$

where σ_f is the fracture strength, ΔT_g is the temperature difference $T_g - T_o$, and V_m is the molar volume of the glassy alloy.

Appendix 2: Concluding Remarks

In this book, we have tried to make use of the free volume theory of Cohen and Turnbull to give an insight into the relaxation phenomena in glassy metals around the glass transition temperature. It was taken into account that the glassy metals are a particular kind of condensed matter, i.e., matter in the amorphous state. The free volume model (FVM) describes the atomic mobility in liquids, undercooled melts, and amorphous solids via the free volume available for a single atom within the material structure. By rapid quenching from the melt into the structure of glass-forming metallic melts, non-equilibrium excess free volume is frozen in. The structural relaxation of a non-equilibrium amorphous structure with concentration of structural defects higher than the quasi-equilibrium concentration of structural defects is identified as structure, striving to reach the quasi-equilibrium for a given temperature structural state, via annihilation or production of free volume. These free volume changes along with the temperature/time changes are, of course, not instantaneous, but possess a definite mechanism and kinetics, about which the free volume theory of Turnbull and Cohen does not propose any indications and which should be specified experimentally.

The authors have endeavored to collect most of the available experimental evidence from the bibliography and from own experimental studies – upon the viscous flow behavior, thermal expansion and specific heat peculiarities, bend stress relaxation behavior, density changes, etc. – of amorphous metallic alloys, showing that the relaxation of glassy metals around the glass transition temperature proceeds via annihilation and/or production of structural defects as carriers of free volume. The kinetics of this process, both under isothermal and non-isothermal experimental conditions, seems to be best described as a kinetics of a bimolecular reaction. The authors believe that their attempt to combine the FVM basic concepts with the proper kinetic equations for description of free volume annihilation and/or production in the amorphous metallic alloys was successful and convincing. It describes well the glassy metal property changes caused by relaxation processes during continuous heating and/or cooling. This approach provides a useful tool for understanding the relaxation phenomena in glassy metals.

The authors recommend to the studious reader to pay a special attention to the developed in this work method for the study of the bend stress relaxation phenomena in ribbonlike glassy metals, especially to the experimental possibility to

K. Russew, L. Stojanova, *Glassy Metals*,
DOI 10.1007/978-3-662-47882-0

separate the fully irreversible bend stress phenomena caused by the viscous flow from the fully reversible anelastic bend stress deformation. The last one can be described very well by the empirical stretch exponent equation of Kohlrausch-Williams-Watts. In this way the possibility arises to open a new window for future studies upon the interrelation between the irreversible relaxation phenomena and the fully reversible relaxation processes in glassy metallic alloys. The nature and mechanism of the reversible anelastic deformation in glassy metals are still not well understood and studied.

Last but not least, the authors of this book would like to point out that they share the belief and conviction of the famous Austrian physicist and philosopher Ludwig Boltzmann, namely, that the task of theoretical modeling consists in constructing a picture of the external world that exists purely internally and must be a guiding star for the researcher:

> …It follows that it cannot be our task find out an absolutely correct theory but rather a picture, which is as simple as possible while representing the phenomena as well as possible…

The FVM of Cohen and Turnbull seems to be such a picture.

The manufacturer's authorised representative in the EU is Springer Nature Customer Service Centre GmbH, Europaplatz 3, 69115 Heidelberg, Germany. If you have any concerns regarding our products, please contact ProductSafety@springernature.com

Printed and bound by CPI Group (UK) Ltd, Croydon, CR0 4YY
29/07/2026
02180566-0008